Distillation and Absorption '92

Institution of Chemical Engineers, Rugby, UK

Hemisphere Publishing Corporation

A Member of the Taylor & Francis Group

New York   Philadelphia   London

Distillation and Absorption '92

Members of the Institution of Chemical Engineers should order as follows:

Worldwide	Institution of Chemical Engineers, Davis Building, 165–171 Railway Terrace, RUGBY, Warwickshire CV21 3HQ, UK

Non-members' orders should be directed as follows:

UK, Eire and Australia	Institution of Chemical Engineers, Davis Building, 165–171 Railway Terrace, RUGBY, Warwickshire CV21 3HQ, UK
or	Taylor & Francis Ltd, Rankine Road, BASINGSTOKE, Hampshire RG24 0PR, UK
USA	Taylor & Francis Inc., 1900 Frost Road, Suite 101, Bristol, PA 19007, USA
Rest of the World	Taylor & Francis Ltd, Rankine Road, BASINGSTOKE, Hampshire RG24 0PR, UK

Library of Congress Cataloging-in-Publication Data
Distillation and absorption '92: a three day symposium organised by the Institution of Chemical Engineers on behalf of the EFCE Working Party on Distillation and Absorption and held at the International Conference Centre, Birmingham, 7–9 September, 1992 / organising committee, K.E. Porter (symposium chairman), J.D. Jenkins, D.W. Reay: technical programme committee, D.W. Reay . . . [et al].
p. cm. — (Symposium series / Institution of Chemical Engineers; no. 128) (EFCE publication series; no. 94) "EFCE event no. 459" ISBN 1–56032–257–8
1. Distillation — Congresses. 2. Absorption — Congresses. I. Porter, K.E. II. Institution of Chemical Engineers (Great Britain) III. EFCE Working Party on Distillation and Absorption. IV. Series. V. Series: Symposium series (Institution of Chemical Engineers (Great Britain)); no. 128.
TP156.D5D58 1993
660'.28425 — dc20 92–34396
CIP

Distillation and Absorption '92

A three day symposium organised by the Institution of Chemical Engineers on behalf of the EFCE Working Party on Distillation and Absorption and held at the International Conference Centre, Birmingham, 7–9 September 1992.

Organising Committee

K.E. Porter (Symposium Chairman)	University of Aston
J.D. Jenkins	University of Aston
D.W. Reay	BP Engineering

Technical Programme Committee

D.W. Reay (Chairman)	BP Engineering
M.W. Biddulph	University of Nottingham
R.J.P. Brierley	Consultant
R.C. Darton	University of Oxford
M.J. Griffiths	BP Chemicals
A.E.O. Jenkins	Distillation Tech Ltd
J.D. Jenkins	University of Aston
J.T. Lavin	BOC Cryoplants
K.E. Porter	University of Aston
Q.M. Siddique	British Gas

INSTITUTION OF CHEMICAL ENGINEERS

SYMPOSIUM SERIES No. 128
EFCE Event No. 459
EFCE Publication No. 94
ISBN 0 85295 281 3

Errata

Paper 36 (page A361): Varioflex valve tray, a high performance mass transfer device (R. Billet).
On page A363, equation (9) should read:

$$\lambda = \frac{F_V}{u_L} \frac{\sqrt{\rho_V}}{\rho_L} \frac{\mu_L}{\mu_V} m \qquad (9)$$

Poster 14 (page B111): Effects of tilt and motion on a packed deaeration column (R.K. Tanner, S.A. Baker and B. Waldie)
Please note that on the following pages 'kg/m s' should be read as 'kg/m^{2}s':
Page B112: 'Experimental Details' section — third para., line 4 and fourth para., line 3.
Page B114: 'Effects of Tilt' section — first para., line 2.
Page B115: 'Scaling-up Effects of Tilt' section — fourth para., line 5.
Page B119: Figure 8.

Printed by The Chameleon Press Limited, 5–25 Burr Road, London SW18 4SG

Preface

Distillation and absorption, like the motor industry, is a mature business which nevertheless continually changes, driven by market forces and led by research and development. This series of conferences (which started at Brighton in 1960) has both recorded changes in practice and shown what needs to be done.

Thus it was during the discussion of the papers in 1960 (at a time when bubble trays were assumed to produce bubbles), that a Mr Asquith commented that on looking through a window in the side of his column, 'It was sometimes difficult if not impossible to see the bubbles, but it was easy to see the drop systems' (spray). Our next two conferences recorded work on flow regimes (bubbly, spray and mixed) and the implications for tray design.

In 1979, at a time when energy saving had a high priority, it was Frank Rush of Du Pont who identified the need to design distillation columns 'leaner and harder'.

One response has been the development of procedures for the safe design of large diameter packed distillation columns, recorded in 1987 and in several of the papers published here.

What new direction will be identified at this conference? Will it be the rate based approach to column design, or the need for more molecular science in predicting phase equilibria or for more fluid mechanics to understand trays? Or, as in 1960, will it be from some unexpected contribution to the discussions that our next new direction will emerge?

K.E. Porter
(Chairman of Organising Committee)

Contents

Papers

Dynamics, measurement, control

Absorption, distillation with chemical reaction

Sequencing and expert systems

Packing 1

Equipment, plants and processes, batch distillation

Packing 2

Trays 1

Thermodynamics, azeotropic and extractive distillation, models

Trays 2

Closing Address

Posters

Dynamics, measurement, control

Absorption, distillation with chemical reaction

Equipment, plants and processes, batch distillation

Trays

Thermodynamics, azeotropic and extractive distillation, models

† Papers printed out of page sequence

Distillation
and Absorption '92

Posters

CONTROLLING BINARY HIGH PURITY DISTILLATION COLUMNS

J. Broll and H. Gelbe
Institute of Process and Plant Design, Technical University of Berlin, Federal Republic of Germany

The control of binary high-purity distillation columns is difficult because of the nonlinearity of the process and the large time delays and constants. This paper presents a proposal for a feedforward control scheme, which corrects the manipulated variables due to measured disturbances in the process loads so that the controlled variables remain as much as possible unaffected. A nonlinear time discrete rigorous model of a distillation column was used to include as much a priori knowledge as possible from the process. It is shown in simulations and experiments that the control behaviour can be improved, if the feedforward model is used in combination with a feedback controller.

INTRODUCTION

During operation of distillation columns, most of the disturbances occur at the input of the controlled system, in the feed flow, the feed concentration or the feed temperature. Distillation columns usually have very large time constants, so disturbances are noted much later in the controlled variables. Especially when using columns for the production of highly purified products in the top and in the bottom, small disturbances may lead to instable operations when using conventional feedback controllers. It would be preferable not to wait for disturbances to occur in the top and bottom concentrates, but to compensate the measurable disturbances in the feed with the aid of a feedforward controlling device.

Figure 1 shows the general control structure. The necessary manipulated variables u_S are calculated by the feedforward control device from the measurable disturbances d. In practice, an additional feedback controller is necessary, because not all disturbances can be detected and there are always differences between the feedback model and the real plant.

Linear or shortcut feedforward control models have been reported repeatedly in the past (Jafarey, McAvoy, 1980; Shinskey, 1984). These models calculate the manipulated variables for one operating point and for selected disturbances, mostly feed quantity and feed concentration. The disadvantage of such models is that due to the strong nonlinearity in high purity columns, small deviations from the operating point may result in unsatisfactory control behaviour. It is therefore essential to employ a nonlinear feedforward control model which describes the behaviour of the column with sufficient accuracy.

THE FEEDFORWARD CONTROL MODEL AND THE REAL COLUMN MODEL

Rigorous dynamic models are based on balances of quantity, energy and momentum on each tray of the column and consist of a set of algebraic and differential equations. The same models may also be used to calculate the required controller output from feed or other disturbances when setting the controlled variables, namely distillate and bottom product concentrations constant. This solution is only available in steady state, because the two controlled variables are directly related to the variables of state on the trays of the column, which are time dependent and can not be put on invariable settings.

The transitional behaviour of the variables on the individual trays can be described by ordinary differential equations of the first order. The dynamic behaviour of every differential equation in every time step is approximated by a quasi steady-state substitute function. For example, the material balance for the tray *n* (trays numbered from bottom to top):

$$H \frac{dx}{dt} = V_{n-1} y_{n-1} + L_{n+1} x_{n+1} - V_n y_n - L_n x_n \tag{1}$$

with $x_n = x$

leads to the substitute function to calculate a new steady-state value x^*:

$$0 = V_{n-1} y_{n-1} + L_{n+1} x_{n+1} - V_n y_n - L_n x^* \tag{2}$$

setting $x_n(t) = x^*(t - H/L_n)$

H/L_n being the holdup time of the liquid flow L_n on the tray. So the dynamic behaviour of the whole column is marked by a noncontinous behaviour, which however becomes smoother with increasing tray numbers. The following assumptions were applied to the feedforward control model:

- negligible vapour holdup;
- constant liquid holdup on trays;
- complete mixing on trays;
- constant pressure drop;
- stationary energy balance;
- constant tray efficiency.

SIMULATIONS

The first example for an application of the feedforward model is a distillation column, which was presented by Wahl and Harriot, (Wahl, Harriot, 1970). The feedforward model and the column model are simple binary models with a relative volatility of α=2.0 and 26 trays. The feedtray number is 13. Figure 2 shows the calculated necessary transient behaviour of the change of the manipulated variables reflux *L* and vapour *V* after a 10% increase in feed concentration at time zero to hold the product concentrations constant.

If the feed concentration increases, the reflux has to decrease and the vapour flow has to increase in steady-state. This behaviour is well known. The feedforward model calculates additionally to the steady-state the dynamic behaviour of the manipulated variables. It is interesting, that the vapour first decreases, before it starts rising up to the new steady state. The behaviour of the reflux flow is similar. The manipulated variables react like an inverse response system.

Figure 3a shows the dynamic behaviour of the manipulated variables ΔL and ΔV after equal changes of the setpoints of the top and bottom concentrations, both to purer products (Figure 3b). The transient behaviour of the controlled variables is given by a linear second order function with the time constants T_1=0.1 h and T_2=0.05 h. It can be seen that the dynamic behaviour of the manipulated variables is different for the same step but in the opposite direction. There is no linear controller whose output may follow this behaviour.

In a second example, the influence of 'bad modelling' of the feedforward control model was analysed and tested with the relevant rigorous model by simulation of a column separating the binary mixture methanol-water with 28 real trays as depicted in Figure 4. Shown is the real laboratory column, which has 100 mm diameter and 4 meters height and which will be used for the validation of the simulation results. The evaporator is a thermosyphon evaporator, the condenser a total condenser. Feed is added on the tenth tray with a methanol concentrate of 31.5 mol% in equilibrium condition. Level control in top and bottom is taken as perfect, as well as the pressure control.

Reflux *L* and vapour *V* are applied as manipulated variables for concentration control of top and bottom. The developed feedforward control model is connected with the column model and calculates on line the manipulated variables reflux *L* and vapour *V* due to the measured or, in this case, defined disturbances of the feed. A feedback control loop has been superimposed. As feedback controller a multivariable controller or two single input single output PID-controllers could be chosen. The robust multivariable controller is based on the LQG/LTR tuning rules (Doyle, Stein, 1982) with an extension, using the positive real system (Broll, Pohlmeier, 1992).

The column model bears far more resemblance to reality than the feedforward model and is validated at the real plant. This column model is based on the following assumptions:

- dynamic energy balances;
- variable holdup;
- variable pressure drop dependent on vapour flow.

Figure 5 shows the course of the distillate concentration from the steady-state value after a 30% decrease in the feed concentration.

The following five control schemes were examined:

1. Column with a feedforward controller only
2. Column with a multivariable feedback controller
3. Column with two conventional single feedback controllers
4. Column with a feedforward and with multivariable feedback controller
5. Column with a feedforward and with two conventional single loop controllers.

All controllers were designed on the same model assumptions mentioned above for the feedforward model, in such a way that the permissible limits for the controlled variables were not exceeded. It can be seen that the control performance of the multivariable controller is much better than that of the single input single output PID controller. The thick solid line shows the course of the controlled variables when the feedforward regulator without feedback is applied. The deviations are produced by the model mismatch between the feedforward model and the more realistic column model.

When, in addition to the feedforward controller, a feedback controller is superimposed, the initial transient of the controlled variables is improved further. The type of the feedback controller has, in the case of only one measured disturbance, no significant influence, even the conventional control strategy achieves adequate results.

In Figure 6 the bottom concentrations are shown after the 30% disturbance of the feed concentration. The bottom concentrations react much harder than the top ones, due to the high relative volatility of methanol in this region. The results achieved with the feedforward controller and the multivariable controller are superior to those achieved with the single loop PID controller.

A coupling of feedforward and feedback controller shows in simulation tests an improvement without causing stability problems. Figures 5 and 6 demonstrate the advantages of the here explained model based feedforward control system over feedback controllers. The period of deviation from the setpoints will be reduced by several hours.

EXPERIMENTS

In order to improve the good simulation results shown above, we implemented the feedforward controller to a real distillation plant, which is identical with the plant used for the simulations and was shown in Figure 3. The pilot plant is used for teaching and research purposes. The feed line consists of two equally equipped pipes. In this way it is possible to disturb the plant with exactly defined steps in feed flow, feed concentration and feed temperature.

The feedforward model is installed on line on a Vax computer which is interlinked with the process control system via a KXJ-11 single board computer. This single board is plugged in the Vax machine and is used for the translation of the different telegrams. The simulation program is SpeedUp, where we used the external data interface (EDI) for communication between process and simulator.

The plant is observed by a nonlinear observer based on tray to tray calculations and temperature correction as an additive term in the component balance (Zeitz, 1977). In this way the distillate and bottom concentrations are continously available. The controller used is a robust controller, based on the positive real system, which is tuned with respect to the constraints of the manipulated variables.

The most severe disturbance for a distillation column is a step change in feed concentration. Figure 7 shows the transient behaviour of the product concentrations. The solid line is the feedforward/feedback system, the dashed line is the feedback controller without feedforward model. After taking the plant in automatic control the feedback control as well as the feedforward/feedback control the column reaches the steady-state operating point. After a 20% step decrease in the methanol concentration the controller without feedforward model (dashed line) is able to correct the deviations in the controlled variables, but it takes a long time. The feedback controller realizes the full magnitude of the disturbance too late. As a result of the nonlinear behaviour the distillation moves to a complete new operation point. It is not possible to tune the controller more aggressive, because the manipulated variables, especially the boilup, have hard constraints.

If the feedforward model is used to calculate the values for the manipulated variables in addition to the feedback controller, concentrations can be seen as shown with solid lines. The feedforward model, which reacts immediately, takes the manipulated variables to the right steady-state value. The concentrations in top and bottom come back to the setpoint in about 20 minutes.

CONCLUSION

In this paper, a new approach to designing a feedforward controller based on nonlinear rigorous models is presented. The use of feedforward controllers should be useful in practice, because conventionally used and well known control configurations do not have to be changed. Operators may accept this more than complicated model-based feedback control schemes. The feedforward algorithm has only to be connected with the controlled variables. But the useful feedforward model must be based on nonlinear rigorous calculations for high purity columns and validated on the specific column conditions. The influence of model faults, process conditions, different manipulated variables, etc are the subject of further investigations and will be discussed in the oral presentation.

NOMENCLATURE FOR EQUATIONS (1) AND (2)

Symbol	Name	Unit
V	Vapour flow	kmol/h
L	Liquid flow	kmol/h
x	Liquid concentration	mol/mol
y	Vapour concentration	mol/mol
H	Liquid holdup	kmol
t	time	h

REFERENCES

1. Broll, J., Pohlmeier, J., 1992, Robust control of a distillation column using positive-real-system controllers with observer, paper to be presented at BICSS'92, 2nd Int Conf on System Simulation and Scientific Computing, Bejing, China, October 20–23.
2. Doyle, J.C., Stein, G., 1981, Multivariable feedback design: concepts for a classical/modern synthesis, *IEEE Trans on Aut Control,* Vol AC-26, No 1.
3. Jafarey, A., McAvoy, T. J., 1980, Steady state feedforward control algorithms for reducing energy costs in distillation, *ISA Trans*, 19 (4)
4. Shinskey, F.G., 1984, *Distillation Control,* 2nd Edition, McGraw-Hill
5. Wahl, E.F. and Harriot, P., 1970, Understanding and prediction of the dynamic behaviour of of distillation columns, *Ind Engng Chem, Process Des Dev* 9: 396
6. Zeitz, M., 1977, Nichtlineare Beobachter für chemische Reaktoren, *Fortschr Ber VDI* Reihe 8, Nr 27, VDI Verlag Düsseldorf

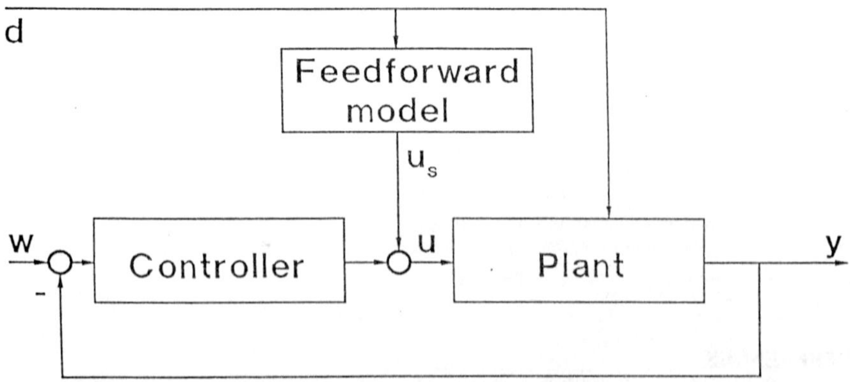

Fig. 1 The feedforward/feedback control structure

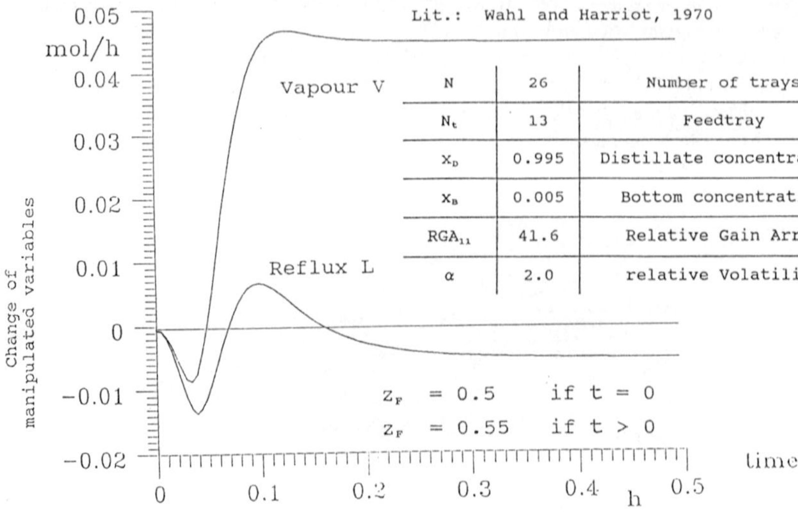

Fig. 2 Dynamic behaviour of the manipulated variables

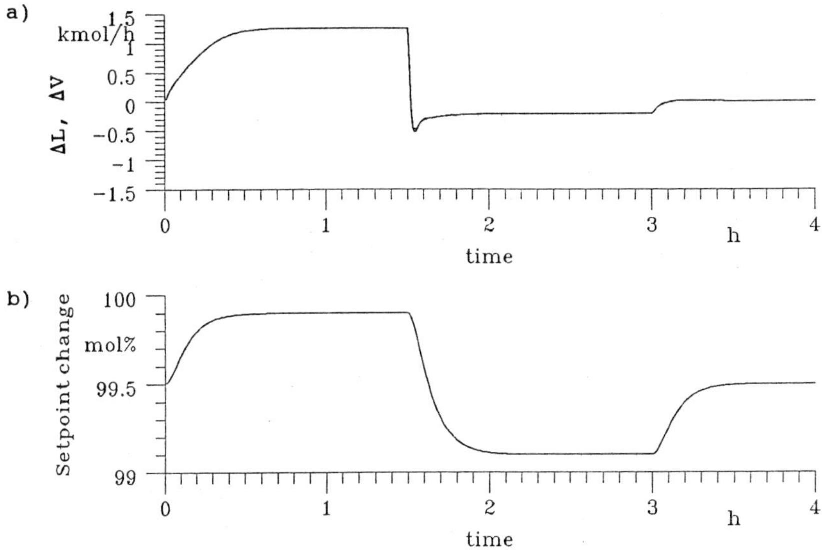

Fig. 3 Dynamic behaviour after setpoint settings

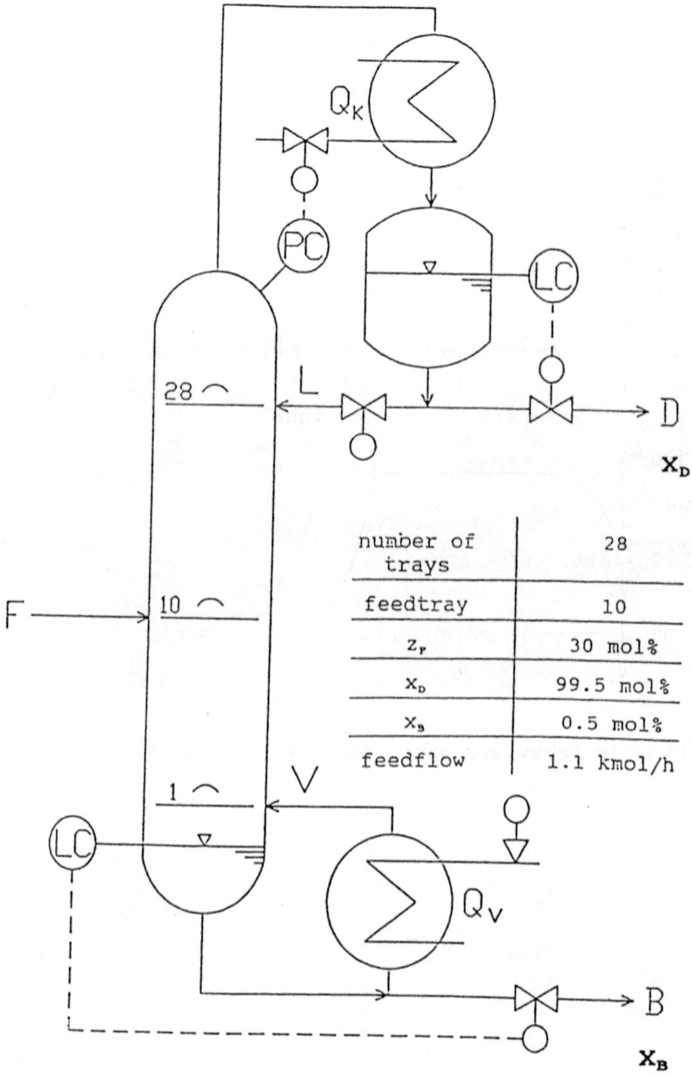

number of trays	28
feedtray	10
z_F	30 mol%
x_D	99.5 mol%
x_B	0.5 mol%
feedflow	1.1 kmol/h

Fig. 4 Distillation column

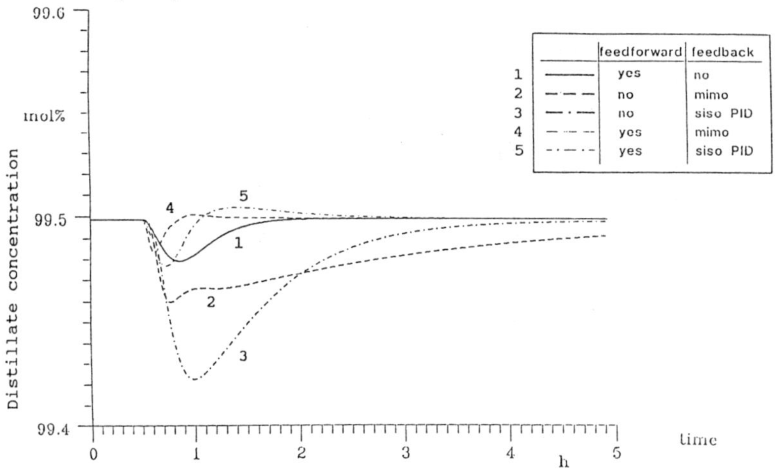

Fig. 5 Distillate concentration x_D, Δz_F = -30% at t= 0.5 h

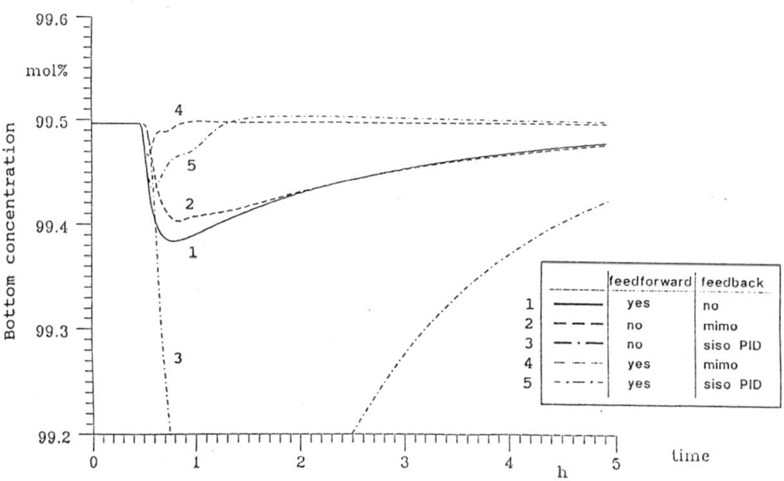

Fig. 6 Bottom concentration x_B, Δz_F = -30% at t= 0.5 h

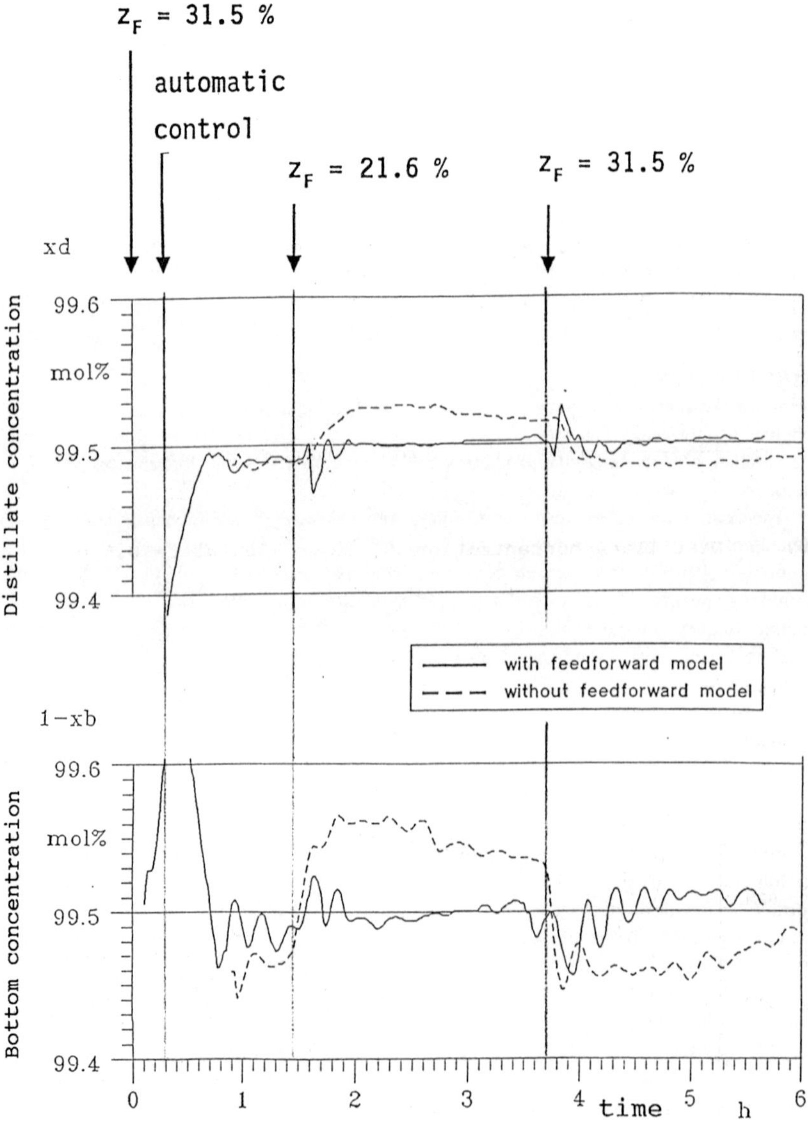

Fig. 7 Experimental data of controlled variables

STEADY STATE AND DYNAMIC SIMULATION OF THREE PHASE DISTILLATION FOR INDUSTRIAL APPLICATION

B. Landwehr, G. Wozny* and H. Hartmann**
* Henkel KGaA, Postfach 1100, D–4000 Düsseldorf 1, Germany
** RWTH Aachen, Wüllnerstr. 5, D–5100 Aachen, Germany

A new steady state and dynamic simulation algorithm for the distillation of highly non-ideal mixtures, consisting of immiscible phases, is presented. First the mathematical model is described. Then literature and experimental results were used to demonstrate the reliability of the steady state algorithm. Finally some dynamic simulation examples are demonstrated.

INTRODUCTION

During the last years several papers concerning three phase distillation were published. Most of them handle only steady state simulations of single columns. A good overview over these steady state simulation methods is given by Cairns and Furzer (1990). Because of the importance of heterogeneous distillation in industrial process, further studies would be desirable.

The existing models are restricted to single columns or the 'conventional structure' of two coupled columns with a common decanter, (eg Bril *et al*, 1976). In practice, one column is commonly not sufficient to separate highly-nonideal mixtures. Multiple columns with several interlinking streams are used to facilitate good separation results. Moreover the 'conventional structure' frequently requires higher reflux ratios (Pham, Doherty, 1990) than other alternatives. Therefore a new steady state model able to describe three phase distillation columns with interlinking streams between any stages was developed.

Steady state solutions are important for the design of industrial columns, but not sufficient to describe operating conditions. Under these circumstances disturbances normally occur. Furthermore, it is known that heterogeneous columns exhibit complex dynamic behaviour and parametric sensitivity. Therefore, models capable to describe dynamic column behaviour are necessary to estimate the influence of disturbances on the processing variables.

Recently two papers considering dynamic three phase distillation were published. Rovaglio and Doherty, 1990, developed the first dynamic model, which was capable to describe some cases of erratic behaviour of heterogeneous distillation columns. Analysing the distillation of a benzene-ethanol-water mixture, they detected such an operating condition, in which only small changes of pressure caused the disappearance of the second liquid phase in the decanter.

The same mixture was examined by Wong *et al*, 1991, whose dynamic model included controllers. The authors found that the appearance and disappearance of a heterogeneous two liquid phase can lead to oscillation or even instability of the operation.

INDUSTRIAL REQUIREMENTS

For industrial purpose, a comfortable user-interface and a connection of different simulation tools is desirable. Our simulation toolbox includes different models, like two and three phase distillation, absorption, batch distillation, rate based approach for distillation and evaporation, determination of immiscibility regions and residue curve maps. A common database provides the different programs with component data (Gorak *et al*, 1991).

Dynamic calculations and control structures have already been implemented in the two phase distillation model (Wozny *et al*, 1987) as well as in the batch distillation.

Our main reason for developing a dynamic three phase distillation model was the necessity to examine dynamic column responses on disturbances in order to design control structures. Furthermore the algorithm should be able to test the chosen structures.

MATHEMATICAL MODEL

The following model is able to describe steady state and dynamic behaviour of single columns and interlinked column systems. The latter case is explained more detailed by Landwehr, Wozny, 1992. The model of a dynamic column with an example control structure is illustrated in Figure 1. The trays are numbered consecutively from the top (condenser) to the bottom (reboiler) of the column. Phase separation and the use of a decanter is allowed at every stage of the column. Figure 2 represents a model showing all possible flows entering or leaving a single stage.

The model equations are derived basing on the following assumptions:

- liquid-liquid equilibrium
- validity of Murphree-tray-efficiency model
- neglect of vapour holdup.

Using the concentrations y, y^*, x^I, x^{II}, the flow rates V, L^I, L^{II}, the temperature T and the liquid holdup HU as system variables, the following equations were formulated:

COMPONENT MATERIAL BALANCE:

$$CB_{k,j} = F_j\, z_{k,j} + V_{j+1}\, y_{k,j+1} + WR_{p,j} y_{k,p} - (V_j + W_j + WR_{j,m})\, y_{k,j}$$

$$+ L^I_{j-1}\, x^I_{k,j-1} + L^{II}_{j-1}\, x^{II}_{k,j-1} + UR^I_{i,j}\, x^I_{k,i} + UR^{II}_{l,j}\, x^{II}_{k,l} \qquad (1)$$

$$- (L^I_j + U^I_j + UR^I_{j,q})\, x^I_{k,j} - (L^{II}_j + U^{II}_j + UR^{II}_{j,r})\, x^{II}_{k,j} = \frac{d\,(HU^L_j\, \bar{x}_{k,j})}{dt}$$

with $k = 1, ..., NK$ and $j = 1, ..., NST$. This range of indices also refers to the following equations.

VAPOUR-LIQUID EQUILIBRIUM:

$$E^{VL}_{k,j} = y^*_{k,j} - K^{VL}_{k,j}\, x^I_{k,j} = 0 \qquad (2)$$

LIQUID-LIQUID-EQUILIBRIUM:

$$E^{LL}_{k,j} = x^I_{k,j} - K^{LL}_{k,j}\, x^{II}_{k,j} = 0 \qquad (3)$$

MURPHREE TRAY EFFICIENCY CORRELATION:

$$MC_{k,j} = y_{k,j} - E_j\, y^*_{k,j} + y_{k,j+1}\,(E_j - 1) = 0 \tag{4}$$

STOICHIOMETRIC EQUATIONS:

$$S1_j = \sum_{k=1}^{NK} y_{k,j} - 1 = 0 \tag{5}$$

$$S2_j = \sum_{k=1}^{NK} x^I_{k,j} - 1 = 0 \tag{6}$$

TOTAL MATERIAL BALANCE:

$$\begin{aligned} MB_j = {} & F_j + V_{j+1} + WR_{p,j} - (V_j + W_j + WR_{j,m}) + L^I_{j-1} + L^{II}_{j-1} + UR^I_{i,j} \\ & + UR^{II}_{l,j} - (L^I_j + U^I_j + UR^I_{j,q}) - (L^{II}_j + U^{II}_j + UR^{II}_{j,r}) = \frac{d\,(HU^L_j)}{dt} \end{aligned} \tag{7a}$$

ENTHALPY BALANCE:

$$\begin{aligned} H_j = {} & F_j\, h^F_j + Q_j + V_{j+1}\, h^V_{j+1} + WR_{p,j}\, h^V_p - (V_j + W_j + WR_{j,m})\, h^V_j \\ & + L^I_{j-1} h^{L^I}_{j-1} + L^{II}_{j-1}\, h^{L^{II}}_{j-1} + UR^I_{i,j}\, h^{L^I}_i + UR^{II}_{l,j}\, h^{L^{II}}_l \\ & - (L^I_j + U^I_j + UR^I_{j,q})\, h^{L^I}_j - (L^{II}_j + U^{II}_j + UR^{II}_{j,r})\, h^{L^{II}}_j = \frac{d\,(HU^L_j\, h^L_j)}{dt} \end{aligned} \tag{8}$$

FRANCIS WEIR FORMULA:

$$FR_j = HU^L_j - \bar{\rho}_{mol} \cdot A_j\,(h_{w,j} + C_j\, \bar{L}^{2/3}) = 0 \tag{9}$$

and controller equations (eg PID-controller).

In the case of a homogeneous liquid phase, the upper equation system would be dependent. Therefore equation (7) is replaced by

$$MB_j = L^{II}_j = 0 \tag{7b}$$

Consequently, we obtain $(4NK + 5)NST$ equations for a column with NST stages and NK components with the same number of variables.

For reboiler and condenser the enthalpy balances were exchanged for specification equations, considering eg a given distillate flow and reflux rate or desired component purities. The correlations D_i describing the controllers were added to the equation system and a corresponding number of new variables Y_i is introduced. This method is explained by Wozny *et al*, 1987 for the two phase system. Special constructions like eg phase separators removing one liquid phase completely are considered in the same way.

SOLUTION ALGORITHM

The presented model consists of a system of algebraic and differential equations. First the latter ones are modified using the implicit Euler's method. Taking the mass balance as an example, we obtain the following equation:

$$\frac{d\,HU_j^L}{dt}\Big|_t = \frac{HU_j^L\,(t+\Delta t) - HU_j^L\cdot(t)}{\Delta t} = MB_j\,(t+\Delta t) \tag{10}$$

In order to apply this method, it is necessary to know starting values at the time $t = 0$. In that manner all differential equations were transformed into algebraic ones and solved by the Newton-Raphson-method.

Defining the following variable vectors

$$\overline{X} = (\overline{X}_1,\ \overline{X}_2\ \ldots\ \overline{X}_{NST}\ ,\ \overline{Y})^T \tag{11}$$

$$\overline{X}_j = (y_{1,j},\ \ldots,\ y_{NK,j},\ y^*_{1,j},\ \ldots,\ x^I_{1,j},\ \ldots,\ x^{II}_{1,j},\ \ldots,\ V_j,\ L^I_j,\ L^{II}_j,\ T_j,\ HU_j)^T$$

and the functional vectors

$$\overline{F} = (\overline{F}_1,\ \overline{F}_2\ \ldots\ \overline{F}_{NST},\ \overline{D}) \tag{12}$$

$$\overline{F}_j = (CB_{1,j}\ \ldots,\ E^{VL}_{1,j},\ \ldots,\ E^{LL}_{1,j},\ \ldots,\ MC_{1,j},\ \ldots,\ S1_j,\ S2_j,\ MB_j,\ H_j,\ FR_j)^T$$

the problem can be written as

$$\overline{F}\,(\overline{X}) = 0 \tag{13}$$

In the case of the steady state problem, that is the time $t = 0$, the equation system consists of equations (1)–(8) without the derivatives. For later times, however, equation (9) is included and the differential terms are simplified according to equation (10). These terms are now written on the other side of the equation, in order to obtain the form of equation (13).

The application of a Taylor series at a point $\overline{X}^o$ truncated after the linear term leads to

$$\overline{J}\cdot(\overline{X} - \overline{X}^o) = -\overline{F} \tag{14}$$

As far as possible, derivatives included in the Jacobian matrix are calculated analytically. For a constant time, the system (14) is solved with a modified Gauss-algorithm for mostly tridiagonal matrices with additional row and column vectors for $\overline{D}$ and $\overline{Y}$ and a new variable vector $\overline{X}$ is determined from the previous one $\overline{X}^{\circ}$. This iteration is repeated until convergence is achieved.

INITIAL VALUES

The first step of the algorithm consists in solving the steady state problem at the time $t = 0$. For this calculation initial values for the Newton-Raphson-procedure must be provided. Our program includes an algorithm for an automatic generation of initial values. This procedure is based on simplifying assumptions like constant molar overflow for the determination of the flow rates. The temperature profile is supposed to be linear between the dewpoint and boiling point of the mean feed concentration. Component balances, and vapour-liquid-equilibrium with constant K-values are used to determine the mean liquid and vapour composition. Murphree tray efficiency is not yet included.

PHASE STABILITY

As mentioned previously, the applied equation system, especially equation (7), depends on the number of liquid phases. Therefore, the phase pattern in the column must be fixed at each iteration step.

For the determination of the phase pattern, the liquid of each tray is examined for a possible phase split. This is carried out by looking for the non-trivial solution of the balances and isoactivity equations. Initial values are calculated by Shah's method (Shah, 1980) and then the equation system is solved by the Newton-Raphson-technique.

An incorrect number of liquid phases has two possible effects on the column simulation: If one or more trays are considered to be homogeneous, although they are heterogeneous, the algorithm finds convergence but a wrong solution. The other way round, the algorithm would diverge (Buzzi Ferraris, Morbidelli, 1981). In order to avoid these mistakes, the phase pattern test is applied between some iteration steps. When a convergent solution is obtained, the phase pattern is checked to confirm the result.

SUMMARY OF THE ALGORITHM

Figure 3 illustrates the most important steps of our algorithm. After the input of the problem describing parameters, initial values are automatically created. The steady state problem is then solved with the Newton-Raphson-technique. Then the time is increased by Δt and the corresponding solution is similarly created. In this way the change of the column variables is determined step by step.

RESULTS

LITERATURE EXAMPLES

In order to test our algorithm, it was applied on several steady state examples of three phase distillation column described in literature. In cases, where the model parameters for NRTL or UNIQUAC were indicated, the results could in general be successfully simulated.

As an example the comparison with the result of Block, Hegner, 1976, simulating a water-propanol-butanol column with 12 stages, is illustrated in Table 1. It can be seen, that our results are in very good agreement with the prior result. Our algorithm using the automatical determination of initial values needs six iterations and 33 s CPU time on a Microvax II, to obtain the convergent result.

EXPERIMENTAL RESULTS

The presented algorithm was also applied to verify some experimental results of an industrial production plant. The analysed example was a high pressure column distillating a four component mixture. The components named in order of increasing boiling points with A, B, C and D, have several binary immiscibilities and strongly different boiling points. A and B have low boiling points, which are almost identical, while C and D have significant higher ones.

In spite of the complex component behaviour, the separation result of the steady state column was well predicted. Figure 4a illustrates the comparison of the calculated concentration profile in the column and the measured top and bottom product purities. The involved temperature profile is illustrated in Figure 4b. Temperatures measured here are indicated, as well. Again good agreement with calculated values was found. The immiscible region is almost identical with the region, where only small changes in concentration and temperature occurred. Detailed results of a steady state sensitivity analysis, that shows regions with high sensitivity, can be found in Kruse (1991).

DYNAMIC SIMULATION

In the following examples the dynamic responses of the column to feed upsets are presented. That kind of sensitivity studies provides important information for the design of control structures, and furthermore facilitates identification and sensor-selection. Furthermore closed-loop responses are demonstrated.

Example 1

In the first example the water-propanol-butanol-column was examined. Data of this example are summarized in Table 2.

The dynamic changes of this column caused by a 40% step increase in feed flow rate was studied with a constant reboiler steam and reflux ratio. The volumetric holdups at the condenser and reboiler were kept constant. This system may be regarded as an open-loop structure. The changes of the temperature and mean liquid composition profiles are shown in Figure 5. The maximum temperature change was detected on tray 8. From that point of view stage 8 would be suitable for a temperature control. The water concentration at the reboiler increases during the first minutes and then decreases slowly to the final value (Figure 7). This behaviour could be interpreted as an effect of washing out by the increased feed flow rate.

A control structure, consisting of two loops was implemented. The loops are tray temperature controls on stage 3 and 8, manipulating the distillate flow and reboiler steam, respectively. The column response to 40% step increase of the feed under closed loop operation is shown in Figure 6. The change of the concentration profile here is much smaller than in the open loop case. In this and other control cases, the liquid at the condenser has to be subcooled in order to avoid overheated states at the column top. This easily appears while regulating the heating vapour at the reboiler with a model supposing constant pressure inside the column.

The phase pattern at the beginning shows phase splitting on stage 8 to 12. In the open loop case, the splitting disappears during the first 1.25 hours while in case of closed control loops it could be kept on stage 9 to 12.

Example 2

A seven stage distillation column for the separation of a water-butanol-butylacetate mixture is analysed. It is similar to the one mentioned in Block, Hegner, 1976. The data are summarised in Table 3.

The examined column is equipped with a decanter at the top, where the water-rich phase is withdrawn. Figure 10 illustrates this structure. If one decanter phase should be refluxed and the other withdrawn, their ratio is fixed by the mean liquid concentration and a degree of freedom is lost. Therefore it is useful to reflux a part of the distillate phase α, in order to gain a degree of freedom for the column control. For high reflux ratios this measure can be necessary to keep the distillate flow smaller than the feed stream.

A step decrease in feed flow of 20% is supposed, keeping a constant reboiler steam and reflux ratio of water-rich phase α. The resulting temperature and concentration response is demonstrated in Figure 8. Big changes inside the column and smaller ones at the top and bottom are stated.

Using a single temperature control loop, that manipulates the waterphase-reflux ratio α, the profile changes are kept near the desired operating point (Figure 9). The reboiler heat steam is still constant, just so the decanter and reboiler volumetric holdup.

The steady state operating point shows phase splitting only in the decanter. Analysing the results of the open-loop operation immiscibility was also only found at the top. The closed-loop response however increases the number of heterogeneous trays up to stage 4. Nevertheless, this does not affect the operation in a negative way, as it could be seen in Figure 9.

SUMMARY

A model for the steady state and dynamic simulation of three phase distillation was presented. The algorithm includes special industrial requirement, shows a high component flexibility and is user-friendly. Only design variables and specification values must be provided. Initial values are generated automatically. Examples of the column responses on step feed changes illustrate, that the algorithm is a powerful tool for the control structure design, identification and sensor choice for the two and three phase distillation. Furthermore it is able to test the chosen control structures afterwards.

NOMENCLATURE

A	effective cross section
B	bottom flow rate
C	height over weir coefficient
D	distillate flow rate
F	feed rate
h	enthalpy
h_w	weir height
HU^L	liquid holdup
K	equilibrium constant
L	liquid flow rate
NK	number of components

NST	total number of stages
Q	heat flow rate
T	temperature
U	liquid side stream
UR	liquid recycle stream
V	vapour flow rate
W	vapour side stream
WR	vapour recycle stream
x	liquid mole fraction
y	vapour mole fraction
z	feed mole fraction
Δ	difference
β	convergence accelerator
ρ	density

SUPERSCRIPT

-	vector
=	matrix
$\bar{}$	medium value
*	at the liquid surface
I	first liquid phase
II	second liquid phase
LL	liquid-liquid
L	liquid
T	transformed
V	vapour
VL	vapour-liquid

SUBSCRIPT

k	component number
j	stage number

REFERENCES

Block, U. and Hegner, B., 1976, *AIChE J*, 5: 582–589.

Bril, Zh.A., Mozzhukhin, A.S., Petlyuk, F.B. and Serafimov, L.A., 1975, *Theo Found Chem Eng*, 6: 761–770.

Buzzy Ferraris, G. and Morbidelli, M., 1981, *AIChE J*, 6: 881–888.

Cairns, B.P. and Furzer, I.A., 1990, *Ind Eng Chem Res*, 1364–1382.

Gorak, A., Toelle, F.-J., Wozny, G., Jeromin, L. and Vogelpohl, A., 1991, *Proc of COPE '91, Barcelona.*

Kruse, C., 1991, Studienarbeit (Universität Clausthal).

Landwehr, B. and Wozny, G., 1992, *European Symposium on Computer Aided Process and Engineering — 1, Elsinore, Denmark*, 395–402.

Pham, H.N. and Doherty, M.F., 1990, *Chem Eng Sci*, 7: 1845–1854.

Rovaglio, M. and Doherty, M.F., 1990, *AIChE J*, 1: 39–52.

Shah, V.B., 1980, *Thesis* (Toledo).

Wong, D.S.H., Jang, S.S. and Chang, C.F., 1991, *Computers Chem Engng*, 5: 325–335.

Wozny, G., Witt, W. and Jeromin, L., 1987, *Chem Eng Technol*, 10: 338–348.

Appendix

Fig. 1: Model of a Column

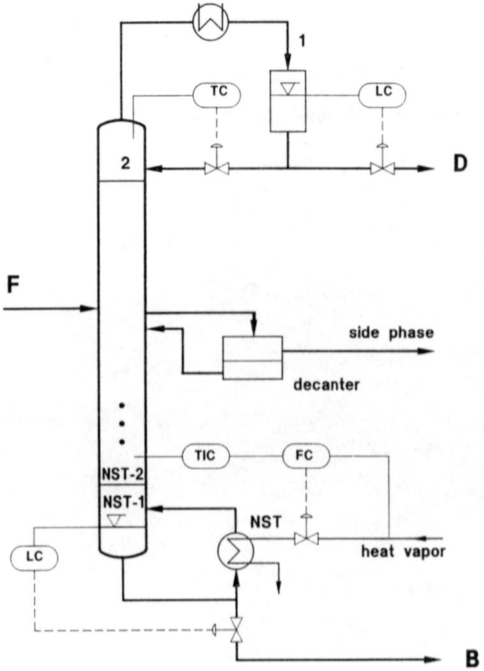

Fig. 2: Model of one Stage

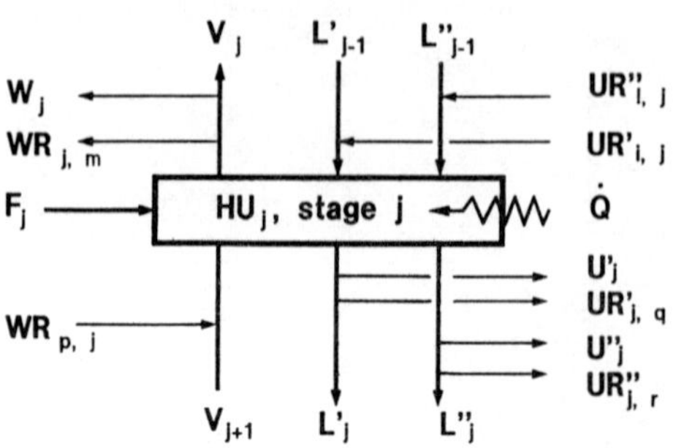

Fig. 3: Flow Diagram for the Dynamic Three Phase Distillation

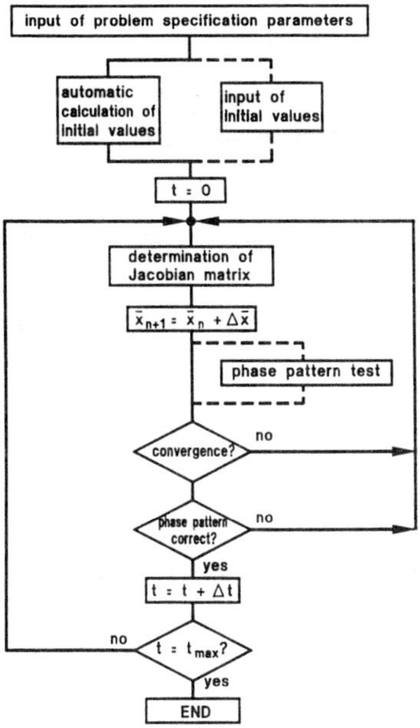

Results of Block, Hegner

stage	T	$\bar{x}(1)$	$\bar{x}(2)$	$\bar{x}(3)$
1	88.4	0.620	0.350	0.031
2	88.9	0.647	0.295	0.058
3	89.5	0.673	0.243	0.085
4	90.1	0.691	0.202	0.107
5	90.5	0.689	0.180	0.130
6	90.9	0.702	0.152	0.146
7	91.3	0.713	0.125	0.162
8	91.7	0.722	0.102	0.177
9	92.0	0.729	0.081	0.190
10	92.2	0.735	0.063	0.202
11	92.4	0.740	0.047	0.212
12	92.6	0.692	0.041	0.267

Our Algorithm

T	$\bar{x}(1)$	$\bar{x}(2)$	$\bar{x}(3)$
88.2	0.618	0.351	0.031
88.8	0.647	0.295	0.058
89.4	0.674	0.241	0.086
90.0	0.692	0.199	0.109
90.5	0.690	0.178	0.132
90.9	0.704	0.149	0.148
91.3	0.714	0.123	0.163
91.7	0.723	0.099	0.178
92.0	0.730	0.078	0.191
92.2	0.736	0.061	0.203
92.4	0.741	0.046	0.213
92.6	0.694	0.039	0.267

Tab. 1: Comparison with Literature Example of the System Water (1) - Propanol (2) - Butanol (3)

Fig. 4: Comparison of Calculated and Experimental Data
a) concentration profile

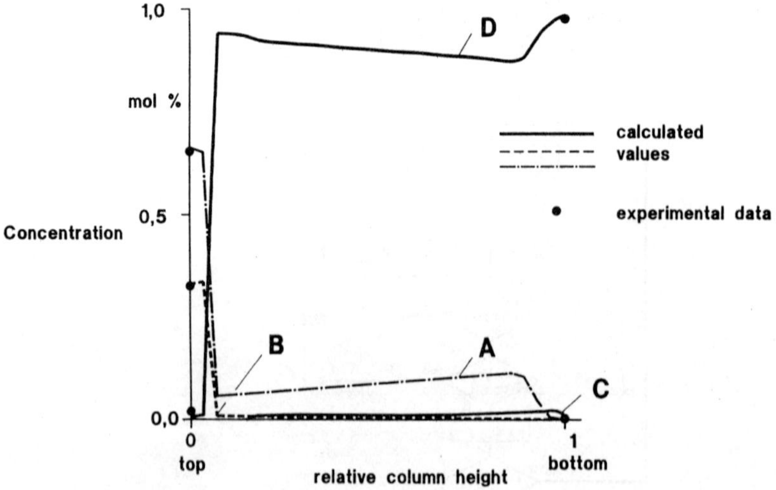

b) temperature profile

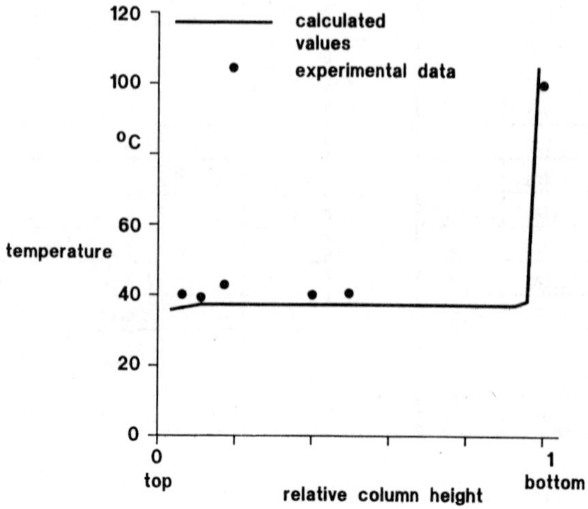

Fig. 5: Open-loop Temperature and Composition Response to a Step Increase in Feed Flow Rate of 40 % (Example 1: Water - 1-Propanol - 1-Butanol)

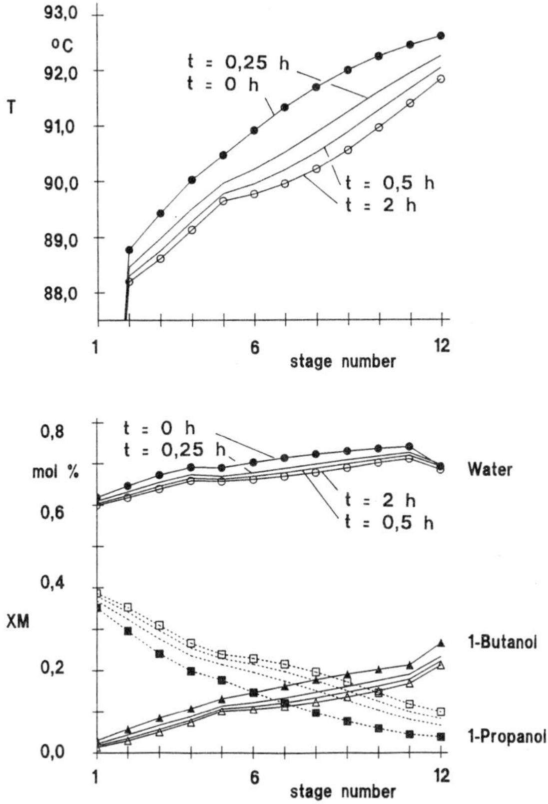

Fig. 6: Closed-loop Response to a Step Increase in Feed Flow Rate of 40 % (Example 1)

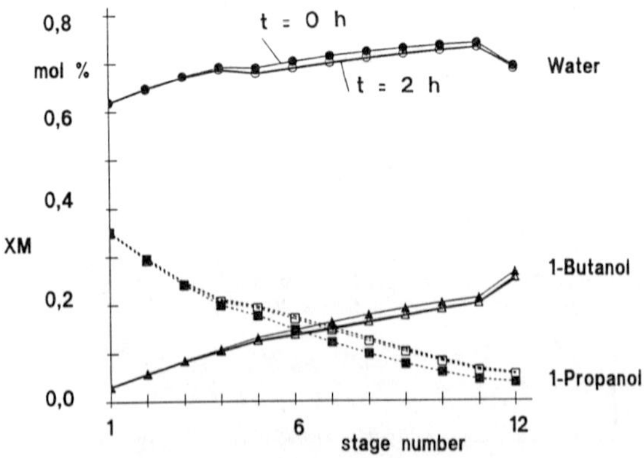

system	1-water 2-propanol 3-butanol
stage number	12
feed stage feed flow rate feed concentr.	5 50 kmol/h 0.65/0.22/0.13
distillate flow	29 kmol/h
reflux rate	3.0
tray parameter: effective area weir heigh coefficient	 A = 0.8 m^2 hw = 0.05 m c = 0.1

Tab. 2: Parameters for Example 1

system	1-butanol 2-water 3-butylacetat
stage number	7
feed stage feed flow rate feed concentr.	2 50 kmol/h 0.24/0.30/0.46
distillate flow	15.1 kmol/h
reflux rate	0.91
tray parameter: effective area weir heigh coefficient	 A = 0.2 m^2 hw = 0.05 m c = 0.5

Tab. 3: Parameters for Example 2

Fig. 7: Dynamic Change of Water Concentration in the Reboiler, for a Step Increase in Feed Flow Rate of 40 % (Example 1)

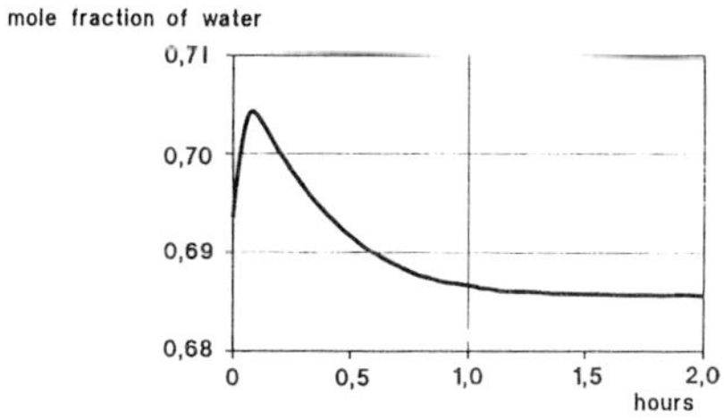

Fig. 8: Open-Loop Temperature and Composition Response to a Step Increase of Feed Flow Rate of 40 % (Example 2)

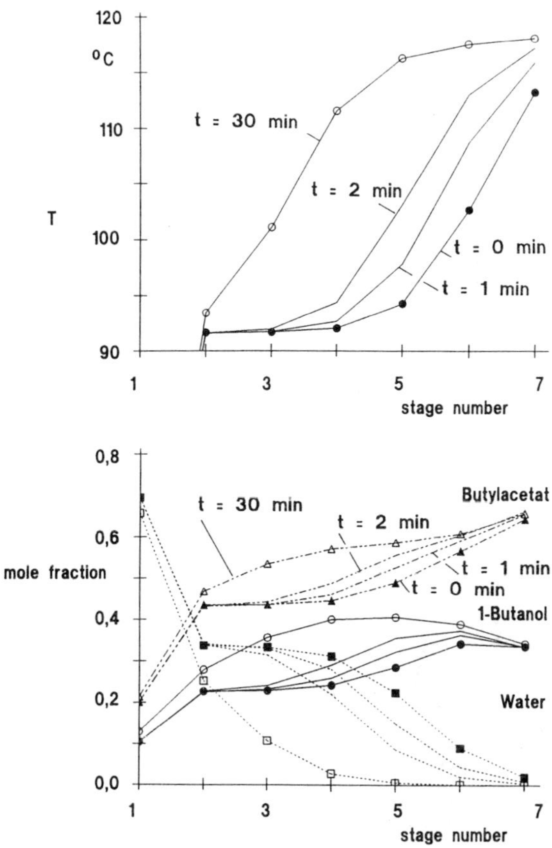

Fig. 9: Closed-loop Temperature and Composition Response to a Step Increase of Feed Flow Rate of 40 % (Example 2)

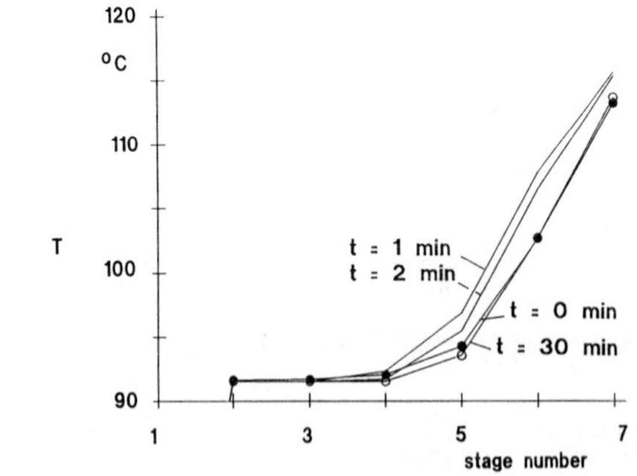

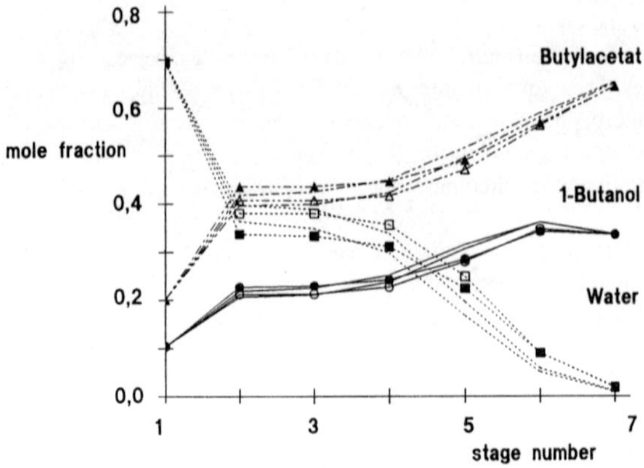

Fig. 10: Model of a Decanter at the Top of the Column

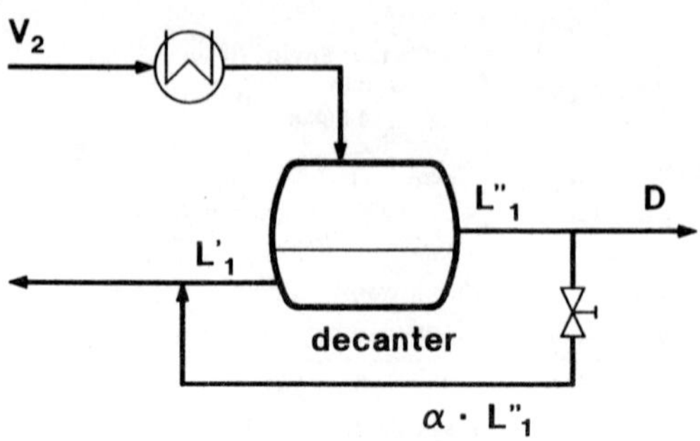

FILM ABSORPTION ACCOMPANIED BY SURFACE CONVECTION. NONLINEAR THEORY.

L.M. Rabinovich and A.V. Vyazmin
Karpov Institute of Physical Chemistry, Ul.Obukha 10, Moscow 103064, Russia

In this paper we used the method of small parameter to obtain a nonlinear evolution equation that describes convection in thin films appearing in absorption and chemisorption processes due to Marangoni effect. Using this equation we investigated the instability of laminar flow. The parametric regions of convection formation are determined and hydrodynamic behaviour of fluid flows are discussed.

Keywords: evolution equation, Marangoni effect, interfacial instability, nonlinear analysis.

INTRODUCTION

The problem of acceleration of interfacial film mass transfer is of principal importance in the intensification of gas absorption and chemisorption processes. Surface convection, that is due to variation of the surface tension with the concentration of the transferred substance[1,2], may highly intensify a lot industrial processes. The hydrodynamic characteristics of convection can be greatly influenced by accompanying heterogeneous and homogeneous chemical reaction[3]. In this case the mass transfer coefficient becomes dependent on the concentration driving force. Surface convection can change the hydrodynamic situation at the interface significantly and cause the intense renewal of the interface[4].

The forces originated from surface tension gradients along the interface bring about convective flows in the vicinity of an interface spreading due to viscosity into the bulks of the phases. According to the effect of nonlinear interactions the growing disturbances evolution may lead either to the formation of regular spatial / temporal structures or to the appearance of chaotic flow with a wide wave spectrum[5].

The most widespread liquid film flow regime is a wavy one. In wave structures development at a liquid film surface the nonlinear effect involving surface tension concentration dependence play an important role.

The paper presents the study of hydrochemical convection influence on the flow and mass transfer in films. A nonlinear evolution equation is obtained by the small parameter method. The stability of laminar flow is investigated. The parametric domains of wavy regimes are defined. The nonlinear analysis is based on the hypothesis about the possibility of surface perturbations stabilization due to nonlinear interactions[6,7]. This permits using the assumption about small amplitudes of stabilized structures and their closeness to harmonic structures.

The nonlinear method used in the present work is a development of the well-known in the theory of nonlinear hydrodynamic stability small parameter method. This method was previously successfully applied to the nonlinear analysis of thermocapillary convection[8], waves in thin liquid films[9,10] and a number of other processes encountered in chemical technology.

EVOLUTION EQUATION DEDUCTION

Consider laminar film flowing down the inclined plane. Free surface of the film contacts with motionless surrounding gas. Liquid consumption in the film is constant and equals Q. Let the incline angle γ, x-axis is directed downstream along the rigid surface and the film is supposed to be semiinfinite ($x \geq 0$); y-axis is directed upwards perpendicular to the inclined plane; $y = 0$ corresponds to the rigid surface. Free surface of the film is described by an unknown function $h = h(x,y)$.

We suppose that there is a mixture of substances above the film one of which is surface-active. Its concentration is constant and equals Co. This substance is soluble and may take part in heterogeneous chemical reaction. The rest of components are neutral and any case they don't influence hydrodynamics, diffusion and chemistry in the system under consideration. Thus we consider only transfer of the soluble component inside the liquid film.

We introduce further dimensionless variables and parameters. We consider the characteristic length of the surface waves l as the longitudinal scale and transversal length scale and velocity scale will be $h_o = (3\nu Q / g \sin\gamma)^{1/3}$ and $U_o = (g \sin\gamma / 3\nu)^{1/3} Q^{2/3}$ respectively, which correspond to the stationary waveless laminar film flow with liquid consumption Q. Introducing an unknown parameter l is widely used in hydrodymamics of thin liquid films[11]. This should emphasize the difference in the longitudinal and transverse length scales inherent to the problem. We will consider further only long waves $l >> h_o$ and consider the parameter $\varepsilon = h_o / l$ to be small: $\varepsilon << 1$.

If we neglect all the terms proportional to ε the problem is reduced to the well known problem of the waveless film creeping flow down the inclined plane without complex effects (concentration-capillary, heterogeneous chemical reaction, etc). In the subsequent analysis we restrict ourselves only to the terms ~ O(ε) and it will allows us to take into account all effects of interest. However, we neglect all the effects connected with the deformation of the free surface during the wavy film flow since we assume those effects to be small in comparison with the other ones. Hence we will not introduce the dimensionless Weber number and will not take into account the corresponding terms in equations and boundary conditions.

We therefore introduce the following dimensionless variables and parameters:

$$t' = \frac{u_o}{l} t, \quad x' = \frac{x}{l}, \quad y' = \frac{y}{h_o}, \quad U' = \frac{U}{u_o}, \quad V' = \frac{V}{u_o}, \quad C' = \frac{C}{C_o}, \quad \varphi = \frac{h - h_o}{h_o}$$

$$P' = \frac{Re}{\rho u_o} P, \quad Re = \frac{u_o h_o}{\nu}, \quad Pe = \frac{u_o h_o}{D}, \quad Ma = \frac{\sigma^o a C_o}{\mu u_o}, \quad Bi = \frac{q h_o}{D}, \quad Th = \frac{K h_o}{D}$$

Further we omit the primes denoting dimensionless variables.

The dimensionless equations are

$$\varepsilon Re \left(\frac{\partial U}{\partial t} + U \frac{\partial U}{\partial x} + V \frac{\partial U}{\partial y} \right) = -\varepsilon \frac{\partial P}{\partial x} + \frac{\partial^2 U}{\partial y^2} + 3 \; ;$$

$$\varepsilon Re \left(\frac{\partial V}{\partial t} + U \frac{\partial V}{\partial x} + V \frac{\partial V}{\partial y} \right) = -\frac{\partial P}{\partial y} + \frac{\partial^2 V}{\partial y^2} - 3\cot\gamma \; ;$$

(1)

$$\varepsilon \mathrm{Pe}\left[\frac{\partial C}{\partial t} + U\frac{\partial C}{\partial x} + V\frac{\partial C}{\partial y}\right] = \frac{\partial^2 C}{\partial y^2} \; ;$$

$$\varepsilon\frac{\partial U}{\partial x} + \frac{\partial V}{\partial y} = 0 \; ; \qquad \frac{\partial \varphi}{\partial t} + \frac{\partial}{\partial x}\int_0^{1+\varphi} U dy = 0$$

It should be noted that inequalities $\mathrm{Re} > \varepsilon$ and $\mathrm{Pe} > \varepsilon$ were used when deriving those equations.

We shall consider further only periodic or quasi-periodic two-dimensional flow. At $y = 0$ the conditions at the rigid boundary are fulfilled:

$$U = V = C = 0 \qquad (2)$$

The last condition (2) may have two physical interpretations. Firstly, the transverse diffusion rate is much less than the downstream flow rate. Secondly, there may be a kind of sink at the rigid surface leading to condition (2).

At the free surface $y = 1 + \varphi$ boundary conditions have to be fulfilled for equations (1) reflecting normal and tangential stress balance and soluble substance balance. This condition reflects the fact that the substance enters the interface from the gas phase, where it is absorbed as a result of heterogeneous reaction. Then it diffuses into the liquid film.

$$P + 2\varepsilon\frac{\partial U}{\partial x} + 2\varepsilon\frac{\partial \varphi}{\partial x}*\frac{\partial U}{\partial y} = 0 \; ; \qquad \frac{\partial U}{\partial y} + \varepsilon\frac{\partial V}{\partial x} = \varepsilon \mathrm{Ma}\left[\frac{\partial C}{\partial x} + \frac{\partial \varphi}{\partial x}*\frac{\partial C}{\partial y}\right] \; ;$$

$$\frac{\partial C}{\partial y} = \mathrm{Bi} - (\mathrm{Th} + \mathrm{Bi})C \; . \qquad (3)$$

We consider here surface tension to depend linearly on concentration. This is valid for small variations of the dissolved substance concentration as compared with its steady state value Co.

In order to solve problem (1–3) we use the small parameter method by means of which it is possible not only to obtain the evolution equation for liquid film thickness but to calculate also an approximate velocity profile and concentration distribution of the surfactant. We already have such a small parameter — ε, and we expand all variables in series on this parameter. Substituting expansion in equations (1) and boundary conditions (2, 3) and holding the terms up to the order of ε we obtain the corresponding problems for the zero and the first approximations.

The solution of the formulated problem for U and C are (if $\mathrm{Th} + \mathrm{Bi} >> 1$)

$$U = 3\left[(1 + \varphi) - \frac{y}{2}\right]y + \frac{\varepsilon \mathrm{MaBi}}{(\mathrm{Th} + \mathrm{Bi})^2 (1 + \varphi)^2}*\frac{\partial \varphi}{\partial x}y - \frac{3}{2}\varepsilon \mathrm{Re}\frac{\partial \varphi}{\partial t}\left[(1 + \varphi)^2 - \frac{y^2}{3}\right]y - 3\varepsilon\cot\gamma\frac{\partial \varphi}{\partial x}\left[(1 + \varphi) - \frac{y}{2}\right]y - \frac{3}{2}\varepsilon \mathrm{Re}\frac{\partial \varphi}{\partial x}(1 + \varphi)\left[(1 + \varphi)^3 - \frac{y^3}{4}\right]y \; ; \qquad (4)$$

$$C = -\frac{Bi\,y}{(Th + Bi)(1+\varphi)} + \frac{\varepsilon PeBi^2}{6(Th + Bi)^2(1+\varphi)^2}*\frac{\partial\varphi}{\partial t}y\left[y^2 - (1+\varphi)^2\right] + \tag{5}$$

$$+ \frac{3\varepsilon PeBi}{8(Th + Bi)(1+\varphi)}*\frac{\partial\varphi}{\partial x}y\left[y^3 - (1+\varphi)^3\right] - \frac{3\varepsilon PeBi^2}{40(Th + Bi)^2(1+\varphi)^2}*\frac{\partial\varphi}{\partial x}y*$$

$$*\left[y^4 - (1+\varphi)^4\right]$$

Substituting (4) into last equation in (1) we obtain the evolution equation. It describes the dependence of the film thickness φ on the longitudinal coordinate x and time t. Finally this equation may be written in the following form:

$$\frac{\partial\varphi}{\partial t} + 3(1+\varphi)^2\frac{\partial\varphi}{\partial x} + \frac{\varepsilon MaBi}{2(Th + Bi)^2}*\frac{\partial^2\varphi}{\partial x^2} - \frac{5}{8}\varepsilon Re(1+\varphi)^3\left[(1+\varphi)\frac{\partial^2\varphi}{\partial t\partial x} + 4\frac{\partial\varphi}{\partial t}*\right.$$

$$\left.*\frac{\partial\varphi}{\partial x}\right] - \varepsilon(1+\varphi)^2\cot\gamma\left[(1+\varphi)\frac{\partial^2\varphi}{\partial x^2} + 3\left(\frac{\partial\varphi}{\partial x}\right)^2\right] - \frac{27}{40}\varepsilon Re(1+\varphi)^5\left[(1+\right. \tag{6}$$

$$\left.+\varphi)\frac{\partial^2\varphi}{\partial x^2} + 6\left(\frac{\partial\varphi}{\partial x}\right)^2\right] = 0$$

Equation (6) is nonlinear and can not be solved analytically. Let's consider the stationary case, $\partial\varphi / \partial t = 0$. It may be easily found from (6) that $\varphi = 0$ is the exact solution. This result follows also from the solution of the evolution equation by means of the low parameter method, where the low parameter is ε. It is interesting to consider further the effect of surfactants on the stability of the film flow when $\varphi = 0$.

LINEAR STABILITY ANALYSIS

Now we proceed to linearization of the evolution equation. We seek the solution of equation (6) as a stationary solution ($\varphi = 0$) plus an infinitely small function of variables x and t. Neglecting the terms of order φ^n ($n \geq 2$) as compared to φ in equation (6) for φ, we obtain the following evolution equation for the waves with infinitely small amplitude:

$$\frac{\partial\varphi}{\partial t} + 3\frac{\partial\varphi}{\partial x} - \frac{5}{8}\varepsilon Re\frac{\partial^2\varphi}{\partial t\partial x} + \left[\frac{\varepsilon MaBi}{2(Th + Bi)^2} - \varepsilon\cot\gamma - \frac{27}{40}\varepsilon Re\right]\frac{\partial^2\varphi}{\partial x^2} = 0 \tag{7}$$

We assume perturbations as harmonic and the unknown function φ as its amplitude multiplied by a periodic function $\exp[i(\omega t - kx)]$. The complex frequency ω can be expressed in terms of the circular frequency Ω and the real increment α, ie $\omega = \Omega - i\,\alpha$. The solution which has an accuracy of the order of linear terms in ε can be expressed in terms of the surface wave phase velocity $V_p = \Omega / k$ as

$$\alpha = \left[\frac{6}{5}\varepsilon \mathrm{Re} - \varepsilon \cot\gamma + \frac{\varepsilon \mathrm{MaBi}}{2(\mathrm{Th} + \mathrm{Bi})^2}\right] k^2 \quad ; \qquad \gamma_p = 3 \tag{8}$$

Note that the surface phase velocity remains constant and is insensitive to physical and chemical properties or regime parameters.

The waveless flow of a liquid film with $\varphi = 0$ is stable when $\alpha < 0$ and unstable when $\alpha > 0$. Equation (8) shows that the sign of α corresponds to the sign of the multiplier enclosed in brackets in the right-hand side of equation (8). In the present case the increase in fluid velocity (Reynolds number) results in hydrodynamic instability of the system, so Reynolds number can be used as a parameter for the problem. The critical Reynolds number Re_{cr} corresponds to the stable-to-unstable mode transition

$$\mathrm{Re}_{cr} = \frac{5}{6}\left[\cot\gamma - \frac{\mathrm{MaBi}}{2(\mathrm{Th} + \mathrm{Bi})^2}\right] \tag{9}$$

This expression for critical Reynolds number correlates well with the results of the paper[11]. Equation (9) shows that critical Reynolds number is directly proportional to the Marangoni number. The basic physical explanation of this phenomenon was given by Levich[12] for the case of unsoluble surface-active substances. The influence of interfacial effects on the liquid film stability decreases monotonically with the heterogeneous chemical reaction rate. The reason is that interfacial chemical reaction level the surface-active substance concentration at the interface and decreases surface tension gradient along the interface.

The influence of the parameter Bi on the system stability is not well defined. When $\mathrm{Bi} \to 0$, the interfacial effects do not control the stability conditions of the liquid film free surface because the mass flux of surface-active substance at the interface diminishes. When $\mathrm{Bi} \gg 1$ the same effect holds, but in this case it is caused by the level of the surface-active substance concentration at the film free surface due to considerable supply rate from the gas phase. The quantity Re_{cr} as a function of Bi has a maximum. The result obtained gives an explanation to the data[13], where a maximum of functional relationship of Re_{cr} from the volumetric concentration of dissolved surface-active substances was found.

NONLINEAR STABILITY ANALYSIS AND CALCULATION OF SURFACE WAVES PROPERTIES

In order to examine the formation of waves on the surface of a falling liquid film, the convective patterns induced by these waves, we consider the case of $\mathrm{Re} > \mathrm{Re}_{cr}$ and $\varphi = 0$ corresponding to unsteady state. We introduce supercriticality $\zeta = \mathrm{Re} - \mathrm{Re}_{cr}$ and rewrite equation (6) in terms of parameter ζ. According to References 8–11, we postulate that nondimensional amplitude of surface waves is of order $\varphi^2 \sim \zeta$.

Let us suppose the dependent variable φ to take a form

$$\varphi = \sum_{n=-\infty}^{\infty} \Phi_n \exp[in(\omega t - kx)] , \qquad \omega = \Omega - i\alpha \tag{10}$$

where k, Ω and α arc real variables. It can be inferred from the condition that φ is real that $\Phi_{-n} = \Phi^*_n$, where '*' means complex conjugate. For periodic weakly nonlinear waves, we may suggest the amplitude of the first harmonic to be much greater than the amplitudes of the others. Leaving the terms of the first order of the quantity $\Delta = \Phi_1 \Phi_{-1} = |\Phi_1|^2 \sim \zeta$, we obtain equations corresponding to harmonics with n = 1 and n = 2. The equation for the second harmonic amplitude gives $\Phi_2 \sim \Delta$. The quantity Φo is determined from the requirement that nondimensional flow rate in the liquid film is equal to its prescribed unperturbed value. This results in $\Phi o = -2\Delta$. Now we obtain the nonlinear dispersion equation from the equation for the first harmonic:

$$\left[i - \frac{5}{8}\varepsilon(Re_{cr} + \zeta)k\right]\omega - 3ik + \varepsilon\left(\frac{15}{8}Re_{cr} + \frac{27}{40}\zeta\right)k^2 + \left\{\frac{5}{4}\varepsilon Re_{cr} k\omega + 9ik - 3\varepsilon\left(\cot\gamma - \frac{27}{40}Re_{cr}\right)k^2 - \frac{45i\varepsilon Re_{cr} k^2\omega - 36k^2 - 54i\varepsilon(\cot\gamma + 27Re_{cr}/40)k^3}{[2i - 2.5\varepsilon(Re_{cr} + \zeta)k]\omega - 6ik + \varepsilon(7.5Re_{cr} + 2.7\zeta)k^2}\right\}\Delta = 0 \tag{11}$$

In the linear theory we have $\zeta \to 0$ (from which it follows that $\Delta \to 0$), and equation (11) gives the values of α and Ω, calculated previously in (8). The nonlinear relations for α and Ω may be presented in the following form: $\alpha = \alpha o(k) + \alpha_1(k)\Delta$ and $\Omega = \Omega o(k) + \Omega_1(k)\Delta$, where $\alpha o(k) = \zeta k^2$ and $\Omega o(k) = 3k$. In the discussion below we restrict ourselves to the qualitative description of the system evolution scenarios at the nonlinear stage of instability rather than making quantitative prediction.

Consider now different types of nonlinear stability loss with respect to the disturbances with fixed wave number in the unstable region. If $\alpha_1(k) < 0$ the role of nonlinear effects becomes more important while the disturbances grow up and these effects stabilize the disturbances when $\Delta = -\alpha o(k) / \alpha_1(k)$. If $\alpha_1(k) > 0$ the nonlinearity leads to further destabilization of the system.

Stabilized periodical waves (and structures) appear only in this case. The steady state condition requires obviously increment α to be equal to zero. The periodic condition (the damping of all disturbances except one characterized by the given k) requires α as the function of k to have maximum at fixed squared amplitudes. Thus we have two equations for k and Δ characterizing the stabilized hydrodynamical wave structure:

$$\alpha_0(k) + \alpha_1(k)\Delta = 0, \qquad \frac{d\alpha_0(k)}{dk} + \frac{d\alpha_1(k)}{dk}\Delta = 0, \qquad \frac{d^2\alpha_0(k)}{dk^2} + \frac{d^2\alpha_1(k)}{dk^2}\Delta < 0$$

The solution k and Δ must be real and positive. Contrarywise the stabilized periodic waves may not exist, ie the instability develops with simultaneous generation of the waves with different wave numbers and interfacial turbulence appears.

It is natural to assume that there exist certain parametric regions where all the mentioned scenario of instability development will be realized. In order to find these regions as well as the characteristics of the surface waves (if they appear) unwieldy calculations have to be made which are the object of an other investigation. If the mentioned conditions are satisfied the solution allows one to calculate the amplitudes of surface waves φ from (10) and velocity and concentration fields from (4) and (5). As a result of calculation we may obtain the interfacial mass flux and enhancement factor of film absorption accompanied by surface convection.

CONCLUSIONS

The following qualitative conclusions may be drawn:

- the rate of the waves amplitude growth is inversely proportional to the supercriticity squared, when the latter is much less than unity;
- dissipative processes may lead to the stabilization of small waves in the supercritical region at the nonlinear stage of the system evolution;
- there exist two types of transition to instability in the systems with heterogeneous chemical reaction — the one resulting in the surface periodical waves, and the one leading to interfacial turbulence;
- this instability generation is also possible for chemical reaction with more complex kinetics.

The study presented seems to be important for development of theoretical foundations of the intensification of the many processes by use of interfacial hydrochemical instability phenomena.

REFERENCES

1. Hozawa, M., Komatsu, N., Imaishi, N. and Fujinawa, K., 1984, *J Chem Eng Japan*, 17: 173.
2. Akselrod, Yu.V., 1985, *Chem Industry*, 4: 237 (in Russian).
3. Buyevich, Yu.A., Vyazmin, A.V., Rabinovich, L.M. and Slin'ko, M.G., 1987, *Proc Acad Sci USSR*, 292: 170 (transl.).
4. Ruckenstein, E., 1968, *Chem Eng Sci*, 23: 363.
5. Buyevich, Yu.A., Vyazmin, A.V. and Rabinovich, L.M., 1989, *Proc Acad Sci USSR*, 307: 584 (transl.).
6. Landau, L.D., 1944, *Dokl Acad Sci*, 44: 311 (in Russian).
7. Hopf, E., 1948, *Comm Pure Appl Math*, 2: 303.
8. Buyevich, Yu.A., 1985, *Eng Phys J*, 48: 230 (in Russian).
9. Barrdahl, R., 1986, *AIChE Journal*, 32: 789.
10. Buyevich, Yu.A. and Kudimov, S.V., 1988, *Eng Phys J*, 54: 406 (in Russian).
11. Boyadjiev, Ch. and Beschkov, V., 1984, *Mass transfer in liquid film flow* (Publ House Bulg Acad Sci, Sofia), 128.
12. Levich, V.G., 1959, *Physico-chemical hydrodynamics* (Phys-Mathgiz, Moscow) (in Russian), 699.
13. Tailby, S.R. and Portalski, S., 1962, *Trans IChemE*, 40: 114.

NOMENCLATURE

a	surface activity
Bi	Biot number
C	concentration
D	diffusion coefficient
g	gravitational acceleration
h	free surface function
K	reaction rate constant
k	wave number
l	longitudinal scale
Ma	Marangoni number

P	pressure
Pe	Peclet number
Q	liquid consumption
q	gas side mass transfer coefficient
Re	Reynolds number
t	time
Th	Thiele number
U, V	components of velocity
V_p	phase velocity
x, y	coordinates
α	real increment
γ	angle
Δ	wave amplitued squared
ε	ho/l
ζ	supercriticality
μ, ν	viscosities of liquid
ρ	density of liquid
σ	interfacial tension
Φ	amplitudes in (11)
Ω	circular frequency
ω	frequency

GAS ABSORPTION WITH MEMBRANE PERMEATION — ACID GAS REMOVAL FROM FLUE GASES BY A PERMABSORPTION METHOD

Susumu Nii and Hiroshi Takeuchi
Department of Chemical Engineering, Nagoya University, Chikusa-ka, Nagoya 464-01, Japan

For removing carbon dioxide and sulphur dioxide from flue gases, a novel operation, 'permabsorption', was presented which is a hybrid operation of the membrane permeation and gas absorption. High permselectivity can be attained by using an appropriate absorbent. The performance of a polydimethylsiloxane hollow-fibre module was comparable to that of a packed column, and the present method can be expected to allow the possibility of operation at low liquid flow rate and therefore, of energy-saving gas separation.

Key words: Hollow fibre membrane, Gas absorption, Permeability, Carbon dioxide, Sulphur dioxide, Polydimethylsiloxane.

INTRODUCTION

One of main causes of global warming is regarded as an increase in carbon dioxide in the atmosphere as a result of human and industrial activities. It would be desirable to reduce the amount of CO_2 as well as sulphur dioxide emitted from the combustion of fossil fuels. Usually, the removal of such acid gases can be achieved by use of conventional gas-liquid contactors, in which high flowrate not only tends to increase overall mass-transfer coefficient but often the contact area as well. High throughput is desirable but limited, as in packed columns, because of flooding.

Membrane process for gas separation can provide lower operating costs, and have process flexibility without losing efficiency at variable operating capacities. In membrane permeation, however, pressure on the permeate (downstream) side must be regulated at a reduced pressure; thus, the membrane demands mechanical stability.

In this article, we have proposed a novel operation mode for membrane modules consisting of homogeneous, nonporous polymeric hollow fibre (HF). If an absorbent liquid flows on the permeate side, then gaseous species to be permeated dissolves into the liquid, subsequent to a reaction with the absorbent, whereby the driving force to mass-transfer across the membrane can be increased without reducing pressure. In addition, sparingly soluble gas cannot permeate through the membrane despite having high permeability. Thus such a hybrid operation mode of gas permeation and gas absorption with a chemical reaction makes it possible to separate a desired gaseous species from its mixture with high selectivity. We call this process 'permabsorption'. This view will be demonstrated in the present study with the object of confirming its applicability to the CO_2 and SO_2 removal from flue gas.

THEORETICAL BASIS FOR EXPERIMENTAL MEASUREMENTS

In permabsorption of gaseous species(A) by use of an HF module, gas and liquid phases flow on the shell and lumen sides of the HF, respectively. A permeant flux across the membrane at a steady state can be expressed as

$$J(A) = k_G(p-p_1)=k_S(\bar{C}_2-\bar{C}_1)=k_L(C_2-C)= (P(A) / \delta)(p_1-p_2) = K(p-p^*) \qquad (1)$$

where the overall mass-transfer coefficient, K, based on the gas phase is given as

$$1/K = 1/ k_G + \delta / P(A) + H / k_L \qquad (2)$$

The membrane permeability, P(A), can be observed by a method to be discussed below. The gas film mass-transfer coefficient for the laminar flow on the shell side is evaluated with a correlation developed by Takeuchi *et al*[7].

$$Sh = 0.85(d_S /d_W)^{0.45} (d_e / L)^{0.25} Re^{1/3} Sc^{1/3} \qquad (3)$$

For laminar flow inside solid tube, the liquid film mass-transfer coefficient can be expressed by the Leveque solution[4]

$$k_L d_i / D = 1.62 [(d_i / L)(Re{\cdot}Sc)]^{1/3} \qquad (4)$$

When a gas dissolves in liquid phase and reacts with a reactant(B) in the liquid, the mass transfer through the liquid film is chemically enhanced. Using the enhancement factor, E, in equation (1), the permeation rate is represented as

$$J(A) = E\, k_L (C_2-C) \qquad (5)$$

For the permabsorption of CO_2 in aqueous alkaline solution, the following reaction is of significance:

$$CO_2 + OH^- \rightleftharpoons HCO_3^- \qquad (6)$$

The E value can be obtained from a classical solution of gas absorption accompanied by a second-order chemical reaction[3]:

$$E = \gamma[1-(E-1)/(rq)]^{1/2} / \tanh \gamma[1-(E-1)/(rq)]^{1/2} \qquad (7)$$

where $r=D_B /D_A$, $q=C_{B0} /C_{A2}$, and $g= [kC_{B0}D_A]^{1/2} / k_L$.

In carbonate solution system, the E value can be evaluated from an analytical solution by Astarita and Savage[1]. The equilibrium partial pressure of CO_2 is also estimated[2] from $p^*=(4H/K_e)\, m\, X^2/(1-X)$, where m is the molarity of carbonate and X is its conversion.

For permabsorption system of SO_2–Na_2SO_3 solution, the SO_2 dissolved in the liquid reacts with the sulphite ion instantaneously:

$$SO_2 + SO_3^{2-} + H_2O \rightleftharpoons 2HSO_3^- \qquad (8)$$

The enhancement factor is given from the solution based on the film theory in the instantaneous reaction regime[3] as

$$E = 1 + r\,q \tag{9}$$

EXPERIMENTAL

Polydimethylsiloxane (PDMS)-HF membrane used in this study was supplied by Nagayanagi Ind Co Ltd (Japan), having the inside diameter of 200 μm and the wall thickness of 60 μm. A membrane module consisted of one hundred HFs, which were fixed with epoxy resin at both ends of teflon tube (1.7 cm ID and 30 cm long). The full details are available elsewhere[5].

The schematic diagram of the experimental apparatus is shown in Figure 1. The completed module was installed vertically. Gas mixture composed of N_2, CO_2 and/or SO_2 was introduced on the shell side at the top of the module. An aqueous solution of NaOH, K_2CO_3, or Na_2SO_3 was pumped on the lumen side at the bottom of the module via a peristaltic pump from a reservoir. After having reached a steady state, concentrations of CO_2 and of SO_2 in the gas phase were determined for the influent and effluent streams from the module on an infra red gas analyser. The permeation rate of A across the membrane, J(A), was calculated on the basis of the concentration difference in the gas phase. In all the experimental runs, the temperature was kept at 299±1K.

RESULTS AND DISCUSSION

GAS PERMEABILITY OF PDMS-HF MEMBRANE

A permabsorption test was conducted using a single PDMS-HF module for pure CO_2 and O_2 with aqueous solution of 1*M* NaOH and pyrogallol as the absorbent, respectively, whereby the amount of gas permabsorbed across the membrane was determined volumetrically using a soap film meter. Permeability of the PDMS-HF membrane for CO_2 and O_2 was determined from $P(A)=V(273/T)\,\delta/(Ftp)$, where T is temperature, V the volume of gas permeated through the membrane, F the contact area, and t the contact time. As for SO_2, the permeability was determined from the slope in a plot of $J(SO_2)$ versus the partial pressure under conditions that the permeation is in membrane controlling: $1/K = \delta/P$ in equation (2). The results obtained for CO_2, SO_2 and O_2 are shown in Table 1, and these values are comparable to literature values obtained by reducing pressure on the permeate side in conventional gas permeators. Such reasonable agreement leads to an expectation that the permabsorption method is available for measuring gas permeability of nonporous HF membranes in a simple operation mode.

For the separation of SO_2 or CO_2 from exhausting gas, the permeation of N_2 through the membrane should be kept as little as possible. In the present experiments, no appreciable permeation of N_2 through the membrane was observed. Therefore, the permabsorption using PDMS-HF membrane has a potential for selective removal of SO_2 or CO_2 from flue gas. The value of a permselectivity defined as $P(SO_2)/P(CO_2)$ is found to be 4 from Table 1.

EFFECT OF NaOH CONCENTRATION ON PERMABSORPTION FLUX

The CO_2 flux, $J(CO_2)$, obtained for various concentrations of NaOH are compared in Figure 2 with the theoretical values, as a plot of $J(CO_2)$ vs the partial pressure of CO_2 together with the respective calculated lines. The observed values for 1*M* NaOH locate on the dashed line with an assumption that the permeation is in membrane controlling. In Figure 2 the values of $J(SO_2)$ for $1M$ Na_2SO_3 are also plotted against $p(SO_2)$, being observed to be controlled solely by the membrane resistance.

Figure 3 shows the effect of liquid velocity, u_L, inside the HF on an apparent value of K with an assumption that $p^*=0$. The K values for the aqueous solutions having smaller C_{B0} tend to decrease drastically with decreasing u_L. It is to be noted that hydroxide ion in the liquid was depleted by the reaction with CO_2 at smaller values of C_{B0} and of u_L. This leads to lowering of the E value given by equation (7) and to rise in the equilibrium pressure of CO_2, p^*, in the liquid bulk (back pressure); thus, the resultant K value can be regarded as apparent.

PERMABSORPTION OF CO_2 BY USE OF CARBONATE SOLUTION

For practical purpose of flue gas decarbonization, it may be sufficient to remove about 70% of the CO_2 as a minimum. Thus, carbonate solution is rather preferable to other absorbent liquid owing to lower energy for regeneration of the rich solution. During the permabsorption of CO_2, carbonate ion is partially converted to bicarbonate. Figure 4 shows the effect of the carbonate-to-bicarbonate ratio on the $J(CO_2)$. When the ratio is less than 1, the flux decreases significantly. It implies that the absorbent liquid must be regenerated at a carbonate conversion of 0.33 under the present conditions.

PERMSELECTIVITY OF SO_2 TO CO_2

Simultaneous permabsorption of SO_2 and CO_2 from a gaseous feed (10%CO_2+0.5%SO_2) was conducted with the variation of u_L. With decreasing liquid velocity, the CO_2 flux decreased remarkably in the same manner as in Figure 3; whereas, the SO_2 flux was not so much influenced by u_L, because the amount of sulfite ion in the liquid was enough in comparison with the SO_2 permeated through the membrane. Thus we can expect that the selective removal of SO_2 to CO_2 is attained when the liquid flowrate is low and the sulphite concentration as well.

Figure 6 shows the effect of the partial pressure of SO_2 on both SO2 and CO2 fluxes for the simultaneous permabsorption in 0.01M Na_2SO_3 at u_L of 2 cm/s, indicating that the presence of SO_2 suppresses the CO_2 flux. Here we define the selectivity of SO_2 to CO_2 as

$$\sigma(SO_2/CO_2) = [J(SO_2) / p(SO_2)] / [J(CO_2) / p(CO_2)] \quad (10)$$

In Figure 6 the σ values are also plotted against the SO_2 partial pressure at p(CO2) of 5, 10 and 17 kPa. The SO_2 selectivity varied with $p(SO_2)$ unusually, depending on p(CO2). Such complicated behaviour may be due to the difference in the membrane resistance to the permeation as well as the reaction kinetics of each gas in the liquid. In the present system, the following reactions are likely to occur in the liquid in addition to the Reaction (8).

$$CO_2 + SO_3^{2-} + H_2O \rightleftharpoons HSO_3^- + HCO_3^- \quad (11)$$

$$SO_2 + HCO_3^- \rightleftharpoons HSO_3^- + CO_2 \quad (12)$$

If there is no membrane between the gas and liquid phases and the gas-phase resistance is of no significance, then the value of s increases with a decrease in $p(SO_2)$ because of the increase in the enhancement factor. At lower $p(SO_2)$, however, the permabsorption of SO_2 is in membrane-diffusion controlling; also the rate of Reaction (11) is very slow. Thus, with decreasing SO_2 in the gas phase, value of σ must approach the permeability ratio $P(SO_2)/P(CO_2)$. On the other hand, as p(SO2) increases, the enhancement factor for SO_2 decreases. This leads to lowering of the σ value. It is to be noted that further increase in $p(SO_2)$ brings about the rise in σ owing to the contribution of Reaction(12) to the stripping of CO_2 as well as the enhancement factor for SO_2.

COMPARISON BETWEEN PERMABSORPTION AND GAS ABSORPTION

To illustrate how the permabsorption by use of HF module is advantageous of the CO_2 removal compared with gas absorption, a comparison of the mass-transfer performance was made between a PDMS-HF module and a packed column with Raschig rings[6]. Table 2 summarizes the result together with the design and operating parameters. An advantage of the permabsorption is appreciable for liquid flowrate: the lower flowrate makes it possible to reduce the energy for the solution circulation through a regenerator as well as the heat of regeneration of the rich solution. Furthermore, the membrane module is free from flooding, unlike the direct contact between gas and liquid phases. In conclusion, the permabsorption process is suitable for treating a large amount of acid gases, including a gas from coal gasification plants and flue gases as well.

CONCLUSION

Permabsorption method using polydimethylsiloxane hollow-fibre membrane was examined for the removal of CO_2 and SO_2 from gas mixtures containing N_2. It was found that an excellent permselectivity can be attained by selecting an absorbent liquid suitable for gaseous species to be removed and by controlling operating conditions. In a practical situations for CO_2 removal from flue gas, the liquid flowrate that required to achieve a desired removal is remarkably low. In addition, the permabsorption method has an advantage of the operability at high gas flowrate because of the indirect contact between gas and liquid flows.

NOMENCLATURE

a	interfacial area (surface area per volume ratio),	1/cm
$\underline{C}$	concentration in liquid phase,	mol/cm^3
$\overline{C}$	concentration in membrane phase,	mol/cm^3
D	diffusion coefficient,	cm^2/s
d_e	hydraulic equivalent diameter,	cm
d_i	inside diameter of hollow fibre,	cm
d_S	inside diameter of shell tube,	cm
d_W	wetted perimeter,	cm
J	permeation flux,	$mol/(cm^2s)$ or $cm^3(STP)/(cm^2s)$
H	Henry's constant,	$cmHgcm^3/mol$
K	overall permeation coefficient or mass-transfer coefficient,	$cm^3(STP)/(cm^2scmHg)$ or $mol/(cm^2skPa)$
K_e	equilibrium constant for the reaction between carbonate ion and CO_2	
k	reaction rate constant for Reaction (6),	$cm^3/(mols)$
k_G	gas film mass-transfer coefficient,	$cm^3(STP)/(cm^2sp)$

k_S	membrane-transfer coefficient,	cm/s
k_L	liquid film mass-transfer coefficient,	cm/s
L	hollow fibre length,	cm
P	permeability of the membrane,	cm^3(STP)cm/(cm^2scmHg)
p	partial pressure of CO_2 or SO_2,	kPa or cmHg
Q	volumetric flow rate,	cm^3/s
u_L	liquid velocity per cross sectional area of hollow fibre,	cm/s
u_G	superficial gas velocity,	cm/s

Greek Letters

δ	membrane thickness,	cm
ε	fractional voidage	

Subscripts

1	membrane surface contacting with gas (outer surface of hollow fibre)
2	membrane surface contacting with liquid (inner surface of hollow fibre)

LITERATURE CITED

1. Astarita, G. and Savage, D.W.,1980, *Chem Eng Sci*, 35, (3): 649.
2. Joshi, S.V., Astarita, G. and Savage, D.W., 1981, *AIChE Symp Series 77*: 202, 63.
3. van Krevelen, D.W. and Hoftyzer, P.J., 1948, *Rec Trav Chim*, 67: 563, 587.
4. Leveque, M.A., 1928, *Ann Mines Rec Mem L'Explotas Mines*, 13: 201.
5. Nii, S., Takeuchi, H. and Takahashi, K., 1992, *J Chem Eng Japan*, 25: 67.
6. Onda, K., Sada, E. and Takeuchi, H., 1968, *J Chem Eng Japan*, 1: 62.
7. Takeuchi, H., Takahashi, K. and Nakano, M., 1990, *Ind Eng Chem Res*, 29: 1471.

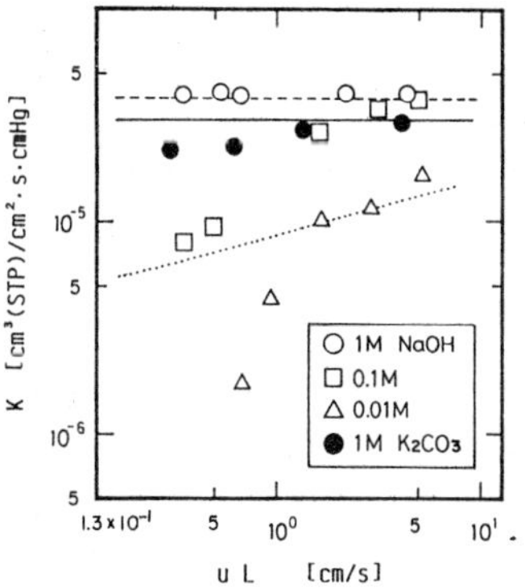

Fig.3. Effect of liquid velocity on overall permeation coefficient. Dashed, solid, and dotted lines represent the calculated values for membrane controlling, 1M K_2CO_3, and water, respectively. Feed stream contains 18% CO_2.

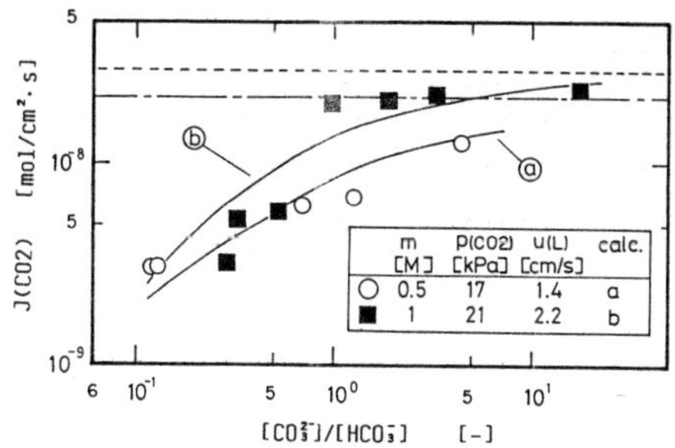

Fig.4. Effect of the carbonate-to-bicarbonate ratio in K_2CO_3 aq. on CO_2 flux. Solid lines are calculated value, dashed and dash-dotted lines being of membrane-controlling for m=0.5 and 1M, respectively.

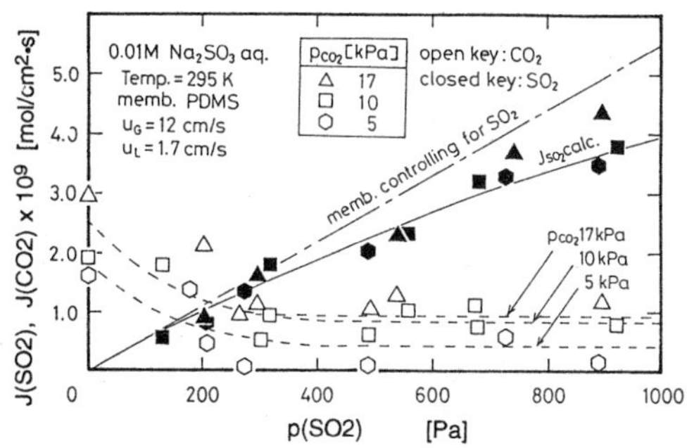

Fig.5. Effect of partial pressure of SO_2 on SO_2 and CO_2 fluxes. Dashed and solid lines represent calculated values on the basis of one reaction plane model. Dash-dotted line is of membrane diffusion controlling.

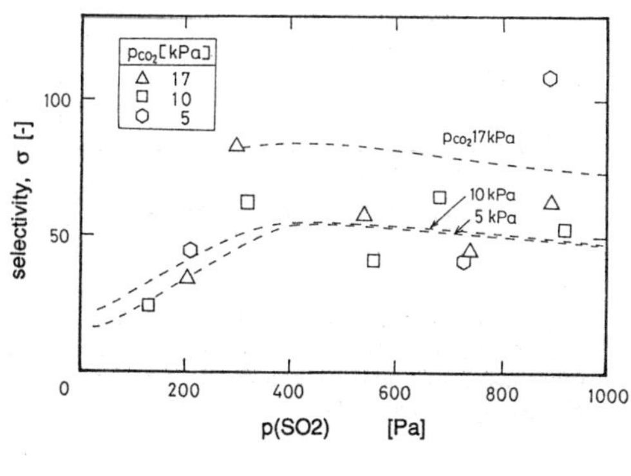

Fig.6. Effect of partial pressure of SO_2 on the selectivity of SO_2 to CO_2 permeation. Dashed line represents calculated value on the basis of one reaction plane model.

Table 1. Gas permeability of PDMS-HF membrane at 299K.

Gas	Permeability	Liter. value
CO_2	2.27×10^{-7}	3.0×10^{-7}*
O_2	3.49×10^{-8}	4.8×10^{-8}*
SO_2	9.8×10^{-8}	-
N_2	-	2.2×10^{-8}**

* From Hwang, S.-T. & K. Kammermeyer, "Membranes in Separations ", Wiley-Interscience, New York(1975).

** From Stern, S.A., F.J.Onorato & C.Libove, AIChE, J., **23**, 567(1977).

Table 2. Comrarison between PDMS-HF membrane module and packed column for CO_2 removal.

	PDMS-HFs membrane module	Packed column
Column diameter, cm	10	10
Packing	15000 PDMS-HF	12mm R-ring
ε	0.97	0.72
Gas velocity, g/(cm²·s)	0.192*	0.25
Liquid velocity, g/(cm²·s)	1.0	2.78
Q, cm³/s	4.71	218
a, 1/cm	20	1.20
K, cm³(STP)/(cm²·s·cmHg)	3.80×10^{-5}	6.80×10^{-4}
K*a*, cm³(STP)/(cm³·s·cmHg)	7.60×10^{-4}	8.19×10^{-4}

* Module length: 50cm, Percent removal of CO_2: 80.

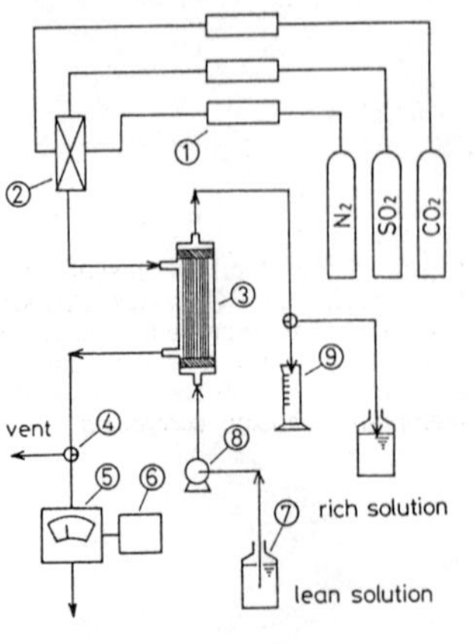

1 flow controller
2 gas mixer
3 module
4 valve
5 analyzer
6 recorder
7 reservoir
8 peristaltic pump
9 measuring cylinder

Fig.1. Schematic diagram of experimental apparatus.

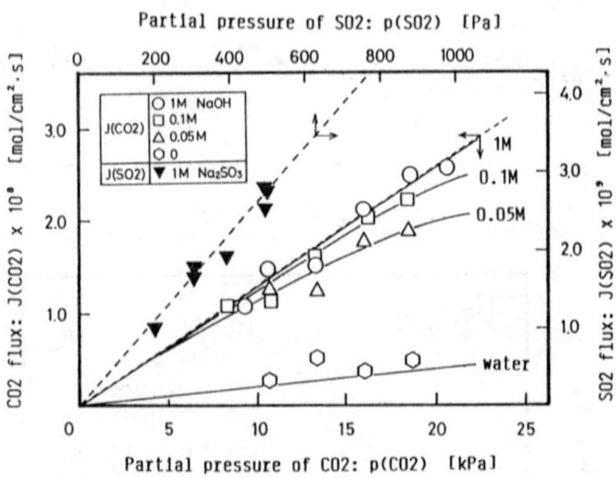

Fig.2. Effects of partial pressure of CO_2 and SO_2 and each permeant flux through the membrane in aqueous solutions of NaOH and Na_2SO_3, respectively.

MASS AND HEAT TRANSFER COEFFICIENTS FOR WATER ABSORPTION BY LITHIUM CHLORIDE

T.-W. Chung, T.K. Ghosh, and A.L. Hines
College of Engineering, University of Missouri-Columbia, Columbia, Missouri 65211

An absorber-stripper system was used to study the absorption of water vapour from moist air by contacting the air with an aqueous solution that contained from 28% to 40% lithium chloride. The absorption system was designed to handle a maximum air flow rate of 1.87 kg/m^{2}s and a maximum liquid flow rate of 15.18 kg/m^{2}s. For 1/2 inch ceramic Berl saddles, which have a surface to volume ratio of 142 ft^2/ft^3 (466 m^2/m^3), the operating air and liquid flow rates were varied from 0.75 to 1.20 kg/m^{2}s and 4.33 to 10.83 kg/m^{2}s, respectively. Mass and heat transfer coefficients were measured over a wide range of flooding conditions. For a solution containing 40% lithium chloride in water, mass transfer coefficients varied from 0.0938 kmol/m^{3}s [mole fraction] at 50% of flooding to 0.1420 kmol/m^{3}s [mole fraction] at 80% of flooding. These correspond to heights of a transfer unit of 0.28 and 0.29 m, respectively. Heat transfer coefficients at the same flooding conditions were 4.48 and 8.03 kw/m^{3}s. Column efficiencies were typically greater than 70%. The experimental data for ceramic Berl saddles are compared with the data for polypropylene Flexi rings.

INTRODUCTION

Liquid desiccant based air conditioning systems are finding increasing applications as humidity control devices in commercial buildings, schools, hospitals and supermarkets. These systems are competitive with traditional vapour refrigeration systems because of their efficient handling of the latent heat load. Air can be cooled and dehumidified to the desired condition by controlling the temperature, flow rate, and concentration of the aqueous liquid desiccant solution. As a result, further subcooling or reheating of the air prior to its use is not required. Another advantage of liquid desiccant systems is their effective utilization of low temperature heat sources for regeneration of the spent desiccant solution.

Aqueous lithium chloride is one of the most frequently used desiccants for dehumidification of air. Usually, a packed bed arrangement is employed to provide contact between the air and liquid. Several types of packing have been employed to provide the contact surface, including cross-corrugated cellulose-base structured packing, polypropylene Flexi rings, ceramic Berl saddles, and glass Raschig rings. However, very little experimental work is available in the open literature relating heat and mass transfer coefficients and heights of a transfer unit. Bichowsky and Kelley (1935) employed a 35% by weight lithium chloride solution in a column that was packed with 2 inch clay Raschig rings. The packing had a specific surface of only 27 ft^2/ft^3 (88.6 m^2/m^3). The mass transfer coefficients obtained in their study were in the range of 0.0145 to 0.02 lb of water/ft^{2}hmmHg (0.0736 to 0.1015 kmol/m^{3}s [mole fraction]) when the air velocity over the packing surface was in the range of 300 to 425 ft/min (1.52 to 2.16 m/s). The flow rate of the lithium chloride solution was maintained at 1.8 gallons per hour per square foot of the packing surface (1.22 L/min m^2, using 1 gallon = 3.8 L). The heat transfer coefficients were in the range of 3 to 5 Btu/hrft^2°F (1.52 to 2.53 kw/m^{3}K) under similar operating conditions. Leboeuf and Lof (1980) investigated the bed height required to achieve 3 tons of cooling as a function of air inlet temperature when employing a 40% lithium chloride solution. Using 1.5 inch Berl saddles, they obtained a mass transfer coefficient of 12,800 kg/m^{3}h[kg water/kg air] (0.32 kmol/m^{3}s [mole fraction]). The spent solution was regenerated by using solar energy. Lof *et al* (1984) used McAdams' correlation to predict the heat transfer coefficient in their system. A large deviation between the experimental data and predicted values was reported. Scalabrin and Scaltriti (1984) employed glass Raschig rings and polypropylene Pall rings as packings to study the dehumidification of air by a 40% lithium chloride solution. The flooding conditions in their column were in the range of 15% to 20%. This is significantly lower than the typical flooding conditions of 50% to 80% used by industry. The height of a transfer unit (HTU) was calculated from data provided by Gandhidasan and Satcunanathan (1983), who used 1.5 inch Berl saddles as a packing and a 41% lithium chloride solution. The HTU was in the range of 0.3 to 0.4 m. However, the packing height to column diameter ratio of their system was less than one. Using a 60 cm column that employed a 35% lithium chloride solution, Chen *et al* (1989) concluded that a polypropylene mesh type packing was better than polystyrene foam balls for dehumidifying air. They noted that the column efficiency increased with an increase in liquid flow rate, but then it reached a plateau at higher liquid rates. Recently, Chung *et al* (1992) studied the dehumidification of air in a column packed with 5/8 inch polypropylene Flexi rings. The column was operated from 50% to 80% of flooding. They noted that the packing surfaces were best utilized under these operating conditions as indicated by the height of a transfer unit.

The objective of the present study was to evaluate the performance of a packed bed absorber for the dehumidification of air when using 1/2 inch Berl saddles as the packing. The parameters that were investigated included flooding conditions, the temperature and humidity of the inlet air, temperature of the lithium chloride solution and its circulation rate, and concentration of the lithium chloride solution. Mass and heat transfer coefficients and the height of a transfer unit were calculated from the experimental data and were compared with the values for polypropylene Flexi rings and the literature data for similar types of packing.

EXPERIMENTAL SYSTEM

A schematic of the absorption-stripping system that was used in the present study is shown in Figure 1. A detailed description of the system was provided by Chung *et al* (1992). The absorber is a 6 inch internal diameter glass column packed with 0.5 inch Berl saddles to a height of 42 cm. To operate the absorption column between 50% to 80% of the flooding velocity, air flow rates in the range of 22.5 to 36 ft^3/min (38 to 61 m^3/h) were required. The humidity and temperature of the inlet air stream were maintained at the desired levels and were reproduced from one run to the next by flowing outside air through a column (referred to in Figure 1 as a pre-dehumidifier), where it was dehumidified by contacting a lithium chloride solution. The air was then humidified to a predetermined level in a Plexiglas chamber. The humidity and temperature of the air stream were controlled by changing the water temperature and amount of water sprayed into the system. After humidifying the air, it was introduced in the bottom of the absorber. The lithium chloride solution was sprayed on top of the packing material. The relative humidities of the inlet and outlet air streams were measured by using a Humicap-moisture sensor manufactured by Vaisala Inc. The temperature was measured by using thermocouples. The spent lithium chloride solution was heated to 180°F in a heat exchanger prior to spraying on top of the packing material in the stripper column. A small quantity of air was injected into the bottom of the column to facilitate the stripping process. The regenerated solution was then cooled and returned to the absorber. The concentrations of the lithium chloride solutions employed in this study varied from 28% to 40% by weight in water.

RESULTS AND DISCUSSION

The theoretical amount of lithium chloride solution required to dehumidify an air stream is usually very low. Based on equilibrium data, the minimum liquid flow rate necessary to dehumidify 35 ft^3/min (59.5 m^3/h) of air at standard conditions from 75% relative humidity (RH) to 40% RH was only 0.0007 gallons per minute (0.0026 L/min). Because this quantity was not sufficient to wet the packing surfaces effectively, the actual operating liquid flow rates were determined from the flooding conditions using the correlation of Eckert (1970). The operating conditions, air and corresponding liquid flow rates, and percent of flooding for each run are given in Table 1.

The performance of the absorption column was determined by evaluating its efficiency and other parameters, including heat and mass transfer coefficients and the heights of a transfer unit. The column efficiency was defined as the ratio of the actual change in moisture content of the air leaving the absorber to the maximum possible change in moisture content under a given set of operating conditions. Therefore, the column efficiency, ε, can be expressed as

$$\varepsilon = \frac{W_{in} - W_{out}}{W_{in} - W_{equ}} \qquad (1)$$

where W_{in} and W_{out} are the water contents of the inlet and outlet air streams, respectively, and W_{equ} is the water content of the air, which is at equilibrium with the lithium chloride solution at a particular concentration and temperature. Column efficiencies are plotted as a function of air and liquid flow rates in Figures 2 and 3, respectively. Lithium chloride solutions of 28% and 40% were used for experiments in which the air flow rate was varied. They were 31% or 37% for experiments in which the liquid flow rates were varied. The efficiency of the column decreased with an increase in air flow rate, but increased when the liquid flow rates were increased. However, in both the cases, the changes were less than 3%. Since the liquid flow rates investigated in this work were several times greater than values dictated by equilibrium considerations, the change in the column efficiency was expected to be a weak function of either flow rate. Similar results were observed with the polypropylene Flexi ring packing when the concentrations of the lithium chloride solution were greater than 35%. However, the efficiency dropped by more than 10% when the solution concentration was 30%.

The overall mass transfer coefficient was calculated from experimental data following the procedure given by Hines and Maddox (1985). A differential section of the packed column can be schematically represented as shown.

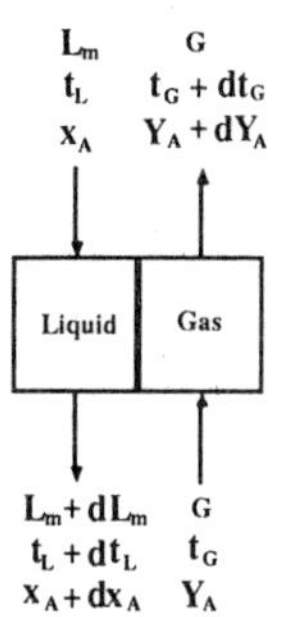

The molar flux of the system can be written as the product of an overall mass transfer coefficient and the difference between the bulk and equilibrium concentrations as

$$N_A = \frac{d(G\, y_A)}{a\, dZ} = \frac{K_{GA}}{\beta_{v-v^*}}(y - y_A^*) \tag{2}$$

where G is the molar flow rate of air. Since it can be assumed that the transfer of mass is taking place from air through a stagnant film of liquid, the bulk flow correction factor β_{v-v*} can be expressed as

$$\beta_{v-v^*} = (1 - y_A)_{*M} = \frac{(1 - y_A^*) - (1 - y_A)}{\ell n\left(\frac{1 - y_A^*}{1 - y_A}\right)} = \frac{y_A - y_A^*}{\ell n\left(\frac{1 - y_A^*}{1 - y_A}\right)} \tag{3}$$

After combining equations (2) and (3) and integrating, the overall mass transfer coefficient can be written as

$$(K_{GA}\, a)_{avg} = \frac{G}{Z}\int_{y_{A,t}}^{y_{A,b}} \frac{(1 - y_A)_{*M}}{(1 - y_A)} \frac{dy_A}{(y_A - y_A^*)} \tag{4}$$

The integration of equation (4) was carried out numerically by using Simpson's integration method. The height of a transfer unit is usually less dependent on liquid and gas flow rates and, therefore, provides a mean to compare systems at different flow rates. It is given as

$$H_{OG} = \frac{G}{K_{GA}\, a} \tag{5}$$

The heat transfer coefficients were calculated by assuming the column operated adiabatically and the interface temperature was equal to the temperature of the bulk liquid. A heat balance over a differential section of the column can be written as

$$GH_G - \{G\,[H_G + dH_G\,] - G\, dY_A\,[c_{pA}\,(t_g - t_o) + Q_{to}]\} = h_G\, a\,(t_G - t_i)\, dZ \tag{6}$$

The enthalpy of air is calculated from the expression

$$H_G = c_{p_{air}}\,(t_G - t_o) + Y_A\,[c_{pA}\,(t_G - t_o) + Q_{to}] \tag{7}$$

After substituting equation (7) into equation (6) and rearranging

$$- G\,(c_{p_{air}} + Y_A\, c_{pA}) \int_{t_{G_{out}}}^{t_{G_{in}}} \frac{dt_G}{t_G - t_i} = h_G\, a \int_0^Z dZ \tag{8}$$

Upon integration of equation (8), the heat transfer coefficient is obtained as

$$h_G a = \frac{G(c_{p_{air}} + Y_A c_{p_A})}{Z} \ln \frac{t_{G_{in}} - t_i}{t_{G_{out}} - t_i} \quad (9)$$

where the interface temperature, t_i, is assumed to be the average of the inlet and outlet liquid temperatures.

The overall mass and heat transfer coefficients calculated from equations (4) and (9) are plotted in Figures 4 through 7 as a function of the inlet air or liquid flow rates. Both transfer coefficients varied linearly with flow rates. The mass transfer coefficients increased with an increase in the concentration of lithium chloride in the solution. The ability to lower the dew point of air is directly proportional to the concentration of lithium chloride, which results in a greater driving force between the gas and liquid phases. Chung *et al* (1992) reported a similar result for polypropylene Flexi rings. The mass transfer coefficients obtained with the 1/2 inch Berl saddles were greater than the ones obtained using polypropylene Flexi rings as the packing. The mass transfer coefficients are compared in Table 2. The values obtained by Leboeuf and Lof (1980), who used 1.5 inch Berl saddles, are higher than those obtained in the present work. The values calculated from the data given by Gandhidasan and Satcunanathan (1983), who used 1.5 inch Berl saddles, were less than those obtained in this study. However, their total packing surface was about eight times greater than that in this study. This suggests that the packing surface of their column was not effectively utilized. The heights of a transfer unit calculated by Chung *et al* (1992), who used polypropylene Flexi rings, were approximately 10% higher than the values obtained for 1/2 inch Berl saddles. These results are shown in Figure 8. It should be noted that the 1/2 inch Berl saddles have a surface to volume ratio that is 1.4 times greater than that of 5/8 inch polypropylene Flexi rings. In this study, the total packing volume was constant. Therefore, the Berl saddles provided a greater surface area for gas-liquid contact, which resulted in a better column performance.

Heat transfer coefficients for two packings are compared in Figures 6 and 7 as a function of the air and liquid flow rates, respectively. The heat transfer coefficients decreased as the lithium chloride concentration increased. The viscosity of the lithium chloride solution increases with an increase of lithium chloride in the solution, which may have reduced the heat transfer rate between the two phases. As the liquid flow rate increased, the heat transfer coefficients also increased due to the higher rate of heat removal from the system. The heat transfer coefficients obtained using Berl saddles were larger. This may be due to the thermal characteristics of the packing. Selected physical characteristics of these two packings are given in Table 3.

Onada *et al* (1968) and Bolles and Fair (1982) developed mass transfer correlations for an absorption-stripping system, but these were intended for use when employing a pure solvent. Therefore, a large error was observed when these correlations were used to predict the mass transfer coefficients for the present system. For mixed solvents, such as lithium chloride-water or calcium chloride-water solutions, the change in the driving force due to the change in the solution concentration should be taken into consideration. Gandhidasan *et al* (1986) developed a correlation using the data from calcium chloride-water-air systems. Predicted values for the lithium chloride-water-air systems were greater than experimentally determined values by a factor of two when their correlation was used.

Lof *et al* (1980) used McAdams' empirical correlation to predict heat transfer coefficients for solar regeneration of an aqueous solution of lithium chloride in a column packed with 1 inch ceramic Raschig rings. Errors in the predicted values were in the range of 2% to 26%, which they attributed partly to experimental error and partly to the failure to maintain adiabatic conditions in the column. The differences between the experimental values and those predicted from the McAdams' correlation for this study were in the range of 3% to 44% with an average error of 14.4%. However, it should be noted that, the correlation of McAdams did not take into account the variation of the concentration of the salt in the desiccant solution. Since, the driving force for mass transfer is related to the concentration difference between the air and liquid phases, the concentration of the desiccant solution will play an important role.

CONCLUSIONS

Although both ceramic Berl saddles and polypropylene Flexi rings were inert to lithium chloride, the column packed with 1/2 inch saddles provided better contact between the two phases as indicated by the lower heights of a transfer unit. However, the total surface available for contact was 1.4 times greater for the Berl saddles than for the polypropylene Flexi rings. Several other factors should be considered when selecting a packing material. Polypropylene Flexi rings are lighter in weight and are less expensive than ceramic Berl saddles. Flexi rings provide a greater bed voidage that results in a lower pressure drop in the column. Also, the 6 inch absorption column used in the present study had an air handling capacity of 65 ft^3/min (110.5 m^3/h) at 80% of flooding when polypropylene Flexi rings were used as the packing. The capacity was only 36 ft^3/min (61 m^3/h) under a similar flooding condition when Berl saddles were employed.

NOMENCLATURE

a	Specific interfacial surface area, m^2/m^3
c_{pA}	Heat capacity of water vapour, kJ/kgK
c_{pair}	Heat capacity of dry air, kJ/kgK
e	Efficiency of the packed bed absorber, %
G	Superficial mass velocity of air, kmol/sm^2
h_G	Heat transfer coefficient, kw/m^3K
K_{GA}	Overall gas phase mass transfer coefficient, kmol/sm^2 [mole fraction]
t_i	Temperature at the gas-liquid interface, K
t_{Gin}	Temperature of air at the inlet, K
t_{Gout}	Temperature of air at the outlet, K
W_{in}	Water content of air at the inlet, kg moisture/kg dry air
W_{out}	Water content of air at the outlet, kg moisture/kg dry air
W_{equ}	Minimum possible water content at the outlet, kg moisture/kg dry air
Y_A	Mole fraction of water vapour in dry air, kmol/kmol dry air
y_A	Mole fraction of water vapour in the bulk phase, kmol/kmol of gas mixture
$y_{A,a}$	Mole fraction of water vapour at the top of the column, kmol/kmol of gas mixture
$y_{A,b}$	Mole fraction of water vapour at the bottom of the column, kmol/kmol of gas mixture
y_A^*	Equilibrium mole fraction of water vapour in the air, kmol/kmol of gas mixture
Z	Height of packing, m

REFERENCES

1. Bichowsky, F.R. and Kelley, G.A., 1935, *Industrial and Engineering Chemistry*, 27 (8): 880.
2. Bolles, W.L. and Fair, J.R., 1982, *Chemical Engineering*, 89: 109.
3. Chen, L.-C., Kuo, C.-L. and Shyu, R.-J., 1989, in *Proceedings of the Eleventh Annual ASME Solar Energy Conference*, Fanney, A.H. ed, San Diego, California, April 2–5, 1989.
4. Chung, T.-W., Ghosh, T.K. and Hines, A.L., 1991, *7th Symposium on Separation Science and Technology for Energy Applications, Tennessee, Oct 20–24*, 1991.
5. Eckert, J.S., 1970, *Chem Eng Prog*, 66 (3): 39.
6. Gandhidasan, P. and Satcunanathan, S., 1983, *Sol World Congr/Proc Bienn Congr Int Sol Energy Soc 8th Meeting*, 3: 1726.
7. Gandhidasan, P., Kettleborough, C.F. and Ullah, M.R., 1986, *ASME J of Solar Energy Engineering*, 108: 123.
8. Hines, A.L. and Maddox, R.N., 1985, *Mass Transfer Fundamentals and Applications*, Prentice-Hall, Inc, New Jersey.
9. Leboeuf, C.M. and Lof, G.O.G., 1980, *Proceedings of the Annual Meeting, American Section of the International Solar Energy Society*, 3 (1): 205.
10. Lof, G.O.G., Lenz, T.G. and Rao, S., 1984, *ASME J of Solar Energy Engineering*, 106: 387.
11. McAdams, W.H., Pohlenz, J.B. and St. John, R.C., 1949, *Chem Eng Processing*, 45 (4): 241.
12. Onda, K., Takeuchi, H. and KuMoto, Y., 1968, *J of Chemical Engineering of Japan*, 1 (1): 56.
13. Scalabrin, G. and Scaltriti, G., 1984, *La Termotecnia*, 38: 87.

Table 1. Experimental Data and Heat and Mass Transfer Coefficients

Percent of Flooding (%)	Air Flow Rate (ft^3/min)	Liquid Flow Rate (gal/min)	Air Inlet Temp (°F)	Air Outlet Temp (°F)	Air Inlet Humidity (kg H_2O/kg dry air)	Air Outlet Humidity (kg H_2O/kg dry air)	Liquid Inlet Temp (°F)	Liquid Outlet Temp (°F)	LiCl Conc (% wt)	Equilibrium Humidity (kg H_2O/lkg dry air)	Packing Height (cm)	Efficiency (%)	Mass Transfer Coefficient (kmol/m^3·s·[mole fraction])	Heat Transfer Coefficient (kw/m^3·K)	Height of Transfer Unit (m)
52	22.5	1.5	80.9	69.2	0.0144	0.0081	67.2	69.9	28	0.0057	42	72.4	0.0791	5.49	0.33
63	27	1.5	80.2	69.2	0.0139	0.0080	67.2	70.4	28	0.0057	42	72.0	0.0952	7.50	0.33
73	31.5	1.5	80.0	69.1	0.0148	0.0083	67.1	69.8	28	0.0057	42	71.4	0.1046	8.87	0.35
84	36	1.5	81.1	69.2	0.0150	0.0084	67.2	70.2	28	0.0057	42	71.0	0.1205	9.59	0.35
52	25	1.0	79.6	69.9	0.0122	0.0075	68.0	69.7	31	0.0058	42	73.4	0.0905	4.82	0.32
58	25	1.5	81.8	69.6	0.0127	0.0076	68.0	69.6	31	0.0058	42	73.9	0.0929	5.78	0.31
71	25	2.0	80.9	69.4	0.0125	0.0075	68.0	69.6	31	0.0058	42	74.6	0.0937	6.22	0.31
83	25	2.5	81.5	69.0	0.0126	0.0075	68.0	69.0	31	0.0058	42	75.0	0.0945	7.20	0.31
50	22.5	1.5	83.3	73.3	0.0185	0.0060	70.3	74.3	40	0.0025	42	78.1	0.0938	4.48	0.28
60	27	1.5	81.3	72.3	0.0182	0.0060	69.8	73.3	40	0.0025	42	77.1	0.01114	5.74	0.28
70	31.5	1.5	82.0	72.3	0.0182	0.0061	70.3	72.4	40	0.0025	42	77.0	0.01269	6.31	0.29
80	36	1.5	83.8	72.1	0.0180	0.0062	70.3	72.2	40	0.0025	42	76.1	0.1420	8.03	0.29
50	25	1.0	76.8	68.7	0.0161	0.0063	66.2	68.7	37	0.0028	42	73.7	0.0915	4.17	0.32
56	25	1.5	73.9	67.7	0.0162	0.0062	66.2	68.3	37	0.0028	42	74.6	0.0947	5.58	0.31
63	25	2.0	74.4	67.5	0.0162	0.0061	66.2	68.0	37	0.0028	42	75.4	0.0960	6.02	0.30
76	25	2.5	76.2	66.8	0.0161	0.0060	66.1	66.8	37	0.0028	42	75.9	0.0979	6.90	0.30

1 ft^3/min (CFM) = 4.72×10^{-4} m^3/s
1 gal/min (GPM) = 3.8 L/min

Table 2. Comparison of Mass Transfer Coefficients with Literature Data

	Packing	LiCl Concentration (%)	Gas flow rate (kg/s·m^2) / Temperature (°C)	Liquid flow rate (kg/s·m^2) / Temperature (°C)	Mass transfer coefficients (kmol/s·m^3·[mole fraction])	Height of a transfer unit (m)
Leboeuf and Lof (1980)	1.5"-Berl saddles	40	1.36 / 50-80	0.27 / 30-50	0.320	0.15
Gandhidasan and Satcunanathan (1983)	1.5"-Berl saddles	41	0.099 / 25-41	0.26 / 24-40	0.013-0.009	0.26-0.38
Chung et al. (1992)	5/8"-Flexi rings	30-40	0.94-1.87 / 22.7-25.5	8.70-15.23 / 12.7-22.7	0.1661-0.0981	0.31-0.49
This work	1/2"-Berl saddles	28-40	0.75-1.20 / 23-28	4.33-10.83 / 16-21	0.1420-0.0791	0.28-0.35

Table 3. Physical Characteristics of Packings

Packing	1/2" Ceramic Berl Saddles	5/8" Polypropylene Flexi Rings
Surface to volume ratio	142 ft^2/ft^3 (466 m^2/m^3)	104 ft^2/ft^3 (341 m^2/m^3)
Weight of packings	54 lb/ft^3 (866 kg/m^3)	7.3 lb/ft^3 (117 kg/m^3)
Free-gas space	62 %	86 %
Packing factor	240 ft^{-1} (788 m^{-1})	100 ft^{-1} (328 m^{-1})

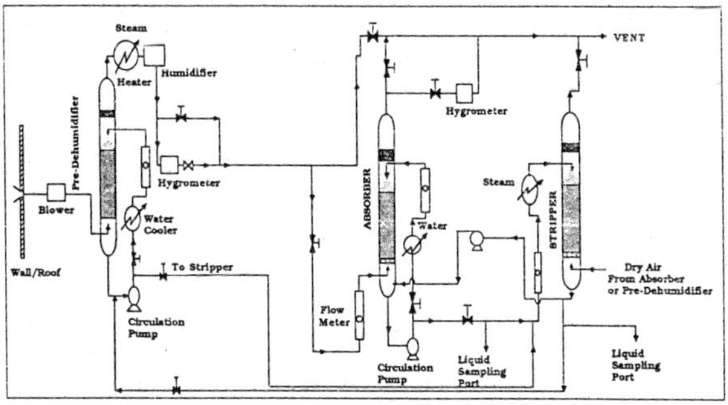

Figure 1. Absorption Stripper System for This Study

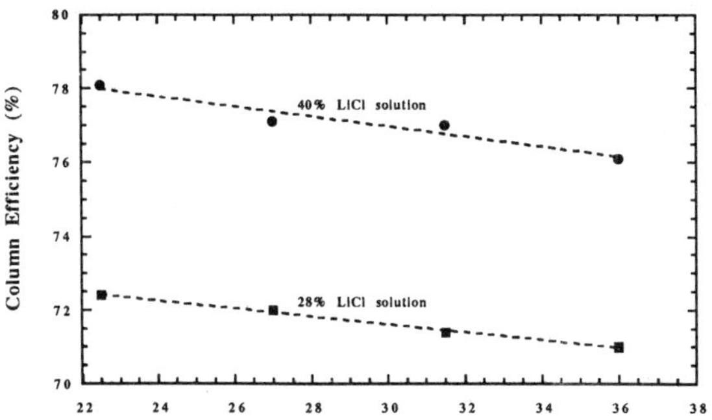

Inlet Air Flow Rate, CFM (1 CFM = 1.7 m^3/h)

Figure 2. Effect of Inlet Air Flow Rate on Column Efficiency

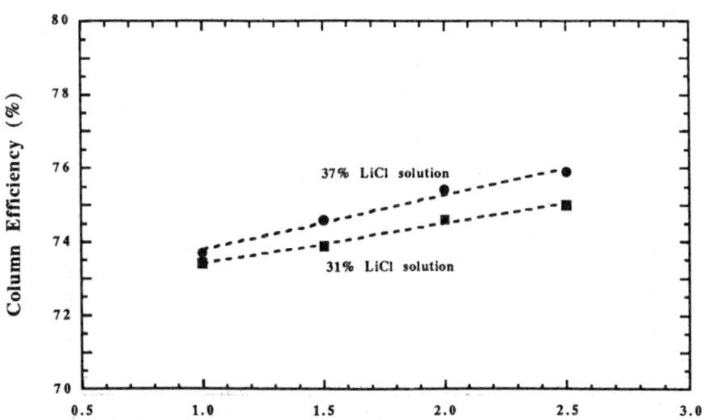

Inlet Liquid Flow Rate, GPM (1 GPM = 3.8 L/min)

Figure 3. Effect of Inlet Liquid Flow Rate on Column Efficiency

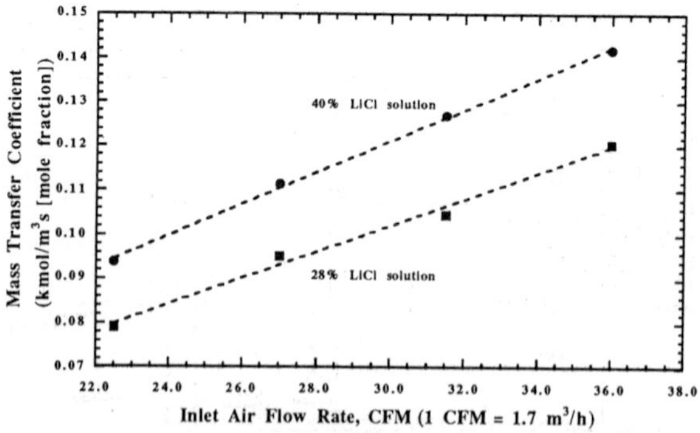

Figure 4. Effect of Inlet Air Flow Rate on the Mass Transfer Coefficient

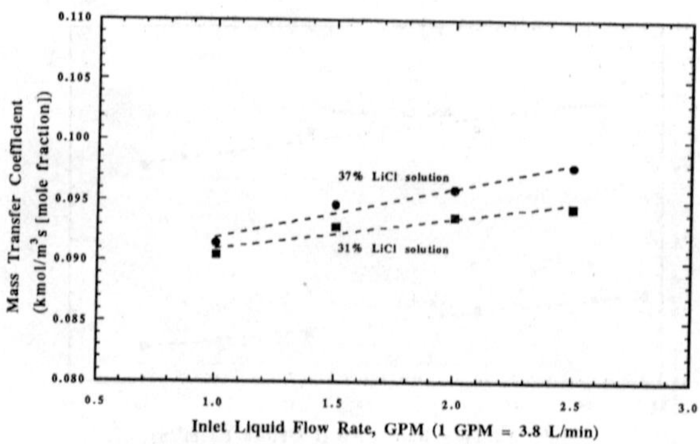

Figure 5. Effect of Inlet Liquid Flow Rate on the Mass Transfer Coefficient

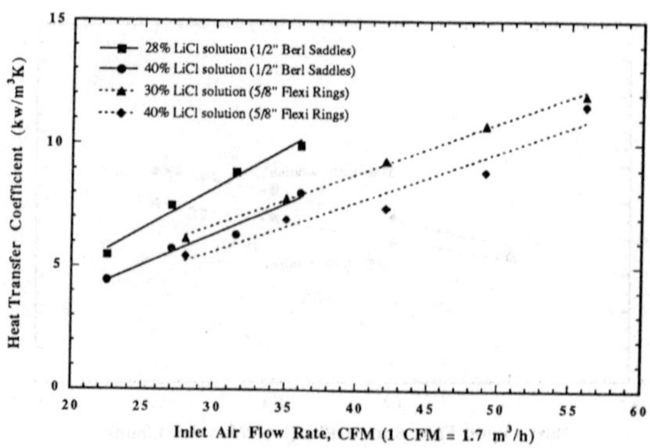

Figure 6. Effect of Inlet Air Flow Rate on the Heat Transfer Coefficient

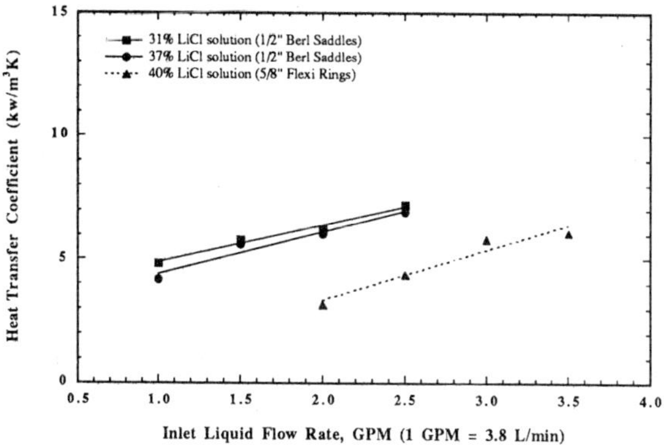

Figure 7. Effect of Inlet Liquid Flow Rate on the Heat Transfer Coefficient

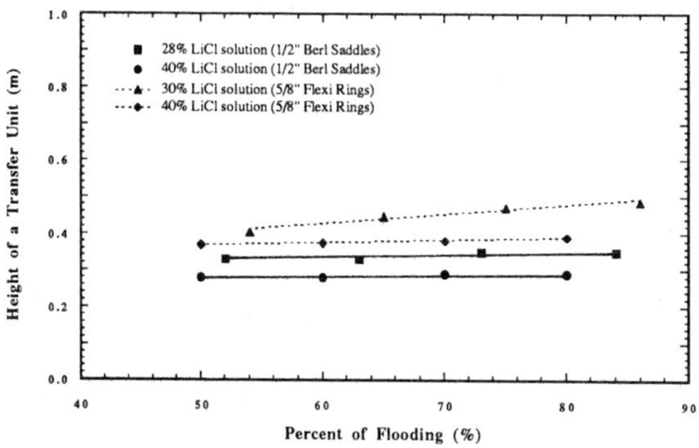

Figure 8. Change in Height of a Transfer Unit as a Function of Flooding

THE CONCEPT OF 'REVERSE-FRACTIONAL-DISTILLATION' EXERGY ANALYSIS AND APPLICATIONS

P. Le Goff and J. Labidi
Laboratoire des Sciences du Génie Chimique, CNRS-ENSIC, NANCY, France

The separation of a mixture by fractional distillation requires the degradation of a heat flux between the reboiler and the condenser of the column. This operation can be described as the transformation of thermal exergy into 'chemical' exergy.

Such an operation can be made to work backwards, that is to transform some chemical exergy into thermal exergy: remixing two separated components in a suitable apparatus will bring heat from a low temperature source, to a higher temperature sink.

We show that such a 'reverse-distillation column' can constitute the main element of a new type of absorption heat pump, able to operate with any working pair, even with an ideal mixture.

INTRODUCTION

In previous papers[1,2,3,4], we have proposed a new type of heat pump, resulting from the combination of a rectification column (fractional distillation) and a 'reverse-rectification' column. In other words, the heat pump is composed of a multi-stage separator combined with a multi-stage mixer, as compared to the conventional absorption heat pump made of single-stage separator and mixer.

The absorption heat pump works as a TYPE-I, or a TYPE-II (heat transformer) whether the operating pressure in the separator (rectification column) is greater or smaller than that in the mixer (reverse-rectification column).

Figure 1 shows the principle of a heat transformer for upgrading heat from 100°C to 198°C, using a mixture of water and ethylene-glycol as the working pair.

In order to understand how such a machine can work, let us first come back to the basic aspects of rectification and its reverse operation.

RECTIFICATION: THE EXERGY BALANCE

Figure 2 shows the operating principle of a standard continuous rectification column for the separation of a 50/50% mole mixture of water and ethylene glycol. This mixture is introduced in the column at about its mid-point.

A water rich distillate is collected at the top of the column and a glycol rich product is collected at the bottom of the column.

If we assume that the column is infinitely efficient and is working under 1 bar pressure then the distillated will be pure water condensed at 100°C and the bottom product will be pure glycol boiling at 198°C.

To carry out this separation, it is necessary to supply a heat flux Q_B at 198°C to the reboiler and to remove a heat flux Q_C at a temperature of 100°C at the condenser. We may say that the operation has converted thermal exergy into 'chemical' exergy.

• Thermal exergy is that resulting from the degradation of the heat flux supplied at 198°C and recovered at 100°C that is:

$$Ex_{th} = \dot{Q}\left[\frac{T_o}{T_C} - \frac{T_o}{T_B}\right] \tag{1}$$

• Chemical exergy is that resulting from the fact that the products which were initially mixed together have been separated:

$$Ex_{chim} = -RT_o\,[x\,Ln\gamma_e x + (1-x)\,Ln\gamma_g(1-x)] + H_m\left(1 - \frac{T_o}{T}\right) \tag{2}$$

where x is the mole fraction of water in the initial mixture, γ_e, γ_g are the activity coefficients of water and glycol in the mixture*.

WHY TO REVERSE RECTIFICATION?

Let us now imagine that it is possible to carry out the reverse operation of mixing, that is the conversion:

(chemical exergy) → (thermal exergy)

This means that an apparatus is available (a 'black box' which for the moment has an internal structure yet to be defined) which can perform the following operations (cf Figure 3):

- input of pure water at a given point in the apparatus;
- input of pure glycol at another point in the apparatus;
- output of a mixture at a third point.

At the same time:

- input of heat flux at a low temperature: that is at 100°C if the operation is at 1 bar;
- output of heat flux at a high temperature (that is at 198°C).

The overall result of such an operation would therefore upgrade a heat flux from 100°C to 198°C.

THE OPTIMUM STRUCTURE OF THE MIXER

Let us first consider the fictitious experiment shown in Figure 4.

• The left hand side of the figure shows a single stage evaporator-condenser working with pure water at 1 bar pressure. The evaporator is supplied with a heat flux Q_{ev} at 100°C to vaporise water and this same heat is recovered at the condenser $Q_c = Q_{ev}$ at the same temperature 100°C.

• The right hand side of the figure shows an identical evaporator-condenser working with pure glycol at 1 bar. Here we supply the heat flux Q_{gv} to the evaporator at 198°C and this same heat Q_{gv} is recovered at the condenser at 198°C.

---

* The simple expression (2) is based on the assumption that the distillate and the bottom product are brought to the temperature of the feed mixture, and therefore that the system includes heat exchangers not shown on Figure 2.

• In Figure 4, we assume that the water condenser is thermally coupled with the glycol evaporator. But this operation is obviously fundamentally impossible. Heat cannot be transferred by simple conduction from a source at 100°C to a sink at 198°C.

• Now let us suppose that this transfer is coupled with a simultaneous operation of mixing between the two fluids as shown schematically in Figure 5. The structure of this mixer should be such that the chemical exergy supplied should be entirely converted into thermal exergy for upgrading the heat flux from 100°C to 198°C.

It can be understood that the structure shown in Figure 5 is not the best one: in order to minimize the exergy loss by mixing, the mixer should be a long column forming a counter-current liquid-vapour contactor, where each phase is in piston flow. Furthermore, the contactor must be equivalent to a cascade of a very large number of theoretical equilibrium stages (see Figure 6).

PRE-HEATING THE GLYCOL

In the system described above, we assumed that water and glycol were introduced as liquids at 100°C and 198°C respectively. In fact, as we intend to show later in the treatment of heat transformers, it is better to assume that the two liquids are introduced at the same temperature of 100°C. A part of the heat produced in the apparatus is then used to pre-heat the glycol from 100°C to 198°C.

Two techniques can be used to do this:

• either by means of an external counter-current heat exchanger (see Figure 6) between the glycol to be heated and a part of the hot heat carrier fluid from the condenser. This solution has the inconvenience of involving two heat transfers and therefore a double creation of entropy;

• or by directly pre-heating the glycol in a preheater-exchanger integrated into the absorption column itself as shown in Figure 7. It is easy to imagine that the creation of entropy is less in this case as there is only one wall for heat transfer.

GENERALISED NOMENCLATURE: MASS AND ENTHALPY BALANCES

In the preceding paragraphs, we assumed that pure glycol and pure water were introduced at the top and the bottom of the column respectively. Now let us generalise the problem (cf Figure 7).

Into the evaporator, at the bottom, we introduce a molar flow rate $\dot{R}$ of a water-rich phase.

Into the top of the column, we introduce a molar flow rate $\dot{P}$ of a water-poor phase.

Two mixtures leave the apparatus:

• an upper mixture leaving the condenser at flow rate $\dot{U}$;

• a lower mixture leaving the foot of the column at a flow rate $\dot{L}$.

If x is the mole fraction of water, the balance equations are as follows:

$$\dot{P} + \dot{R} = \dot{U} + \dot{L} \tag{3}$$

$$\dot{P}\, x_P + \dot{R}\, x_R = \dot{U}\, x_U + \dot{L}\, x_L \tag{4}$$

In the same way, the enthalpy balance is:

$$\dot{P}\,h_P + \dot{R}\,h_R + \dot{Q}_e = \dot{U}\,h_U + \dot{L}\,h_L + \dot{Q}_a \tag{5}$$

where $\dot{Q}_e$ and $\dot{Q}_a$ are the heat fluxes to the evaporator and from the condenser respectively.

REPRESENTATION ON A PONCHON DIAGRAM

The points marked on the diagram 8 (enthalpy/mole fraction of water) represent the various input and output fluxes:

- point [R] is the water rich liquid introduced into the evaporator at boiling point;
- point [R_v] is the saturated vapour leaving the evaporator;
- point [P] is the water poor liquid introduced at the head of the column at the same temperature as [R];
- point [U] is the upper liquid mixture taken to be at boiling point;
- point [U_v] is the vapour at the same composition as U at its inlet to the condenser;
- point [L] is the lower liquid mixture taken to be at its boiling point.

Now let us re-write equation (5) as follows:

$$\dot{P}\,h_P + \dot{R}\left[h_R + \frac{\dot{Q}_e}{\dot{R}}\right] = \dot{U}\left[h_U + \frac{\dot{Q}_a}{\dot{U}}\right] + \dot{L}\,h_L \tag{6}$$

This shows that expressions (4) and (6) are scalar projections of a vector equation which represents the standard 'lever rule':

$$\dot{P}.\overrightarrow{AP} + \dot{R}.\overrightarrow{AR_V} = \dot{U}.\overrightarrow{AU_V} + \dot{L}.\overrightarrow{AL} \tag{7}$$

where point [A], the intersection of the two straight lines $P.R_V$ and $L.U_V$, is the 'addition point' of P and R_V and also of U_V and L, as defined by the following symbolic addition:

$$P + R = U + L = A \tag{8}$$

THE LIMITING CASE OF A QUASI-REVERSIBLE MIXER

Point $(UV)_\infty$ on Figure 9 is a saturated vapour in thermodynamic equilibrium with the water-poor liquid injected at the head of the column. This point is therefore the left hand limit of point U_V, which is free to move along the dew point isobar. The right hand limit of this point U_V is just point R_V which corresponds to the case where absorption column was not present.

In the same way, the point L_∞ on Figure 9 is a liquid which would be in thermodynamic equilibrium with the vapour R_V leaving the evaporator. This point is the right hand limit of point L which is free to move along the bubble point isobar.

The left hand limit of point L is the point [J] located at the intersection of the straight line $P.R_V$ and the bubble point isobar. In fact, in this limiting case, the addition point [A] would also be located at [J] and, by the lever rule, we would have U = 0. That is there would be no upper mixture produced.

Point [A_{opt}] on Figure 9 designates the optimum operating conditions for which there is both:

- thermodynamic equilibrium at the head of the column between the water poor liquid and the vapour U of the upper outlet;
- thermodynamic equilibrium at the foot of the column between the water rich vapour R_V and the liquid L of the lower outlet.

This limiting case corresponds to a quasi-reversible system with no creation of entropy which is obtained with a column equivalent to a double infinity of theoretical equilibrium stages.

THE COMPLETE HEAT TRANSFORMER

Figure 10 shows the complcte flowsheet of a heat transformer based on the previous principle. The separation of the mixture is made in a conventional rectification column of high efficiency operating under a pressure of 0.033 bar. The bottom product is quasi-pure glycol (water content: 0.1%). The distillate is composed of water with 1% glycol.

The reverse-rectification column, operating under a pressure of 1 bar is such that the vapour leaving the head of the column is in quasi-equilibrium with the 'poor' liquid injected at the top: its water mole fraction is $y_a = 0.02$ and its temperature: 195°C.

The Coefficient of Performance (COP) of the whole heat transformer is directly dependent from the enthalpy efficiency of the reverse-rectification column, which is defined as the ratio of the amount of heat removed at the condenser to the amount of heat supplied to the evaporator.

TO MAXIMIZE THE ENTHALPY EFFICIENCY OF THE SYSTEM

Let us return to the schematic diagram of Figure 6:

- the water rich phase is pure water: $x_R = 1$;
- the water poor phase is pure glycol: $x_P = 0$;
- the 'upper' mixture is also pure glycol which means that thermodynamic equilibrium is attained at the top of the column: $x_U = 1$.

It can be seen that the flux of pure glycol is separated into two streams: one part (f) is used in the mixer and leaves at the foot of the column mixed with water. The other part (1 – f) continues to leave at the foot of the upper condenser in a pure state.

The upgraded heat flux taken at the condenser is:

$$\dot{Q}_a = \dot{P}(1-f) \, . \, \Delta H_{aV} \tag{9}$$

where ΔH_{aV} is the latent enthalpy of vaporisation of glycol at 198°C.

The heat flux supplied to the evaporator is:

$$\dot{Q}_e = \dot{R} \, . \, \Delta H_{eV} \tag{10}$$

where ΔH_{eV} is the latent enthalpy of vaporisation of water at 100°C.

The enthalpy efficiency is therefore:

$$\eta_T = \frac{\dot{P}}{\dot{R}}(1-f) \, . \, \frac{\Delta H_{aV}}{\Delta H_{eV}} = \left(\frac{\dot{P}}{\dot{R}} + 1 - \frac{1}{x_L}\right)\frac{\Delta H_{aV}}{\Delta H_{eV}} \tag{11}$$

where x_L is the mole fraction of water in the lower mixture L.

Figure 11a shows the variation of $(\dot{P}/\dot{R})$, and of f, when the mole fraction x_L varies between its maximum value 1, and its minimum value 0.15, for which the upper mixture flow rate is equal to zero $\dot{U} = 0$). As a consequence, the enthalpy efficiency increases from zero (for $x_{LO} = x_S$) till to its maximum value $(\eta_{th})_{max} = 0.85$ for $x_L = 1$.

RELATION BETWEEN THE ENTHALPY EFFICIENCY AND THE EXERGY EFFICIENCY

The chemical exergy liberated by the mixing of the flow rate $\dot{R}$ of water rich phase and the flow rate $(f.\dot{P})$ of the water poor phase is:

$$\dot{E}x_{chim} = [\dot{R} + f.\dot{P}] . \left[RT_o \, [x_L \, Log \, \gamma_e x_L + (1 - x_L) \, Log \, \gamma_g \, (1 - x_L)] + H_m \left(1 - \frac{T_o}{T}\right)\right] \quad (12)$$

with

$$x_L = \frac{\dot{R}}{\dot{R} + f.\dot{P}} = \frac{1}{1 + f \, . \, \frac{\dot{P}}{\dot{R}}} \quad (13)$$

and H_m is the enthalpy of mixing.

In addition, the thermal exergy produced is:

$$\dot{E}x_{ther} = \dot{Q}_a \left[1 - \frac{T_o}{T_a}\right] - \dot{Q}_e \left[1 - \frac{T_o}{T_e}\right] \quad (14)$$

If η_{ex} is the exergy efficiency of the conversion, we have:

$$Ex_{ther} = \eta_{ex} \, . \, Ex_{chim}$$

As any temperature T_o can be used as reference; we choose to use $T_o = T_e = 100°C = 373$ K and to calculate variations in exergy at that temperature. Combining the above expressions leads to the following general equation:

$$\eta_{ex} = \frac{\dot{U}}{\dot{L}} \cdot \frac{[1 - (T_e/T_a)].\Delta H_{aV}}{T_e S(x_L) + H_m[1 - (T_e/T_L)]}$$

with

$$S(x_L) \equiv R \, [x_L \, Log\gamma_e x_L + (1 - x_L) \, Log\gamma_g(1 - x_L)] \quad (15)$$

Figure 11b shows that the exergy efficiency reaches its maximum value of 1 when the chemical exergy supplied by the mixture, [represented by the denominator of equation (15)] becomes just sufficient for providing the thermal exergy necessary to upgrade the heat flux $\dot{Q}_a$ [represented by the numerator of equation (15)].

Calculation shows that the corresponding maximum value of x_L is: $x_{L\,max} = 0.45$.

Finally, when the composition of the lower mixture x_L varies from $x_J = 0.15$ to $x_{L\,max} = 0.45$:

- the exergy efficiency increases from 0 to 1;
- the enthalpy efficiency increases from 0 to $(\eta_T)_{max} = 0.70$;
- the ratio of the flowrates $(\dot{P}/\dot{R})$ decreases from 5.7 to 1.75;
- the fraction f of the glycol flow rate used in the mixing operation decreases from 1 to 0.70.

CONCLUSION: THE OPTIMUM HEAT TRANSFORMER

The calculation given above shows that it is possible to upgrade a heat flux from 100°C to 198°C with an excellent enthalpy efficiency $\eta_T = 0.7$ and exergy efficiency $\eta_{ex} = 1$ by using an efficient countercurrent gas-liquid contactor with integral pre-heater. The flow rate of pure liquid glycol injected at the head of the column is 5.7 times the flow rate of water vapour injected at the foot of the column. The mixture containing 45% water removed at the bottom of the column must then be re-separated in a standard rectification column.

NOMENCLATURE

Ex	exergy, kJ/kmol
f	fraction of glycol used in the mixing operation
h	enthalpy, kJ/kmol
H_m	enthalpy of mixing, kJ/kmol
$\dot{L}$	molar flow rate of lower mixture, mol/s
$\dot{P}$	molar flow rate of poor phase, mol/s
$\dot{Q}$	heat flux, kJ/s
$\dot{R}$	molar flow rate of rich phase, mol/s
S	entropy, kJ/mol.K
$\dot{U}$	molar flow rate of upper mixture, mol/s
x	mole fraction in water,
γ_e, γ_g	activity coefficient of water and glycol in the mixture
η	efficiency
ΔH_V	latent enthalpy of vaporisation, kJ/mol
ΔT	temperature lift, K

SUBSCRIPTS

a	condenser
b	reboiler
chim	chemical
e	evaporator
ex	exergy
g	glycol

L	lower mixture
m	mixing, mixture
P	poor phase
R	rich phase
ther	thermal
U	upper mixture
V	vapour
o	reference

LITERATURE CITED

1. Le Goff, P., Ranger, P.M., Jeday, M.R. and Ramadane, A., 1988, *Symposium International* 'Pompes à chaleur chimiques de haute performance', Sept. 1988, *Récents Progrès en Génie des Procédés*, V. 2, p. 118–127.
2. Le Goff, P., Ranger, P.M. and Jeday, M.R., 1988, *2nd International Workshop on Research Activities on Advanced Heat Pumps*, Graz, Sept. 1988, p. 187–197.
3. Ranger, P.M., Matsuda, H., Le Goff, P. and Jeday, M.R., 1989, *2nd Congrès Français de Génie des Procédés*, Toulouse, Sept. 1989, *Récents Progrès en Génie des Procédés*, V. 8a, p. 708–713.
4. Ranger, P.M., Matsuda, H. and Le Goff, P., 1990, *J Chem Eng Japan*, 23 (5): 530–542.

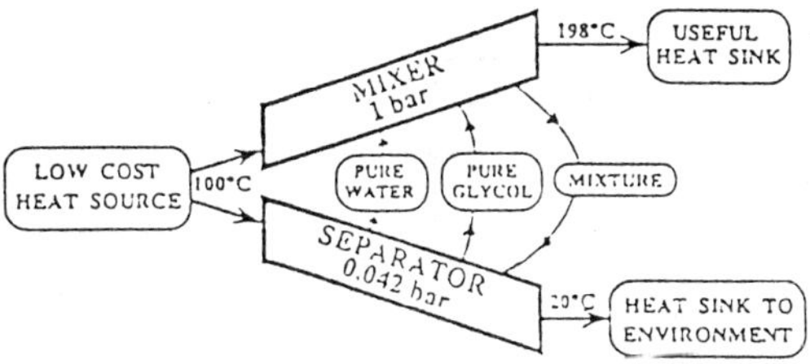

Figure 1. A heat transformer composed of a mixer associated to a separator.

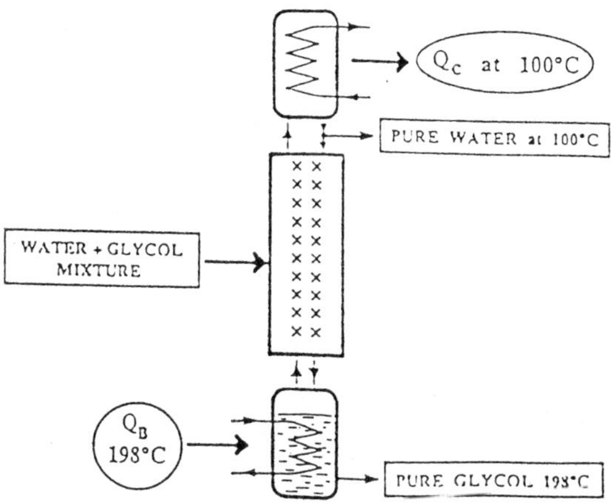

Figure 2. A classic rectification (fractional distillation) column for the separation of a glycol/water mixture by the degradation of a heat flux from the reboiler to the condenser.

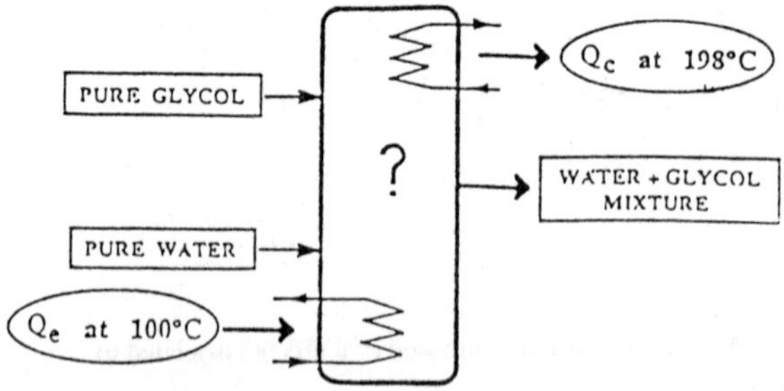

Figure 3. A 'reverse-rectification' column seen as a black box upgrading a heat flux at a higher temperature level, by the mixing of glycol and water.

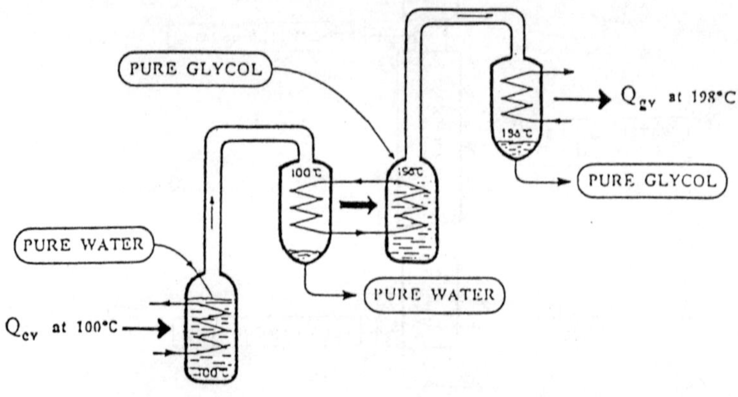

Figure 4. A fictitious coupling of a condenser at 100°C heating a boiler at 198°C.

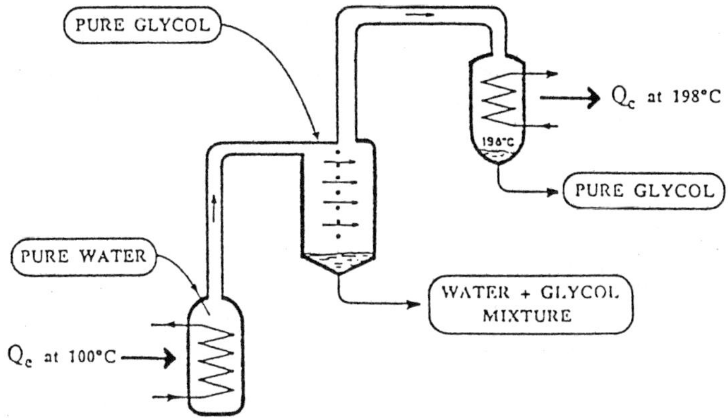

Figure 5. The coupling of (Figure 4) is now obtained by mixing the water vapour and the liquid glycol.

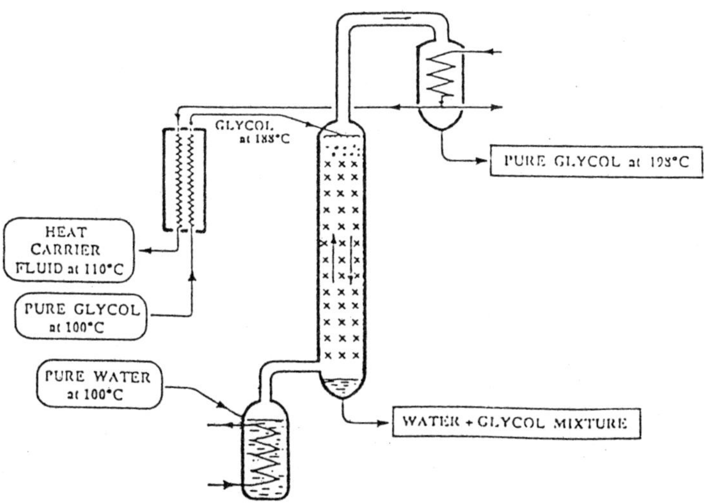

Figure 6. External preheating of the liquid glycol injected at the head of the column.

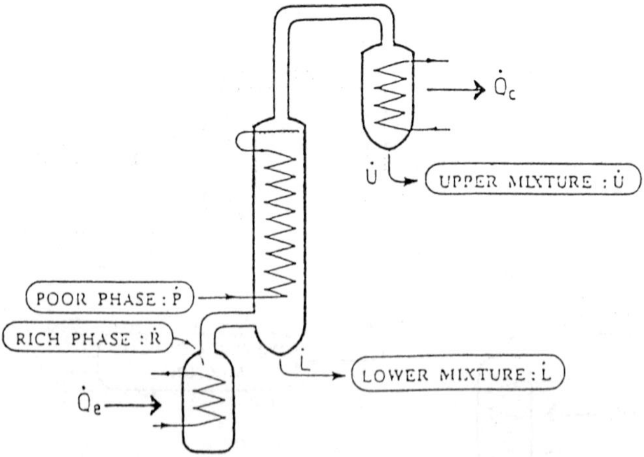

Figure 7. A generalized scheme, where a 'rich phase' and a 'poor phase' are mixed and form an 'upper mixture' and a 'lower mixture', the poor phase being pre-heated in an exchanger integrated into the absorption column.

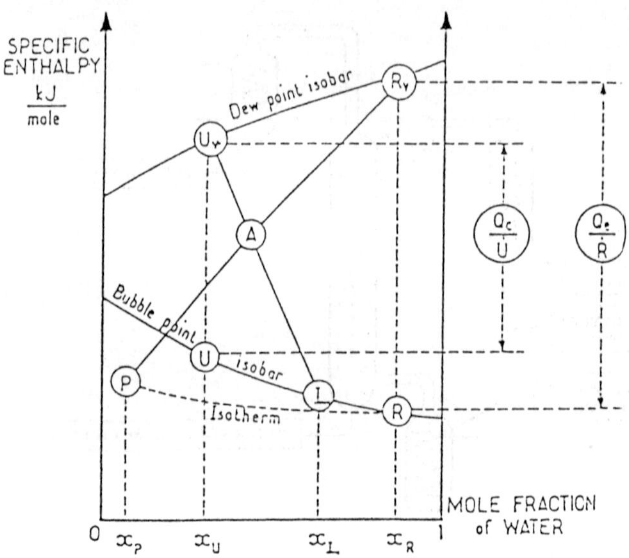

Figure 8. Representation of the mixing operation on the PONCHON diagram, with the help of the 'Lever rule'.

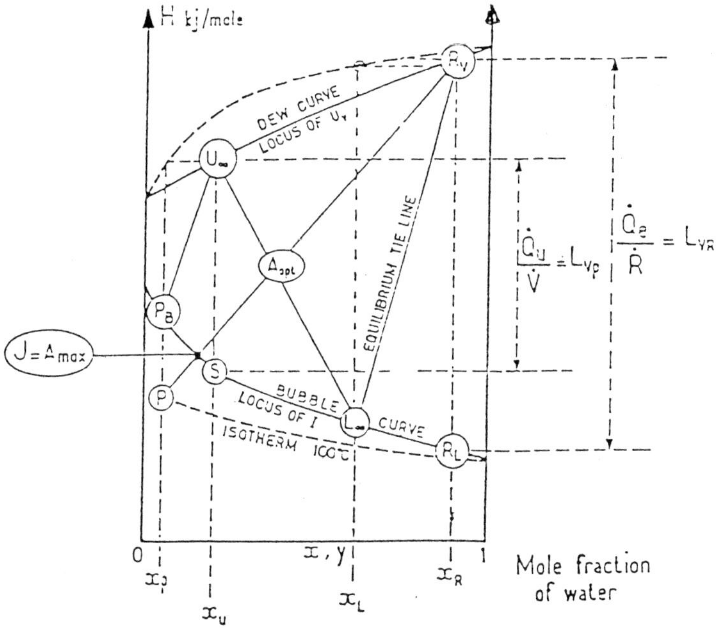

Figure 9. The limiting cases corresponding to a quasi-reversible system, with no creation of entropy.

HEAT PUMPING IN ENERGY-INTEGRATED DISTILLATION — A PARAMETRIC STUDY

P. Mizsey and Z. Fonyo
Technisch-Chemisches Laboratorium, Swiss Federal Institute of Technology, ETH-Zentrum, CH-8092 Zurich, Switzerland

Firstly, the influence of relevant parameters on the economics of distillation plants involving closed heat pump cycles is scrutinized and the results are compared to the features of mechanical vapour recompression schemes. The results show that from an economic viewpoint the closed heat pump cycle has a similar effect to vapour recompression, but the closed cycle is less competitive. However, the closed cycle scheme can still be attractive due to safety and operability reasons. Preliminary screening for the selection of the most economic variant also shows that the application of the heat pumps outside the field of stand-alone columns is restricted to separating close boiling mixtures in a column of small pressure drop, since heat integration should be carried out before heat pumping.

Secondly, a study of an industrial separation plant is demonstrated in a retrofitting situation. If adequate heat integration cannot be realized in the case of retrofitting situations, due to the inherent features of the plant, heat pumping can significantly improve the process performance.

INTRODUCTION

The significant energy requirements of distillation processes can often be reduced by both recovering and recycling energy. Since heat integration is usually more economical than heat pumping, it has been widely recommended to first exhaust the heat integration options if a background process exists before exploring heat pumping (Linnhoff *et al*, 1983, Meszaros and Fonyo, 1986, Mizsey and Fonyo, 1990). Therefore, heat pumping is still questionable and somewhat less studied than the various heat integration methods.

Heat pumping is, however, competitive if no background process exists and seems to be one of the more promising energy saving techniques. It has also to be considered that heat integration often reduces flexibility in addition to making operation more complex.

PREVIOUS WORK

Many different ways of heat pumping have already been studied but, in the case of a distillation column, electrically driven solutions are preferred. They are vapour recompression, bottom flashing, and closed cycle systems. Ranade and Chao (1990) even recommend guidelines of where and when to use the different types of heat pumps. Vapour recompression seems to be the most economical variant. That is why it has been investigated much more than the other options (eg Lüdecker, 1985).

On the other hand, the simplest way of introducing a heat pump into a process is the closed system with a working fluid (Omideyi *et al*, 1984), which has less technical risk than the other solutions. The behaviour of a closed cycle heat pumping process can be simply generalized and, by the help of a safe working fluid, it is much easier to realize heat transfer between two or more equipment items. These features make it an appropriate and preferred alternative, especially in the case of retrofit design, where the heat recovery matches are usually constrained by the mechanical restrictions of the already existing equipment.

The published industrial examples show that, in the case of retrofitting, the closed cycle heat pump is a suitable method to recover energy. Benstead and Sharman (1990) design a closed cycle heat pump to develop an existing whisky distillation plant. The payback time proves to be less than one year. Farhanieh and Sunden (1990) also present a retrofit case study of an existing heat exchanger network. The maximum saving is obtained by also designing a closed cycle heat pump system.

GOALS OF THIS RESEARCH

Despite these many attractive features of the closed cycle heat pump system there is no general parametric study reported to understand fully its behaviour and to predict the circumstances of its best use. In the first part of this paper the influence of relative volatility, feed flow rate and tray pressure drop on the economics of distillation plants involving closed heat pump cycles is investigated. The results are compared to the behaviour of mechanical recompression.

In the second part, an industrial example demonstrates that in the case of an existing plant the efficiency of energy utilization can still be improved by the use of heat pumping.

INVESTIGATION OF CLOSED CYCLE HEAT PUMP PROCESSES

In our parametric study two computer programs are written and used for the simulation of heat pump assisted distillation and heat integration assisted by closed cycle heat pumping, in order to calculate the economics and the typical process parameters. The payback time of excess capital (PBT) is selected as the objective function:

$$PBT = \frac{CCOMP + CHX}{UCONV - UHP}$$

It is considered that in the closed cycle process the areas of the condenser and reboiler are different to those of the conventional process and the costs of cooling and heating can be also saved.

As a reference working fluid R718 is selected because, for generalization, it has proved to be the most suitable one (Holland, 1982, Gopichand *et al*, 1984).

STAND-ALONE DISTILLATION COLUMN

The stand-alone distillation column with and without heat pump is simulated by short-cut methods. For the capital cost calculation the procedure of Douglas is applied (Douglas, 1988) and for the estimation of the efficiency of the adiabatic compression of the working fluid Dimoplon's equations are used (Dimoplon, 1978). Swiss energy prices are supposed (Table 1). The project life is taken as 10 years. Besides the payback time, the optimal temperature difference in the heat exchangers (reboiler, condenser) is determined. If necessary a trim reboiler or condenser is designed.

The relative volatility, the feed flow rate and the distillation column type are selected as variables. The components to be separated are selected so that the relative volatility ranges from

1.12 to 2.19 and the top and bottom temperatures of the distillation column would be higher than 100°C at 1 bar, respectively. The investigated range of the feed is 50–1000 kmol/h. The feed is binary, with equimolar composition. Two column types are selected, a common distillation column with a tray pressure drop of 3 torr and a packed column with the possible smallest pressure drop, 0.1 torr/HETP (eg Fabry and Manczinger, 1987, Sulzer Mellapack). Figure 1 shows the temperature difference of the top and bottom of the distillation column for three different column types. For comparison the number of theoretical trays is also plotted. Due to the different tray pressure drops investigated, the number of theoretical trays might have a deviation of about three percent at the smaller relative volatility values. In Figure 1, however, the average number is plotted.

It is to be noted that, for the packed columns of low pressure drop, there exists no minimum of the difference of the column bottom and top temperatures as a function of the relative volatility, a phenomena observed for tray columns of higher pressure drop (Meszaros and Fonyo, 1983, Lüdecker, 1985).

Figures 2–5 show the payback time as a function of the relative volatility and the feed rate for the tray column ($\Delta p = 3.0$ torr/tray) and one packed column ($\Delta p = 0.1$ torr/tray). The tendencies are similar to those of a vapour recompression process (eg Lüdecker, 1985). As it is to be expected, at smaller relative volatility, higher feed, and smaller column pressure drop the systems with heat pump prove to be more economical. The optimal temperature difference driving forces across the condenser and evaporator using the closed cycle heat pump process are represented in Figures 6 and 7. They range practically from 4°C to 6°C. (Gopichand et al (1982) have determined about 9°C for ethanol-water separation.)

The closed cycle process and vapour recompression are compared in Figures 8 and 9. The payback times of the two different processes as functions of the relative volatility and the column feed rate show that the vapour recompression is always more economical than the closed cycle process. The difference, however, in the case of packed columns is smaller than for tray columns.

HEAT INTEGRATION ASSISTED BY CLOSED CYCLE HEAT PUMPING

Heat transfer between two arbitrary points of a process, eg condenser and reboiler of two different columns, is also possible if the temperature of the heat source is lower than the sink's temperature and the heat source is in a vapour phase. In this case, by the help of a heat pump, the temperature of the heat source can be increased and the match will be possible.

The return on capital employed of this type of match assisted by heat pumping is calculated on the same basis as the stand-alone columns. The optimal temperature differences between the working fluid and the heat source and sink temperatures are determined and found to be in the same range as stand-alone columns.

For this kind of heat integration two cases are possible. If there are no heat exchangers at the selected points of the technology, they are designed and their prices are included, otherwise only the necessary supplementary heat exchangers and their costs are considered.

Figure 10 shows the economy of heat integrations assisted by closed heat pump cycles, which is a good guide for the designer to estimate the expected payback time of a heat pumped match. For example: the temperature difference between the sink and the source is 15°C, the possible heat transfer is 106 kJ/h, then the expected payback time is 5 years if there are existing heat exchangers and about 8 years if there are none. The economy heat transfer diagram can be used for stand-alone columns as well.

RETROFITTING CASE STUDY

An industrial case study represents this problem. In a three column distillation system light hydrocarbons (C3–C5) are separated (Figure 11). There is a possibility for energy integration, column 1 could boil up column 2 but their pressures would have to be shifted. The pressure of column 2 cannot be decreased because the top temperature would be too low for a normal condenser using cooling water. The only choice is to increase the pressure of column 1. But the pressure limit of this equipment is 20 bar due to mechanical restrictions. So this column match could not be realized.

But if we use an intermediate heat pump to increase the pressure as well as the temperature of the vapour top stream of column 1, then the column match would be realizable in spite of the mechanical restriction (Figure 12). The vapour top stream in our case is propane, which can be compressed without any technical risk, so in this special case direct vapour recompression can be used otherwise a closed heat pump cycle is recommended.

The pressure of column 1 should be set on the allowable highest, that is 20 bar. The output pressure of the heat pump is also a design parameter. If the output pressure is low, the temperature difference is small and an enormously large heat exchanger is necessary for the match. If the output pressure is high, the heat pumping is expensive. Figure 13 shows the trade-off. The nonintegrated base case and a hypothetical energy integrated version neglecting the mechanical pressure limitations are also denoted. The utility and capital costs are shown in Figure 14.

This industrial case study also demonstrates that the heat pump can be competitive if energy integration is impossible. Thus, in the retrofitting case heat pumping should be additionally considered after energy integration. Based on the industrial case study above it is represented that at the retrofitting situation if adequate heat integration cannot be realized due to the inherent features of the plant, heat pumping can significantly improve the efficiency of the energy utilization. In this case the total cost saving is about 15 %.

It is to be noted here, that this statement seems to be in contrast with the statement of Null (1976), who states that the range of application of a heat pump is severely diminished in replacement situations. According to his investigations, the utility cost saving depends on the local utilization of the utility saved by heat pumping.

In our study, however, we suppose that the energy saving means utility cost reduction which is independent of whether the utility saved is utilized in the process or elsewhere.

CONCLUSIONS

The parametric study shows that the economic factors of the closed cycle heat pump assisted distillation are similar to the direct vapour recompression. The difference in economic figures is small if the heat flow transferred is high and the temperature difference is small. In the case of a distillation column it means big feed rate, difficult separation and small column pressure drop.

A generalized nomogram is presented for the quick estimation of the payback time for heat integration by closed cycle heat pumping.

Finally, it is shown that in the case of retrofit design, heat pumping can help to save energy that could not be otherwise realizable due to mechanical constraints of the equipment.

NOMENCLATURE

CCOMP	compressor capital cost, $
CHX	additional heat exchanger capital cost, $
PBT	payback time of excess capital
UCONV	utility cost of conventional process, $/year
UHP	utility cost of heat pumped process, $/year

REFERENCES

1. Benstead, R. and Sharman, F. W., 1990, Heat pumps and pinch technology, *J. Heat Recovery Systems & CHP*, 10 (4): 387–398.
2. Dimoplon, W., 1978, *Hydrocarbon Processing*, 57 (5): 221.
3. Douglas, J. M., 1988, *Conceptual Design of Chemical Processes*, McGraw-Hill, New York.
4. Farhanieh, B. and Sunden, B., 1990, Case study, Analysis of an existing heat exchanger network and effects of heat pump installations, *J. Heat Recovery Systems & CHP*, 10 (3): 285–296.
5. Gopichand, S., Omideyi, T. O., Kasprzyaki, J. and Devotta, S., 1984, The economic of heat pump assisted distillation systems II, *J. Heat Recovery Systems*, 4 (4): 271–280.
6. Holland, F. A., Watson, F. A. and Devotta, S., 1982, *Thermodynamic design data for heat pump systems*, Pergamon Press Ltd., Oxford.
7. Linnhoff, B. *et al*, 1982, *User Guide on Process Integration for the Efficient Use of Energy*, Institution of Chemical Engineers, London, United Kingdom.
8. Lüdecker H., 1985, Wirtschaftliche Einsatzbereiche von Wärmepumpen- und Mehrstufenschaltungen in der Rektifiziertechnik, PhD. Dissertation, Berlin.
9. Meszaros, I. and Fonyo, Z., 1986, Design strategy for heat pump assisted distillation systems, *J. Heat Recovery Systems*, 6, 469–476.
10. Mizsey, P. and Fonyo, Z., 1990, Toward a more realistic process synthesis — The combined approach, *Comput Chem Engng*, 14 (11): 1213–1236.
11. Null H. R., 1976, Heat pumps in distillation, *Chem Eng Progress*, July 58–64.
12. Omideyi, T. O.,Kasprzycki, J. and Watson, F. A., 1984, The economics of heat pump assisted distillation systems — I. A design and economic model, *J. Heat Recovery Systems*, 4 (3): 187–200.
13. Ranade, S. M. and Chao, Y. T., 1990, Industrial heat pumps: Where and when?, *Hydrocarbon Processing*, October, 71–73.

TABLE I

Utility prices

Cooling water	:	1.2 $ / GJ
Low pressure steam	:	4.9 $ / GJ
High pressure steam	:	15.0 $ / GJ
Electricity	:	0.1 $ / GJ

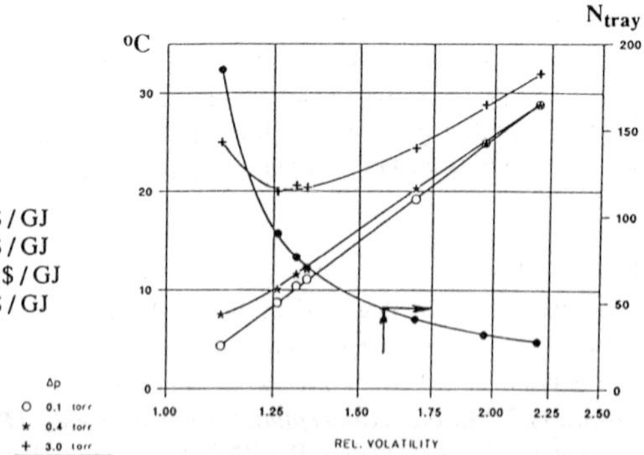

Figure 1 Temperature difference of column top and bottom Number of theoretical trays vs. relative volatility

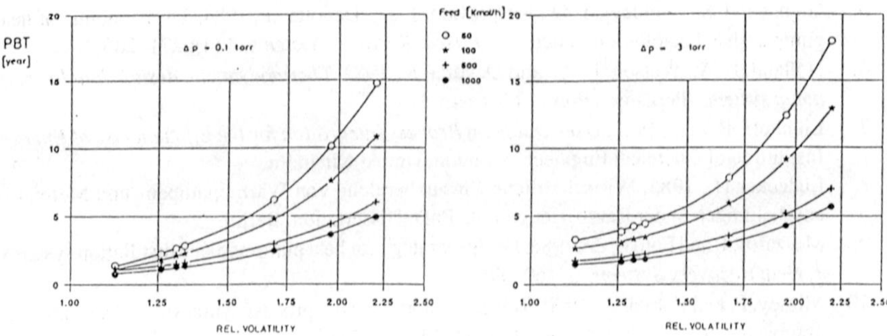

Figure 2-3 Payback time vs. relative volatility

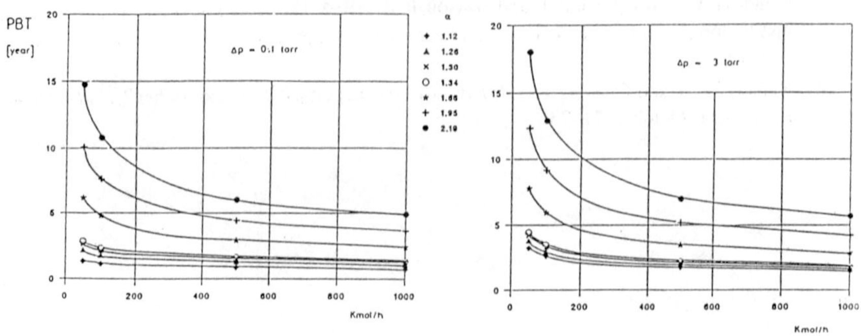

Figure 4-5 Payback time vs. column feed

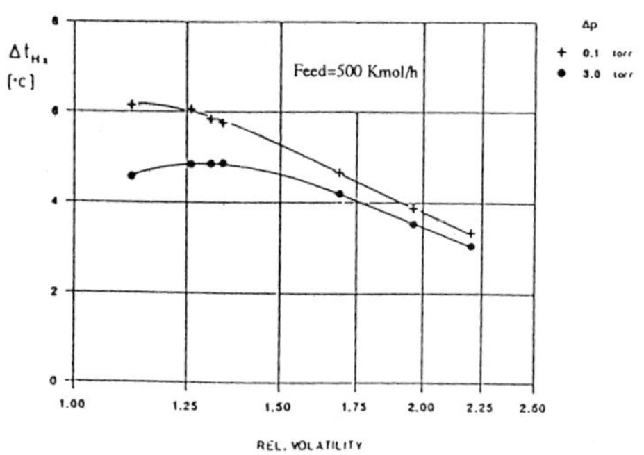

Figure 6 Temp. diff. in the heat exchangers vs. rel. volatility

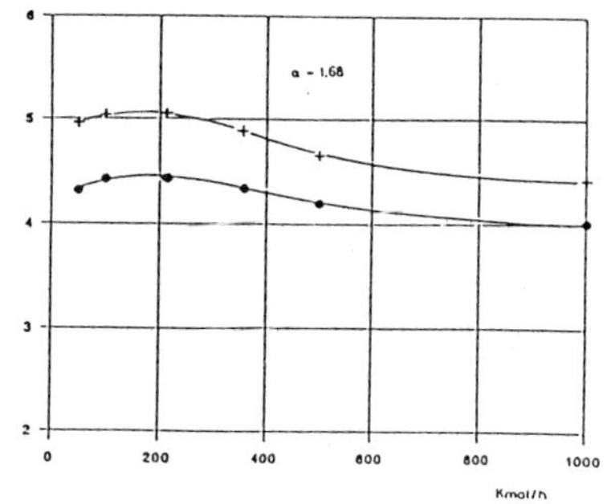

Figure 7 Temp. diff. in the heat exchangers vs. column feed

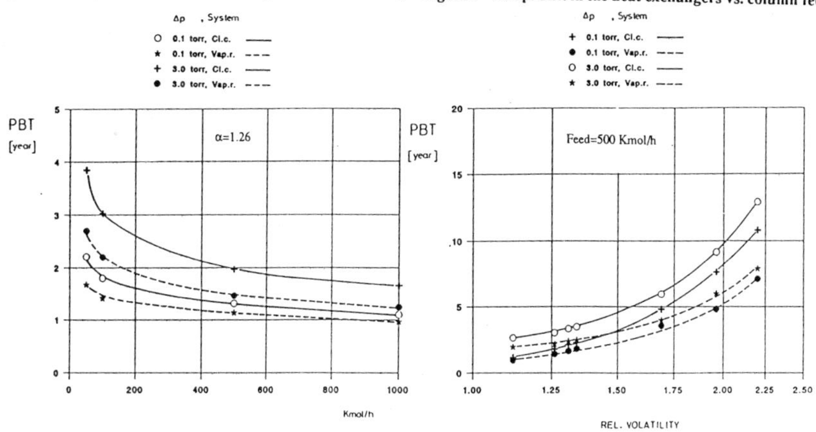

Figure 8-9 Comparison of the closed cycle process and the vapour recompression

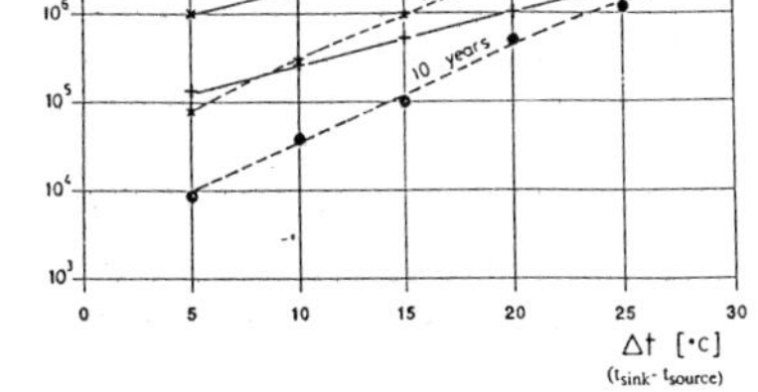

Figure 10 Economy of heat integration assisted by closed heat pump cycle

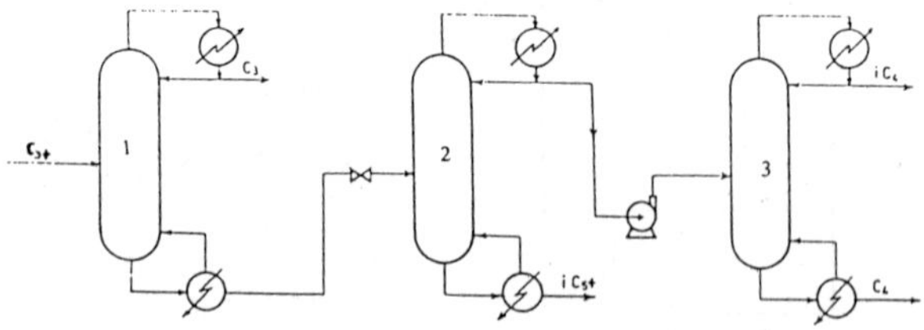

Pressure	bar	16	20	25	4.5	5
T_{TOP}	°C	40	51	62	40	38
T_{BOTTOM}	°C	110	122	135	90	52
R		3.6	4.2	5.1	1.5	5.7
Q_{COND}	GJ/H	15.0	15.5	15.8	7.7	8.5
Q_{REB}	GJ/H	16.3	17.6	19.0	5.1	8.7

Figure 11 Separation of light hydrocarbons

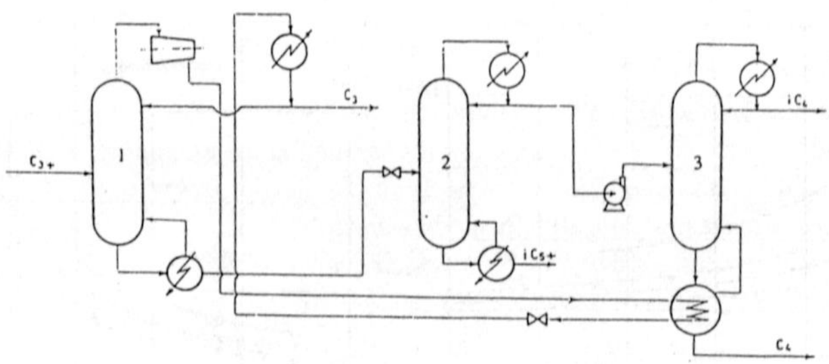

Figure 12, Column match assisted by heat pumping

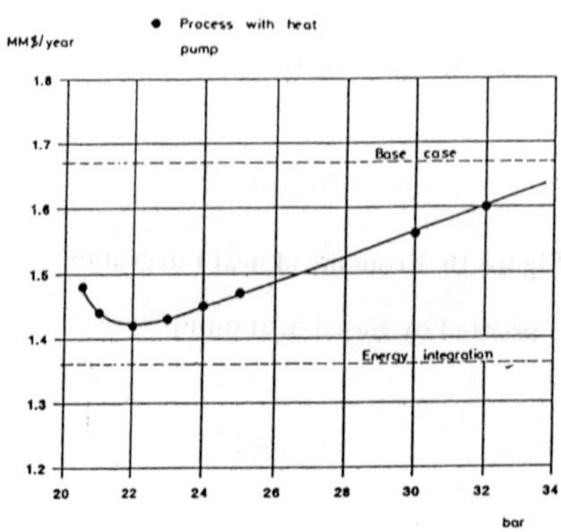

Figure 13 Total costs of the industrial case study

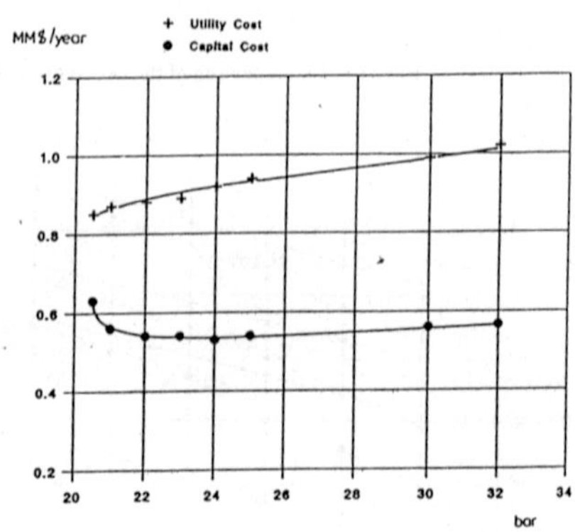

Figure 14 Utility and capital costs of process with heat pump

STRUCTURED COLUMN PACKINGS IN THE OLEOCHEMICAL INDUSTRY

Wilhelm Johannisbauer and Lutz Jeromin
Henkel KGaA, Verfahrenstechnik, 4000 Düsseldorf, Postfach 11 00, Germany

Oleochemicals such as glycerol, fatty acids, fatty alcohols and wax esters are refined by distillation and deodorization. As these products are extremely heat sensitive it is absolutely imperative that they are distilled at low temperatures, hence the low vacuum and short residence time.

Consequently processes were developed that applied structured packings as column internals. Processes for the distillation of glycerol, fatty acids and for deodorizing esters for cosmetic products will be presented.

Data concerning pressure drop and the mass transfer efficiency of industrial scale columns for the distillation of fatty alcohols are compared with pilot plant experiments that use test systems.

INTRODUCTION

Renewable raw materials which include natural fats and oils are becoming increasingly important in the chemical industry. Today about 8 million t per annum, being approx. 15% of the entire production of natural fats and oils, are processed into chemical products[1].

In order to utilize them within different products such as paints, varnishes, plastics, detergents and cosmetics, raw materials are presently processed in many ways (Figure 1).

For this purpose, distillation processes are used to a large extent for the separation of reaction products.

One special characteristic of fatty chemical substances is that they change under thermal stress. Colour, odour and taste of the products are unpleasantly intensified and can exert a negative effect with regard to the utilization of detergents, in cosmetics and the food sector.

DECOMPOSITION PRODUCTS OF FATS AND OILS

In spite of numerous investigations, no clear statement can yet be made as to which substances evoke changes. Fatty chemical products generally have the same secondary products, which come into being during the oxidation of the fat itself. Here, with oxygen taken from the atmosphere, a chain reaction takes place which comprises of several reaction steps which transform the fatty acid into anhydride, ketone and aldehyde. In this case, highly unsaturated fatty acids are oxidized more quickly than single unsaturated fatty acids[2]. Additionally, at higher temperatures, polymerization, due to the catalytic influence of metal traces, takes place.

Tests were carried out during the distillation of erucic acid, a C22-fatty acid with a double bond, in order to determine how much decomposition depends on temperature and residence time. The decomposition rate was defined as being the increase in the proportion of unsaponifiable material related to the initial value. The tests were carried out in a falling film evaporator with total condensation. The apparatus was made out of glass in order that catalytic influences could be eliminated[3].

With a residence time of < 120 min and a temperature of 280°C, the increase of UV can be kept at < 1%. An increase in temperature of 20°C leads to a 3-fold higher decomposition rate (Figure 2).

SELECTION OF COLUMN EQUIPMENT

In order to keep the thermal stress as small as possible during distillation or rectification of fatty chemical products, devices must be chosen which ensure a brief residence time and low temperature.

Here, in addition to the utilization of falling film or thin film evaporators as reboilers[4], attention must be especially paid to the selection and dimensioning of columns and column equipment. In order that work can be carried out under the lowest possible temperatures, the pressure drop per stage of the rectification column, the hold-up and the residence time, must be low. Therefore, in the last 10 years, regularly structured packings prevailed in the fatty chemical industry. Due to a greater number of manufacturers and increased sales the prices for such packings could be considerably reduced.

In the meantime there are 'tailor-made' packings, which through the variation of specific surfaces allow an optimal adaptation to each actual application. If a high separation stage number is necessary a fine-structured packing will rather be chosen or with respect to a low pressure drop, or in the case of fouling problems, a coarse-structured packing is more suitable.

Compared to random packings and trays the improved characteristics of these packings resulted in new distillation processes in the oleochemical industry.

Today considerably higher purities can thus be achieved. Through the extension of columns and equipment with sidestreams, several distillation steps can be combined within a plant.

SELECTED EXAMPLES FOR THE UTILIZATION OF STRUCTURED PACKINGS

GLYCEROL DISTILLATION

Glycerol from natural raw materials is a byproduct when fats and oils are saponified, hydrolysed and transesterified. Glycerol is used today to a great extent in the pharmaceutical, cosmetic and food industry[5]. In addition to the removal of salt and organic high-boiling impurities secondary products that are only contained in the range of 0,1%. also have to be separated. Predominantly 1,3-propandiol, glycerol monomethyl ether, ethyleneglycol and 1,2-propandiol were identified as secondary products. In addition, the demand for odour and colour of the final product is very high.

Extensive experimental tests on the laboratory and pilot scale resulted in a patented process[6], (Figure 3) in which the latest distillation technique was consistently used.

For a milder separation of the high-boiling components, where glycerol losses have to be kept at a minimum, a thin layer evaporator was selected into which dehydrated crude glycerol is fed. The glycerol vapour, with lower and higher boiling impurities, is separated in a rectification column which is filled with a structured packing. Here, a preliminary fraction is obtained at the top of the column where the lowboiling fractions accumulate up to 10%. High-boiling components are removed from the bottom of the column. The main fraction, the glycerol, is drawn off as a sidestream. The glycerol usually meets the pharmaceutical quality demands without the requirement for further treatment. If the crude glycerol is of a low quality, further bleaching with active carbon is necessary in order that a colourless product can be obtained.

Only by using packings with a low pressure drop was a bottom temperature of max. 165°C maintained, during which, as far as is known, no condensation polymerization of glycerol occurs.

The combination of precut separation and glycerol distillation in a column by means of a sidestream also allows the removal of secondary products that might have been formed during distillation.

FATTY ACID DISTILLATION

A similar process was developed for the working-up by distillation of fatty acids (Figure 4)[7].

Fatty acids resulting from the hydrolysis of fats and oils also contain triglycerides and other secondary products which produce a dark colour and an unpleasant odour in the crude fatty acid. Since fatty acids are also utilized as starting materials for detergents, cosmetic esters and glycerides in the food industry, they must meet very high standards with regard to purity and especially colour and odour.

The crude fatty acid is distilled in two stages in falling film evaporators; about 80% of the fatty acid is vaporized in the first stage under relatively mild conditions, and only a part of the remaining 20% is distilled at a higher temperature from the high-boiling residue. In the rectification column the vapours are separated first into a preliminary fraction at the top which contains the odour carriers and a part of the colour carriers, secondly into the main fraction as a sidestream and thirdly into a residue which contains the main part of the colour carriers.

DEODORIZING

Continuous deodorizing represents a third example for the application of structured packing in the fatty chemical industry.

During deodorization only a small quantity of more volatile components, normally below 1%, is usually separated from a high-boiling product. The mass transfer resistance lies in the liquid phase, thus it is important that the liquid is distributed as a thin film with high turbulence. Since the product in general is not treated further, deodorizing being the ultimate phase of a process, no darkening particularly or change in taste of the product should occur.

The process presented (Figure 5) has been used for some years on an industrial scale in order to deodorize ester for cosmetics and food[8]. By means of a preheater with a low hold-up the crude product is fed on the head of a column which is filled with a structured packing. In counter-current, super-heated steam, 5 to 10% of the feed, is blown into the bottom in order to lower the partial pressure of the impurities, which have to be separated, and their boiling temperature. In the vacuum pumps these substances are compressed together with steam and then condensed. The deodorized product is drawn off at the bottom of the column by means of a cooler. A bottom evaporator is not necessary, since the quantity of heat required for the distillation of the small quantity of impurities can be given by the latent heat of the feed.

Structured packings ideally meet the demands of this process and lead to considerably lower energy costs when compared to other deodorizing processes.

DIMENSIONING OF PACKED COLUMNS

With the examples presented for the application of structured columns in oleochemistry it is very clear that a computation of the column dimensions is difficult.

The impurities which influence colour and odour are often not completely detected in their composition. Equilibrium data concerning the impurities and the products are also only partially known. Therefore it is necessary to determine the required number of stages and process conditions with the aid of some possibly representative impurity examples, or with pseudo components determined with thermogravimetric measurement.

Distillation tests are still often necessary in these cases in order to check and to safeguard the calculated results. However, even with a task that can be well defined, certain deviations from the expected data result.

The separation of a fatty alcohol mixture in different production plants yielded the data specified in Table 1. By comparing the data with manufacturers specifications, efficiencies were measured which were lower by 10–20% and pressure drops which were higher by 10–20%.

Both the manufacturers of such packings and some institutes have measured the efficiency and the pressure drop on test mixtures and pilot columns (eg see References 9 and 10). The number of theoretical stages and the pressure drop are generally specified with reference to the packing height. Since the tests are carried out with a total reflux, the liquid load is also changed when the vapour load (F-factor) is varied.

Especially when working under vacuum, a low liquid load of below 1 m^3/m^2h was often obtained, resulting in the incomplete wetting of the packing, thus the liquid is unevenly spread. This results in a drastic drop in efficiency. Therefore further tests on packings are very important with regard to low liquid loads and high vapour loads.

When calculating rectification columns with structured or random packings a fundamental problem is the use of a plate model with suppose equilibrium on each theoretical plate. A model which describes the continuous mass and heat transfer at every location of the packing presents a more realistic approach. Gorak[11] revealed, that the concentration and temperature profiles calculated according to the mass transfer model considerably differs from the one calculated according to the plate model. This consideration is especially important for columns with sidestreams and for the control of these columns if temperatures or concentrations along the column are used as an control parameter.

CONCLUSIONS

With structured packings the oleochemical industry found a tool by which distillation processes can be realized without thermally damaging the products.

High purities can be obtained, several columns can be combined and an extensive energy integration is feasible. Great throughputs can be achieved in a plant.

The further development of process models and the test of mass transfer in packed columns will ensure a more precise dimensioning of the equipment and the optimization of control concepts.

REFERENCES

1. Dieckelmann, G. and Heinz, H.J., 1988, *The Basis of Industrial Oleochemistry* (Henkel KGaA, Düsseldorf).
2. Stage, H., 1987, *Fat Sci Technol*, 89: 9.
3. Peukert, E., Henkel KGaA, Internal Report.
4. Johannisbauer, W. and Marzenke, J., 1980, *Fette, Seifen, Anstrichmittel*, 82: 8.
5. Steinberner, U. and Preuß, W., 1987, *Fat Sci Technol*, 89: 8.

6. Brockmann, R., Jeromin, L., Johannisbauer, W., Meyer, H., Michel, O. and Plachenka, J., 1987, *EP 0 141 358.*

7. Jeromin, L., Johannisbauer, W., Thorausch, K., Scrapac, F., Hartmann, H. and Michel, O., 1986, *EP 0 132 570.*

8. Jeromin, L. and Johannisbauer, W., 1990, *DE 3 839 017.*

9. Spiegel, L. and Meier, W., 1987, *Proceedings Distillation and Absorption Conference 1987* (Institution of Chemical Engineers), A203 ff.

10. Fair, J.R. and Bravo, J.L., 1987, *Proceedings Distillation and Absorption Conference 1987* (Institution of Chemical Engineers), A183 ff.

11. Gorak, A., Wozny, G. and Jeromin, L., 1991, Industrial application of the rate-based approach for multicomponent distillation simulation, *Proceedings of Chemical Engineering, Karlsruhe, 1991.*

packing type	column dimensions	F-factor $ms^{-1} (kg \cdot m^{-3})^{1/2}$		liquid load $m^3 \cdot m^{-2} \cdot h^{-1}$		number of** theoretical stages m^{-1}		pressure drop mbar m^{-1}	
		top	bottom	top	bottom	practice	test	practice	test
Sulzer Mellapak 250 Y	diameter d = 1,1 m	1,60	1,66	2,00	5,23	2,3	3,0	0,9	0,8
		1,85	1,39	6,05	5,70	2,3	2,8	1,1	0,9
	height h = 12 m	1,86	2,46	4,50	10,3	2,1	2,7	1,3	1,1
Montz B1 250	d = 1,0 m h = 10 m	0,5	1,9	2,5	14,8	2,5	2,9	0,6	0,4
	d = 1,5 m h = 12 m	2,0	1,2	1,4	4,3	2,5	3,0	1,4	1,1
	d = 1,25 m h = 4 m	2,3	-	1,3	-	1,5*	3,0	1,8	1,8

table 1: results from production plants of fatty alcohol

* probable due to cold reflux

** practise: production of fatty alcohol
test : test data of the packing producer

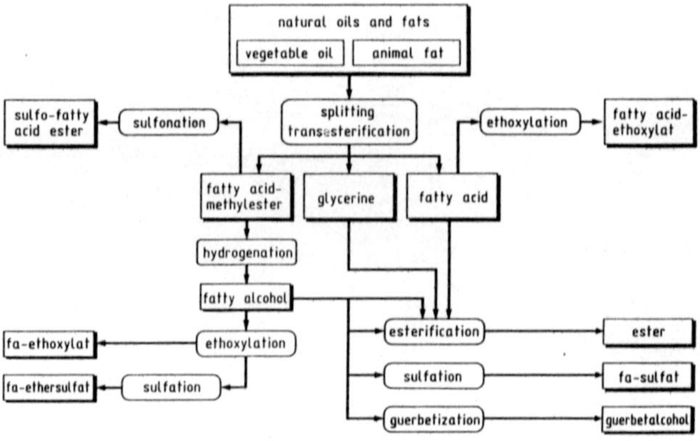

Figure 1: Oleochemistry, reactions and products

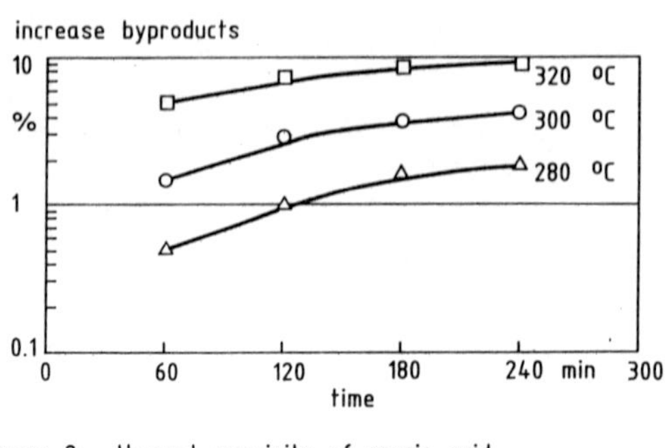

Figure 2: thermal sensivity of erucic acid

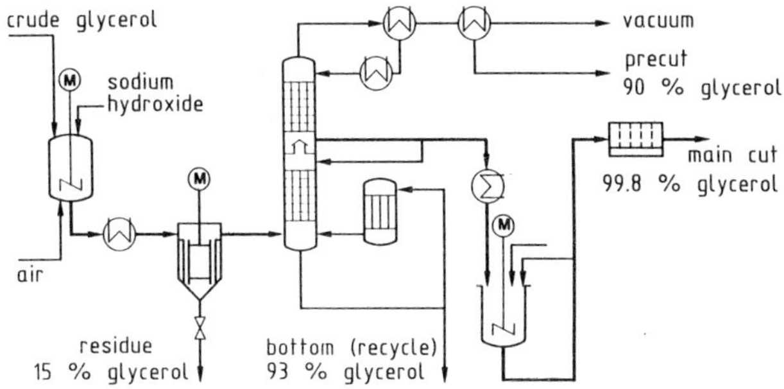

Figure 3: distillation of glycerol

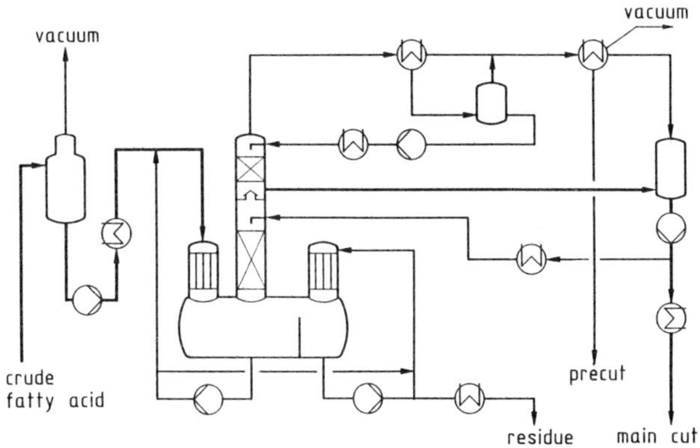

Figure 4: distillation of fatty acids

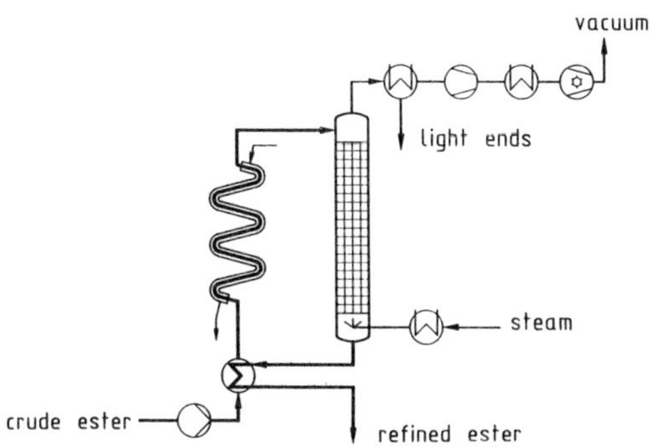

Figure 5: continuous deodorisation

A GENERALIZED PRESSURE DROP MODEL FOR STRUCTURED PACKINGS

L. Spiegel and W. Meier
SULZER Bros Ltd, Separation Columns, CH-8401 Winterthur

INTRODUCTION

The pressure drop model for structured packings of Spiegel and Meier[1] is valid for low liquid loads (<20 m^3/m^2h) only. For many years structured packings have been used not only in vacuum distillation at low liquid loads, but also in applications with very high liquid loads like strippers (>200 m^3/m^2h) or high pressure distillation[2]. Therefore the need to generalize the pressure drop calculation was obvious.

In the last five years a number of publications on pressure drop prediction for structured packings appeared. Bravo et al[3] developed a model valid in the region below the loading point, Stichlmair et al[4] used a more fundamental approach based on holdup and dry pressure drop to predict the wet pressure drop up to flooding with some success. Billet[5] and Mackowiak[6] argued their pressure drop equation developed for random packings to be valid for structured packings as well. These authors based their analysis on their own experimental data or data published by packing vendors. In most cases the highest liquid loads did not exceed 70 m^3/m^2h.

It is the purpose of this work to present a generalized pressure drop model with an extended range of validity with respect to liquid load, liquid viscosity and surface tension.

PRESSURE DROP MODEL

As an example, the pressure drop of MELLAPAK 250.X for air/water at ambient conditions is plotted against the gas load (in terms of the F factor F_V) at constant specific liquid load l in Figure 1.

The model is sketched in Figure 2. Its basic characteristics can best be seen in a double logarithmic representation of pressure drop versus gas load. The point A (pressure drop loading point) divides the operating range in two different regions:

REGION 1

The wet pressure drop dp/dz is parallel to the dry pressure drop $(dp/dz)_o$ for not too high liquid loads.

REGION 2

The pressure drop increases with a higher slope than the dry pressure drop and passes through the point C (100%–capacity).

DRY PRESSURE DROP

The dry pressure drop of MELLAPAK 125.X, 250.X, 250.Y and 500.Y is given in Figure 3. The pressure drop is well represented in the log-log diagram by a straight line with slope 1.8 for all four packings.

CAPACITY

The capacity — defined as point of operation where the local pressure drop is 12 mbar/m (point C in Figure 2) — of MELLAPAK 125.X, 250.X, 250.Y and 500.Y is given in Figure 4 in form of a Wallis diagram. It is a plot of $c_G 0.5$ versus $c_L 0.5$, where

c_G gas capacity factor (= $F_V/(\rho_L-\rho_G)^{0.5}$) (m/s)
c_L liquid capacity factor (= $(l/3600)(\rho_L/(\rho_L-\rho_G))^{0.5}$) (m/s)
ρ_G, ρ_L gas, liquid density (kg/m^3)

According to Spiegel and Meier[1] the capacity data can be represented by a set of two straight lines for each packing. These capacity lines are used to calculate point C. The influence of surface tension on capacity is accounted for in the model as well.

LIQUID HOLDUP

The liquid holdup h_L is only a function of the liquid load below its loading point (point B in Figure 2). In Figure 5 the liquid holdup of various MELLAPAK types is plotted in function of liquid load for the air/water system at ambient conditions. According to reference 7 holdup is proportional to $\mu_L 0.25$, where μ_L is the liquid viscosity.

WET PRESSURE DROP IN REGION 1

Two effects increase the wet pressure drop compared to the dry pressure drop:

1. Due to the space occupied by the liquid the space for the gas is constricted, therefore the pressure drop increases. The increase can be well predicted by the channel model where the pressure drop depends on the hydraulic diameter and the effective velocity of the gas. The increase in pressure drop is therefore given by the following factor:

$$f_1 = (1 - h_L)^{-2.5} \qquad (1)$$

The derivation of this formula is based on structured packings, where a void fraction of 1 may be asssumed.

2. Due to the downward flow of the liquid phase the relative velocity between the liquid and gas phase is not negligible. The pressure drop due to friction is based on the relative velocity rather than the absolute veloctiy of the gas phase. Therefore the increase in pressure drop has to be calculated by the following factor:

$$f_2 = \left(\frac{w_L' + w_G'}{w_G'}\right)^{1.8} \qquad (2)$$

where

$w_G' = F\,V/(\rho_G 0.5(1-h_L))$ (m/s)
$w_L' = l/(3600\,h_L)$ (m/s)

w_G', w_L' are the effective vertical velocity components of the two phases.

The wet pressure drop dp/dz is calculated from the dry pressure drop using the two factors:

$$dp/dz = f_1 f_2 (dp/dz)_o \tag{3}$$

These correction factors give a good description of the wet pressure drop up to 80–100 m^3/m^2h with the air/water system.

For higher liquid loads a third correction factor f_3 is necessary. In Figure 6 the factor f_3 is plotted against a transformed holdup

$$h_L^* = \frac{h_L \sigma^{0.8}}{\left(a_I^{0.75} \mu_L^{0.25}\right)} \tag{4}$$

where

σ surface tension (mN/m)

a_I geometric surface area (m^2/m^3)

μ_L liquid viscosity (cP)

The transformed holdup allows a universal description of the factor f3 for all packings, different liquid viscosities and surface tensions. For h_L* 0.0046 f_3 is unity, indicating the validity of the combination of the two model assumptions. Above 0.0046 it increases drastically. The reason is not yet clear, but it could be explained by a change in the flow regime (from channel flow to dispersed flow).

For values of h_L* 0.0046 the f_3 is calculated by a linear function fitted to the data.

REGION 2

Between point A and C a cubic polynomial is used to represent the pressure drop curve:

$$\log(dp/dz) = ax^3 + bx^2 + cx + d \tag{5}$$

where

$$x = \frac{\log(F_V/F_{V,max}) - \log(0.45)}{-\log(0.45)}$$

$F_{V,max}$ = F factor at 100% capacity

This definition of x limits its values between 0 and 1.

The polynomial starts in point A (at 45% of capacity based on constant liquid load) with the same slope m as in region 1 and passes through point C. These boundary conditions fix the parameters of the polynomial:

a = $\log(12) - m - \log(dp/dz)_A$

b = 0

c = m

d = $\log(dp/dz)_A$ wet pressure drop in point A

VALIDATION OF THE MODEL

The complete pressure drop model was implemented into a computer program and tested on our database containing more than 400 data points. The results are given in form of parity plots of measured versus calculated pressure drop for MELLAPAK 125.X, 250.X, 250.Y and 500.Y (Figure 7). For most of the data points the deviation between observed and calculated pressure drop is less than ±20%. For a few points, especially in the range above 3 mbar/m, the difference is greater. This may be partially explained by experimental errors in the determination of the gas load, liquid load or pressure drop that are in the range of 5%.

CONCLUSION

A model was developed to predict pressure drop of structured packings from low to very high specific liquid loads, considering liquid viscosity and surface tension. The model was tested with experimental data (more than 400 data points) obtained from several air/liquid systems at atmospheric conditions in a column of 1 m internal diameter. The accuracy of the prediction is for liquid loads below 100 m^3/m^2h at least ±20%. The model is available through our PC-program SULPAK 4.0 (1991).

NOMENCLATURE

a_I	geometrical surface area of packing (m^2/m^3)
c_G, c_L	gas, liquid capacity factor (m/s)
F_V	F factor (Pa0.5)
$F_{V,max}$	maximum F factor (at 100% capacity) (Pa0.5)
h_L	liquid holdup (% of packing volume)
h_L*	transformed holdup (= $h_L\sigma0.8/(a_I0.75\mu_L0.25)$)
l	specific liquid load (m^3/m^2h)
m	slope of wet pressure drop
dp/dz	wet pressure drop (mbar/m)
$(dp/dz)_o$	dry pressure drop (mbar/m)
w_G'	effective vertical velocity of gas phase (m/s)
w_L'	effective vertical velocity of liquid phase (m/s)
μ_L	liquid viscosity (cP)
ρ_G, ρ_L	gas, liquid density (kg/m^3)
σ	surface tension (mN/m)

LITERATURE

1 Spiegel, L. and Meier, W., 1987, Performance characteristics of the various MELLAPAK types, *IChemE Symp Ser 104*: A203.

2 Sulzer Brothers, Winterthur, Switzerland, 1991, *Separation columns for distillation and absorption*, Publication No. 22.13.06.

3 Bravo, J.L., Rocha, J.A. and Fair, J.R., 1986, *Hydrocarb Proc*, 65: 45–49

4 Stichlmair, J., Bravo, J.L.and Fair, J.R., 1989, *Gas Sep Purif*, 3: 19–28

5 Billet, R., 1989, *Packed column design and analysis*, Ruhr University, Bochum.

6 Mackowiak, J., 1991, *Fluiddynamik von Kolonnen mit modernen Füllkörpern und Packungen für Gas/Flüssigkeitssysteme*, Salle, Franfurt am Main.

7 Suess, P. and Spiegel, L., 1992, Holdup of MELLAPAK structured packings, *Chem Eng and Proc*, 31, (2): 119–124.

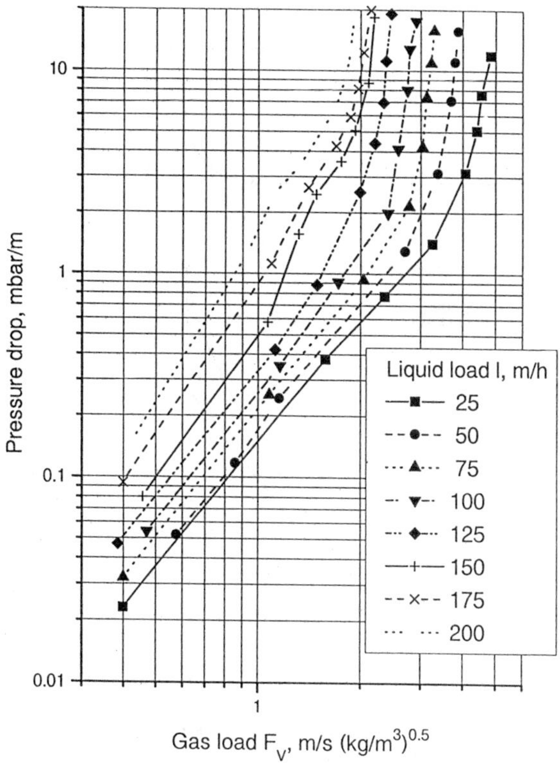

Fig. 1 Pressure drop of MELLAPAK 250.X at various liquid loads. Air/water at 30°C, atmospheric pressure column diameter = 1 m

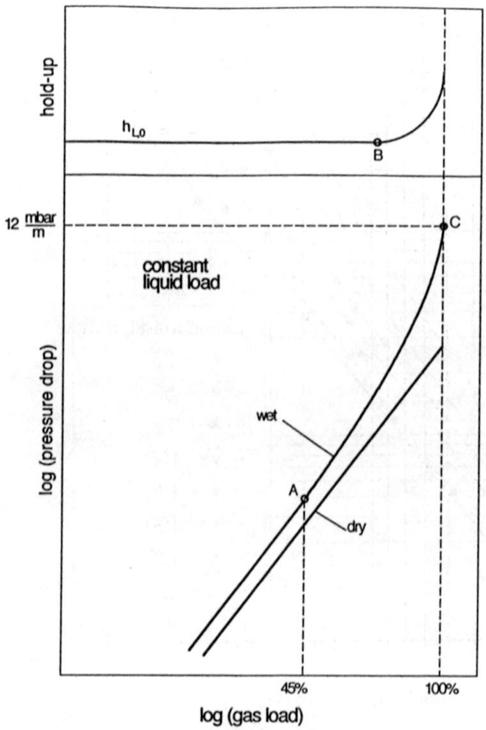

Fig. 2 Behaviour of pressure drop and holdup with gas load at constant liquid load
A: pressure drop loading point (45% capacity)
B: holdup loading point
C: 100% capacity point

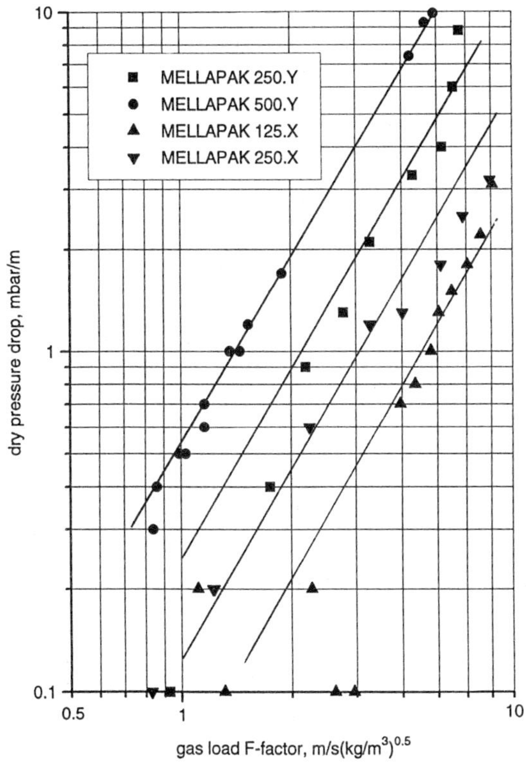

Fig. 3 Dry pressure drop of MELLAPAK 125.X, 250.X, 250.Y, 500.Y
atmospheric pressure, column diameter = 1m

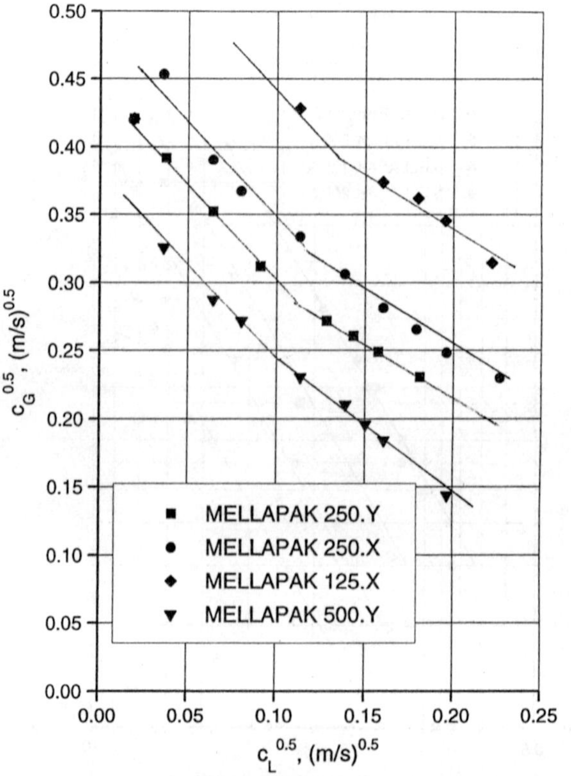

Fig. 4 Capacity of MELLAPAK 125.X, 250.X, 250.Y, 500.Y
Air/water at 30°C, atmospheric pressure,
column diameter = 1m

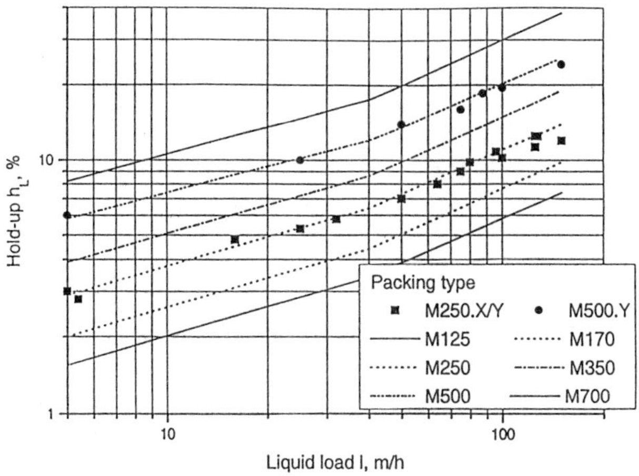

Fig. 5 Liquid holdup below loading point of various MELLAPAK types
Water at 30°C, atmospheric pressure,
column diameter = 1 m,
(data taken from [7])

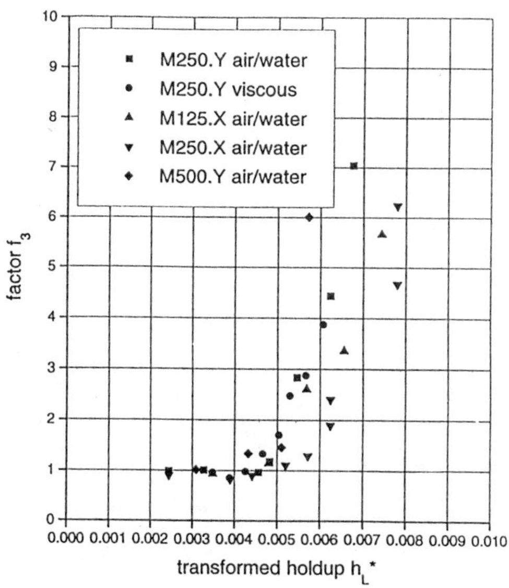

Fig. 6 Factor f_3 versus transformed holdup for MELLAPAK 125.X, 250.X, 250.Y and 500.Y, different air/liquid systems

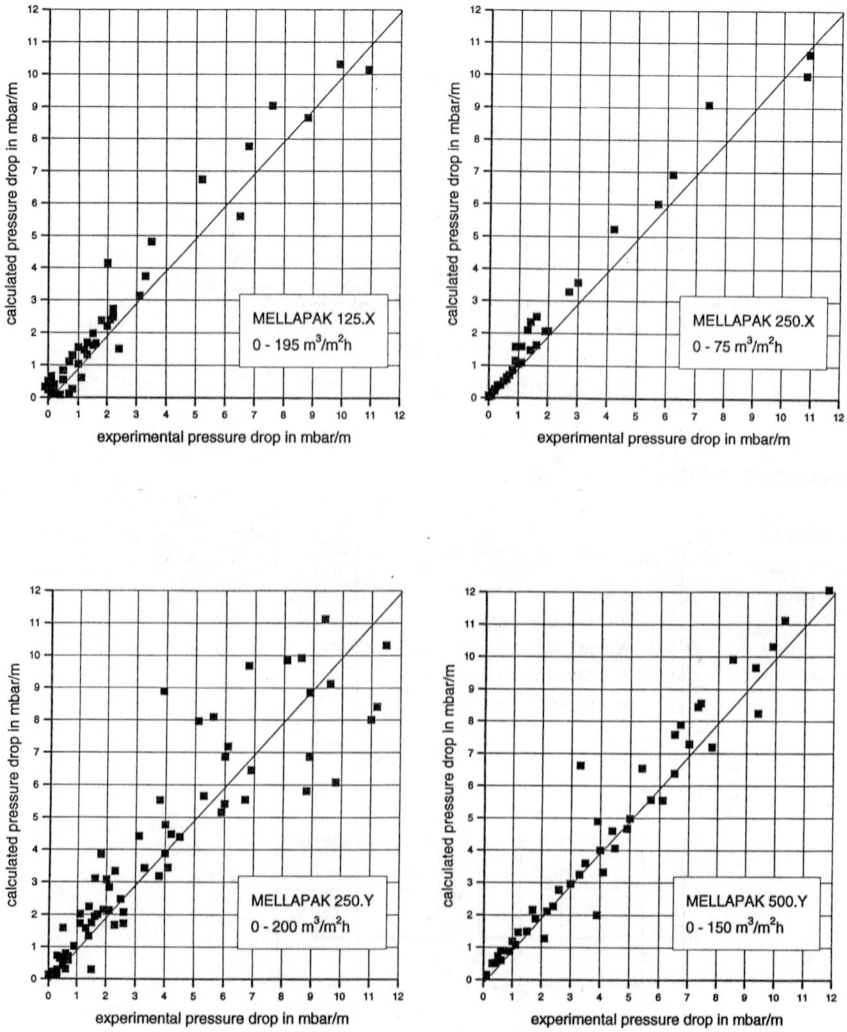

Fig. 7 Parity plot of experimental and calculated pressure drop of MELLAPAK 125.X, 250.X, 250.Y and 500.Y

DIRECT CONTACT HEAT TRANSFER IN PACKED COLUMNS

M. Laso* and P. Bomio#
* ETH-Zentrum, CH–8092 Zürich, Switzerland
\# Absorption Technology, TRK 0614, Sulzer Bros., CH–8041 Winterthur, Switzerland

A thermodynamically rigorous model for simultaneous heat and mass transfer in countercurrent gas-liquid contacting columns is presented. The model is based on two-film theory rate equations. It accounts for non-ideal gas behaviour and includes high-flux corrections to heat and mass transfer. The main strength of the model lies in its ability to deal with gas-liquid systems for which the Lewis number is very different from one and can vary along the column axis, as well as with high-temperature and/or high-pressure applications in which the equilibrium condensables content of the gas phase at the outlet can be very high (sometimes as high as 0.5 mol fraction). For a Lewis number of unity and for low saturation vapor contents the model correctly simplifies to the standard design procedure based on enthalpy as the driving force. A comparison between model predictions and experimental data from a synthesis gas saturation column at 52.3 bar and 200°C is presented.

INTRODUCTION

Countercurrent gas-liquid contacting devices for simultaneous heat and mass transfer belong to the traditional chemical engineering unit operations. The cooling effect obtained by evaporating a fraction of a liquid stream to be cooled into a surrounding unsaturated gas has been exploited for a long time. Similarly, a hot liquid stream can be used to heat and saturate a dry gas stream. Another application is quenching of a hot process gas stream (typically from a reactor or a furnace) by means of a liquid. Countless such units have been built and are widespread in the chemical process industry, in heat and/or power generation plants, in air-conditioning, refrigeration and cryogenic services, etc. In a large fraction of these cases, the gas-liquid system of interest is air-water.

The design of Simultaneous Heat and Mass Transfer (SHMT) devices has progressed from purely empirical procedures to more sophisticated ones, the fundamentals of which were already well established in the fifties[1,2,3,4]. During the last four decades the vast majority of SHMT devices has been designed following such procedures. They succeeded because they were mainly applied in the interchange of heat and mass for the air-water system, for which the wet bulb curves are practically coincident with the adiabatic cooling curves. This allows lumping the driving forces for heat and mass transfer into a single enthalpic driving force. This reduction to a single driving force simplifies the calculation and makes the problem amenable to relatively simple numerical or even graphical methods. For some specific applications like cooling-tower design, little more than table look-up is required in order to perform a reliable design.

There are however many applications of practical industrial relevance in which the Lewis relation does not hold. In such cases the use of a specific enthalpy driving force ceases to be valid. The purpose of this work is to present a thermodynamically consistent design method for countercurrent columns under simultaneous heat and mass transfer conditions in which:

(a) the Lewis number may deviate markedly from unity,
(b) the concentration of condensables in the gas phase may be very high,
(c) the gas phase may behave non-ideally,
(d) the liquid may be a solution containing dissolved non-volatile components displaying large heats of dilution,
(e) there may be an arbitrary number of components in each phase, but only one component may be transferred between the gas and the liquid phases.

These conditions are encountered in numerous applications of the petrochemical (quenching towers in ethylene plants), chemical (quenching of reactor effluent, saturation of synthesis gases prior to shift reaction) and cryogenic industries.

It is important to emphasize that the objective of the paper is to establish a thermodynamically consistent model for the design of SHMT columns of general validity. The model has to be supplemented with closure relationships describing the gas-liquid system (phase equilibria, diffusivities) and column internals (heat and mass transfer coefficients, effective interfacial area). These closure relationships are obviously case specific. It is not the purpose of this contribution to evaluate or recommend calculation procedures for phase equilibria, physical properties of packing characteristics. In spite of this, and as an illustration of the method, two SHMT columns are simulated using specific closure equations which seem to reproduce the experimental results satisfactorily. The application of the present model to other systems and/or column internals requires knowledge of the specific characteristics of both.

THE DESIGN EQUATIONS

In order to set up the design equations, we consider the differential height element across the column as shown in Figure 1, in which the different mass and heat streams are indicated. The sign convention for heat and mass transfer is such that the transfer of a differential amount of heat or mass ($d\dot{Q}$ or $d\dot{N}$) is positive when the transfer takes place from the liquid into the gas phase. We split this differential element in the three control volumes ①, ② and ③ shown in Figures 2 through 4 and set up the corresponding mass and energy balance equations. Control volume ①:

$$d\dot{N}_L - d\dot{N}_A = 0 \tag{1}$$

$$d(\dot{N}_L\bar{H}_L) - H_{ALPh}d\dot{N}_A - d\dot{Q}_L = 0 \tag{2}$$

Control volume ②:

$$d\dot{Q}_L + (H_{ALPh} - H_{AVPh})d\dot{N}_A - d\dot{Q}_{mPh} = 0 \tag{3}$$

where H_{AVPh} is the partial molar enthalpy of component A in the gas phase at the interface.

Control volume ③:

$$-d\dot{N}_m + d\dot{N}_A = 0 \tag{4}$$

$$-d(\dot{N}_m\bar{H}_m) + H_{AVPh}d\dot{N}_A + d\dot{Q}_{mPh} = 0 \tag{5}$$

The mass balance equation for the transferring component A and the overall energy balance yield:

$$- d(\dot{N}_m y_A) + d\dot{N}_A = 0 \qquad d(\dot{N}_L \bar{H}_L) - d(\dot{N}_m \bar{H}_m) = 0 \tag{6}$$

The mol fraction of vapor in the gas phase and the molar flowrate of the gas phase can be obtained from mass balances between the top of the column and a point at height z and between the bottom of the column and the same point z, respectively:

$$x_A = 1 - \dot{N}_{Ltop}(1 - x_{Atop})/\dot{N}_L \tag{7}$$

$$\dot{N}_m = \dot{N}_{mbot}(1 - y_{Abot})/(1 - y_A) \tag{8}$$

Equations (1) through (8) are of general validity and, through the use of partial molar enthalpies, thermodynamically correct. Equations (7) and (8) have been formulated for the case of a gas saturation column, in which the known column feeds are the gas at the bottom and the liquid at the top. Analogous balances for other configurations can be set up trivially. In order to derive working equations from the formal relationships (1) through (8) we will now introduce the necessary closure equations.

In the following development we will consider no liquid-side resistance to mass and heat transfer. This approximation is by no means essential and can be easily incorporated in the present model if required; it makes the numerical problem easier to solve, since it leads to equality of temperature at the interface and in the bulk liquid, thus eliminating the iterative determination of T_L at each integration step. Besides, it implies the equality of the partial molar enthalpies of A at the interface and in the bulk liquid. The extension of Equations (1) to (8) to the case of finite liquid-side resistance is straightforward[9]. The inclusion of liquid-side resistance in preliminary runs was found not to be necessary to reproduce the experimental temperature and concentration profiles. This may be due to the special mass-transfer characteristics of the structured packing Mellapak used in the experiments[11], although it should be a reasonable approximation for most internal types at low to moderate specific liquid loads.

Two other exact closure relations are obtained from the total differentials of the molar enthalpy streams of gas and liquid for the case in which only the component A is transferred:

$$d\bar{H}_m = \left(\frac{\partial \bar{H}_m}{\partial T_m}\right)_{P,n_j} dT_m + (H_{mA} - \bar{H}_m)\frac{d\dot{N}_A}{N_m} \tag{9}$$

$$d\bar{H}_L = \left(\frac{\partial \bar{H}_L}{\partial T_L}\right)_{P,n_j} dT_L + (H_{AL} - \bar{H}_L)\frac{d\dot{N}_L}{N_L} \tag{10}$$

The remaining closure equations are now only dependent on the system we have to deal with, on the operating conditions and on the specific details of the SHMT column, that is, we need:

(i) a description of the phase equilibrium for the gas-liquid system;

(ii) a description of the kinetics controlling heat and mass transfer.

In the way the model is formulated, it can accommodate any form of (i) and (ii). For concreteness, we will make use of the standard Sherwood and Nusselt relationships:

$$Sh = C_1 Re^{C_2} Sc^{C_3} \qquad Nu = C_4 Re^{C_5} Pr^{C_6} \tag{11}$$

Division of these equations under the assumption of analogy between heat and mass transfer yields:

$$\frac{\alpha}{\beta} = c\bar{C}_p\left(\frac{a}{D_{Am}}\right)^{1-n} = c\bar{C}_p Le^{1-n} \tag{12}$$

It is important to emphasize that the separate treatment of heat and mass transfer leading to Equation 12 is only valid when the two transport processes do not influence each other, which is correct at low fluxes. For high transfer rates it can be shown[4] that correction terms of the following form have to be included:

$$\frac{\alpha^+}{\beta^+} = c\bar{C}_p\left(\frac{a}{D_{Am}}\right)^{1-n} \frac{e^{\varepsilon}}{e^{\varepsilon}-1} \frac{(y_{AI}-y_{Am})}{\ln\left(\frac{1-y_A}{1-y_{AI}}\right)} \tag{13}$$

where

$$\varepsilon = \frac{\bar{C}_{pAV}\ln\left(\frac{1-y_{Am}}{1-y_{AI}}\right)}{\bar{C}_{pm}\left(\frac{a}{D_{Am}}\right)^{1-n}} \tag{14}$$

Substituting the kinetic expressions:

$$d\dot{Q}_{Ph} = \alpha^+(T_L - T_m)A_q a_{eff}\, dz \qquad d\dot{N}_A = c\beta^+(y_{AI} - y_{Am})A_q a_{eff}\, dz \tag{15}$$

and Equations (9) through (14) into the balances (1) through (6) and after some algebraic manipulation we obtain the final set composed of the two algebraic equations (7) and (8) and the following four differential equations:

$$\frac{dT_L}{dz} = \frac{A_q a_{eff}\,\beta^+}{\dot{N}_L\left(\frac{\partial \bar{H}_L}{\partial T_L}\right)}\left[\frac{\alpha}{\beta}(T_I-T_m)+c(y_{AI}-y_{Am})(H_{AVPh}-H_{AL})\right] \tag{16}$$

$$\frac{dT_m}{dz} = \frac{A_q a_{eff}\,\beta^+}{\dot{N}_m\left(\frac{\partial \bar{H}_m}{\partial T_m}\right)}\left[\frac{\alpha}{\beta}(T_I-T_m)+c(y_{AI}-y_{Am})(H_{AVI}-H_{Am})\right] \tag{17}$$

$$\frac{dy_A}{dz} = c\beta^+(y_{AI}-y_{Am})(1-y_{Am})\frac{A_q a_{eff}}{\dot{N}_m} \tag{18}$$

$$\frac{d\dot{N}_L}{dz} = c\beta^+(y_{AI}-y_{Am})A_q a_{eff} \tag{19}$$

This set of equations together with the known conditions of the feed streams represent a two point boundary value problem whose solution gives the profiles of the different variables (temperatures and flowrates of each component) along the column. Thus, for known input streams it is possible to predict the conditions at any point in the column as well as the outlets. These design equations are general and thanks to the formulation in partial molar enthalpies can be extended to highly non-ideal systems, Lewis number less than or greater than one and arbitrarily high molar fraction of vapor in the gas phase. Extension to finite liquid-side resistance and to co-current contacting devices (for which the model reduces to a mathematically much easier initial value problem) are trivial. The model can be shown to reduce exactly to the conventional transfer unit approach if the corresponding simplifying assumptions[5] are made.

The model has been implemented as a user-defined subroutine within the PROCESS®[6] flowsheet simulator. Calculation of physical properties and phase equilibria are performed within the PROCESS environment while the calculation of transfer coefficients, partial molar quantities, balance equations, diffusivities as well as the whole numerical solution procedure are performed in user-supplied subroutines. The boundary value problem is solved by means of a 'shooting'[7] technique combining a fourth-order, variable step Runge-Kutta with the third-order variable-step method of Michelsen for the cases in which the model equations turn 'stiff'. In high-flux situations and/or with large temperature and concentrations gradients the model equations can be highly nonlinear, making convergence of the shooting procedure problematic. Therefore, a sophisticated convergence-handling routine has been developed, that upon detection of stiffness and/or non-convergence of the shooting procedure switches to the appropriate integrator or, if necessary, can perform a heuristic search in the manifold of starting values and restart the algorithm until convergence is achieved.

In order to estimate the reliability and efficiency of the model, the 32 benchmark cases of Poppe[8] were run using the present model and Poppe's correlations for mass transfer coefficients and diffusivity. The average deviation between Poppe's results and ours was 0.83% (absolute: 0.18 K) for the gas outlet temperature and 3.3% (absolute 0.0006 kg vapour/kg dry air) for the outlet gas water content. These small discrepancies stem from our using slightly different vapor-liquid equilibrium data. Similarly good agreement was found in all other benchmarks for which an NTU-HTU-based hand calculation was possible.

Since the real strength of the model lies in its ability to treat highly non-ideal cases, special importance was given to its testing under such conditions. However, there are virtually no experimental data and/or numerical examples in the open literature which could be used for a comparison. For this reason the model was compared with experimental values obtained in (A) a low pressure air-water column as well as in (B) an industrial synthesis gas (52% vol. CO, 48% vol. H_2) saturator operating at 52.3 bar and temperatures between 511 K (top) and 298 K (bottom). Full details about both columns and the experimental program can be found elsewhere[9,10]. In both cases, the only transferring component is water. Its diffusivity was estimated using the method of Wilke and Lee[13] and including corrections for high pressure and finite dilution whenever necessary[12,14]. Mass and heat transfer coefficients and effective area were obtained from empirical correlations especially developed for structured packing[15,16,17,11].

Column (A) operated close to the boiling point of water; the air stream at the outlet thus reached water contents as high as 0.73 mol fraction (to be compared with the 0.01–0.05 typical range in which the NTU-HTU approach can be used). The model predictions for outlet temperatures and water content were found to deviate on average by 6% (1σ).

Column (B) is an even more stringent test of the model because of (a) the high pressure and temperature in the column, (b) the presence of large amounts of hydrogen, leading to a Lewis number of between 2.1 and 2.7 (top and bottom respectively) and (c) the drastic change in water vapor concentration in the gas phase (from virtually zero to more than 0.50 mol fraction). The comparison between model predictions and experimental data is summarized in the following table (for a description of the column and the calculation details see[9]):

	Measured	Calculated	
Outlet water temperature	445.2	446.7	K
Outlet gas temperature	498.2	493.6	K
Outlet gas steam content	0.515	0.510	mol fraction
Steam production	1.39×10^3	1.35×10^3	kg/h

Thus we see that the model can successfully handle highly non-ideal situations for which virtually no design procedures exist. The discrepancy of 5 K in the gas outlet temperature is due to supersaturation of the gas phase within the column, leading to condensation of water vapor as droplets within the gas stream. The heat of condensation of the droplets further heats the gas stream. This effect has also been included in the model[9] and its consideration has been shown to significantly reduce the discrepancy.

CONCLUSIONS

A design model for direct heat and mass transfer columns has been presented. It is based on a thermodynamically rigorous formulation of the balance and kinetic equations employing partial molar quantities. It can successfully deal with high-pressure, high-temperature applications and with systems for which the Lewis number differs greatly from unity and can even be variable along the column axis. High relative saturations can be handled properly. The performance of the model on Poppe's benchmarks is excellent. A comparison with a real industrial synthesis gas saturator shows that the model can correctly reproduce the operation of a highly non-ideal system at high saturation values.

ACKNOWLEDGEMENT

We would like to thank an anonymous referee for his numerous constructive remarks.

REFERENCES

1. Treybal, R.E, 1955, *Mass Transfer Operations* (McGraw-Hill Book Co., New York).
2. Brown, G.G., 1950, *Unit Operations* (John Wiley & Sons, Inc., New York).
3. Mickley, H.S., 1949, *Chem Eng Progr*, 45: 739.
4. Ackermann, G., 1937, Wärmeübergang und molekulare Stoffübertragung im gleichen Feld bei grossen Temperatur- und Partialdruckdifferenzen, *VDI-Forschungsheft 382*, (VDI-Verlag, Düsseldorf).
5. Fair, J.R., 1972, *Chem Eng Progr Symp Ser, No. 118*, 68 (1).
6. PROCESS is a registered trademark of Simulation Sciences.
7. Press, W.H., Flannery, B.P., Teukolsky, S.A. and Vetterling, W.T., 1988, *Numerical Recipes; The Art of Scientific Computing* (Cambridge University Press).
8. Poppe, M., 1973, Wärme- und Stoffübertragung bei der Verdunstungskühlung im Gegen-

und Kreuzstrom, *VDI-Forschungsheft 560* (VDI-Verlag, Düsseldorf).
9. Winkler, S., 1989, Verdunstung bei höheren Drücken, *MSc Thesis* (University of Stuttgart).
10. Jaschinski, K., 1989, Stoff- und Wärmeübertragung in einer Gegenstromkolonne, *MSc Thesis* (University Hamburg).
11. Brito, M., 1991, *PhD Thesis No. 984* (Federal Polytechnic Institute Lausanne).
12. Reid, R.C., Sherwood, T.K. and Prausnitz, J.M., 1987, *Properties of Gases and Liquids* (McGraw-Hill).
13. Wilke, C.R. and Lee, C.Y., 1955, *Ind Eng Chem*, 47, 1253.
14. API, 1987, American Petroleum Institute Tables.
15. Spiegel, L. and Bomio, P., 1987, *Chem-Ing Tech*, 59 (2): 130.
16. Spiegel, L. and Meier, W., 1987, *4th Int Symp Dist Abs, Brighton.*
17. Bomio, P., 1977, *Chem-Ing Tech*, 11: 895.

NOMENCLATURE

z	Coordinate along vertical column axis, m
a_{eff}	Effective specific area of packing, m^2/m^3
A_q	Column cross section, m^2
T	Absolute temperature, K
N	Molar flowrate, kgmols/s
Q	Rate of heat flow, KJ/s
y,x	Molar fraction in the gas or liquid phase
Nu	Nusselt number
Sh	Sherwood number
Re	Reynolds number
Sc	Schmidt number
Pr	Prandtl number
Le	Lewis number
α, α^+	Heat transfer coefficients, kW/m^2K
β, β^+	Mass transfer coefficients, m/s
c	Molar density, $kmol/m^3$
$\tilde{C}p$	Specific molar heat capacity, kJ/kmol K
H	Specific molar enthalpy, kJ/kmol
a	Thermal diffusivity, m^2/s
D_{Am}	Diffusion coefficient, m^2/s
C_1 to C_6, n	Numerical constants

SUBINDICES

L	Liquid phase
m	Bulk gas phase
A	Transferring component
I,Ph	Interphase
V	Vapor
bot	Column bottom
top	Column top

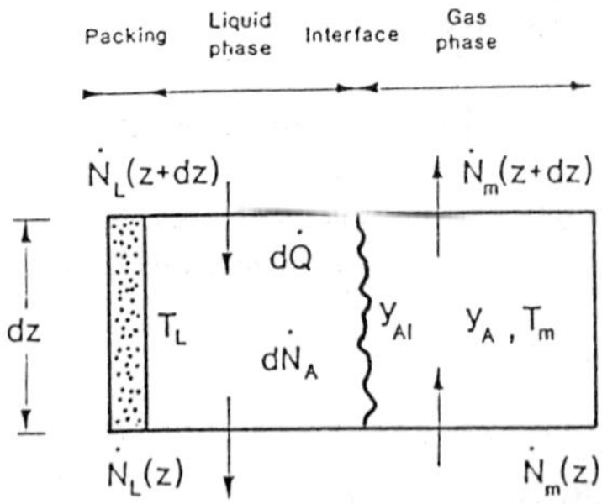

Fig. 1 : Differential column element

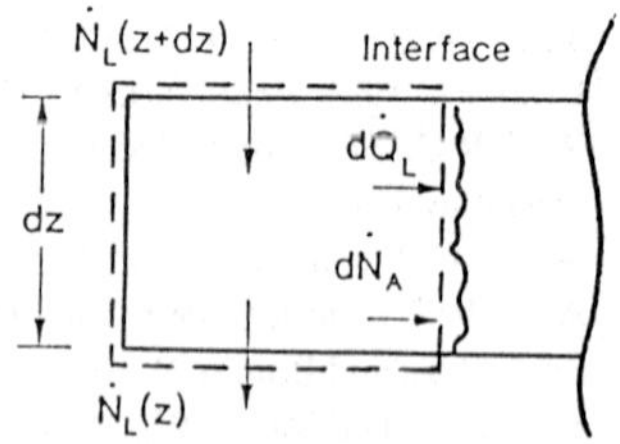

Fig. 2: Control volume 1

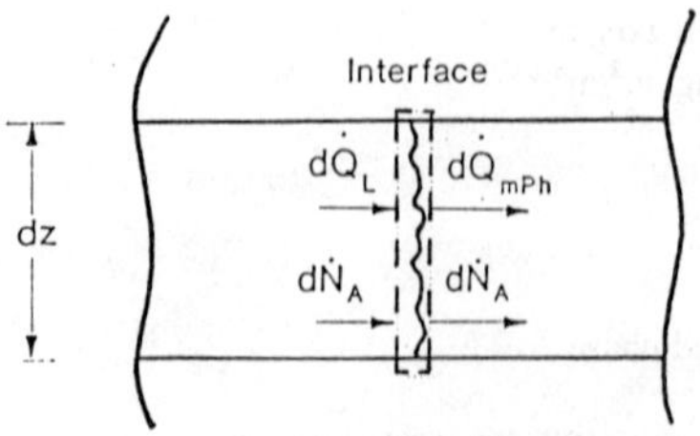

Fig. 3 : Control volume 2

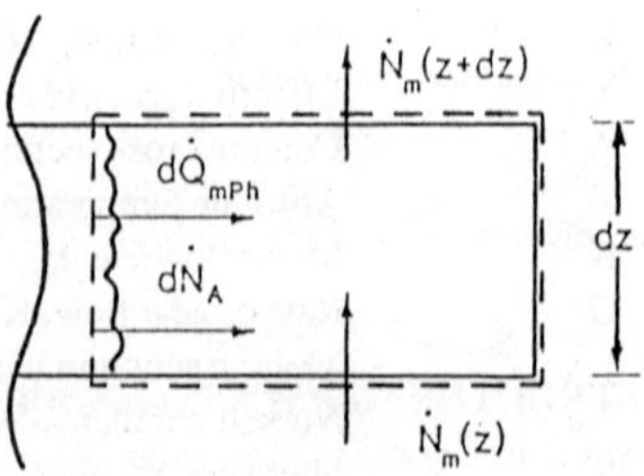

Fig. 4 : Control volume 3

STRUCTURED PACKING IN PRESSURE COLUMNS

L. Ghelfi*, H. Kreis**, J.A. Alvarez*** and R. Hunkeler*
* Sulzer Bros Ltd, CH-8401 Winterthur, Switzerland
** Linde AG, D-8023 Hoellriegelskreuth, Germany
*** Petroquimica Bahia Blanca, Argentina

The capacity of trayed columns can be increased in various ways: by installing better trays or by a column revamp with dumped or structured packing. In this paper the experiences with structured packing are described and a demethanizer revamp is presented.

INTRODUCTION

Improving the cracker in an ethylene plant had the consequence that the capacity of the demethanizer had to be increased by 15%. The tower was equipped with 2-pass valve trays, which due to jet flooding and downcomer backup were not able any more to cope with the higher loads. Thanks to an optimistic design of the packing manufacturer structured packing has been chosen to revamp the 30 bar column.

STRUCTURED PACKING IN A PRESSURE COLUMN?

Structured packing, for good wettability by capillary forces consisting of gauze, has been developed thirty years ago for very small specific liquid loads in heavy water separation plants. Later, for price reasons, the gauze was replaced by metal sheet and the packing was used more and more also in moderate vacuum and in atmospheric columns. In the last few years also pressure columns are equipped with structured packing. The hydrodynamic behaviour of vapour and liquid is different in vacuum towers and in pressure columns. In a vacuum tower the vapour velocity is high (F-Factor 3) and the specific liquid load is relatively small. The liquid flows along the packing surface in a film and the vapour is radially mixed by the structure of the packing. In a pressure column the vapour velocity is in the range of centimeters per second (F-Factor less than 0.5) but the specific liquid load is now up to 100 m^3/m^2h. Obvious the liquid is not flowing down as a liquid film (what for is then the packing surface used?) and the slowly rising vapour is in danger to be carried down by the mass of liquid (vapour backmixing). These considerations are not very favourable for packing in pressure columns. But better than a philosophy about hydrodynamics it is to measure the packing performances which are important, namely efficiency and capacity.

PERFORMANCE OF STRUCTURED PACKING IN PRESSURE COLUMNS

The following perfomances have been measured with structured packing in pressure columns:

EFFICIENCY

Application	Pressure (bar)	Diameter (m)	Number of stages per meter of packing
C3-Splitter	16	0.35	4-5
C3-Splitter	18	3.0,3.4	1.2-2.0
Stabilizer	12.5	2.6/3.5	1.0
C3/C4	16-23	1.3-2.5	1.3-2.2
i-/n-C4	6.6	1.53	2.6
De-C1	32	1.2/2.28	1.6
De-C2	20	0.7	2.5
C2-Stripper	13-15	1.3-2.0	1.0

CAPACITY

The possible capacity of structured packing is well described by a Souders-type correlation. The small surface tension in pressure systems has to be considered. Compared to trays in all cases the capacity was increased with structured packing but only in small columns the number of stages increased by a Mellapak revamp too.

REVAMPING A DEMETHANIZER WITH STRUCTURED PACKING

The existing demethanizer was equipped with 47 valve trays. Spacing in the lower column part was 800 mm and in the upper part 600 mm. For the new design loads (Table 1) these trays are not any more suited. The Mellapak design is given in Table 2 and a sketch of the tower and internals after the revamp in Figure 1 and Figure 2. During the test run the column was operating with 12% higher internal loads than in the design and with a 14% higher reflux rate the specifications could be met. Further, an additional test was recently carried out: the feed rate to the column was 21% higher than design. The specifications were also met with a reflux ratio slightly higher than in the previous test.

CONCLUSIONS

For new pressure columns structured packing is the first choice only if the column diameter is small (less than 1 meter). Exceptions are C4- and C5 columns, especially if they are heat pumped. Revamping of existing tray columns with structured packing is favourable if the capacity has to be increased. If after revamp better separation is needed structured packings need still some development or a modern tray has to be chosen.

Table 1. Demethanizer Expansion Design

COLUMN BLOCK T-10 DEMETHANIZER

COLUMN SUMMARY. FLOWS ARE IN KGMOL/HR

STAGE	TEMP C	PRESS KGCM	FLOW FROM STAGE VAPOR	FLOW FROM STAGE LIQUID	FEED	PRODUCT	HEAT ADDED MM WATT
35	-114.9	30.80		412.8	1345.0	1286.3 VAP	-1.944
34	-37.4	30.80	354.0	655.1			
33	-27.1	30.80	596.3	693.5			
32	-23.2	30.80	634.7	708.2			
31	-21.5	30.81	649.5	714.5			
30	-20.7	30.81	655.7	716.7			
29	-20.2	30.81	658.0	717.0			
28	-19.8	30.81	658.2	715.8			
27	-19.4	30.81	657.1	713.4			
26	-18.7	30.81	654.7	709.5			
25	-17.7	30.81	650.7	1317.5	377.4		
24	-14.6	30.82	881.3	1328.7			
23	-11.3	30.82	892.5	3012.4	183.7		
					457.6		
					591.6		
22	-8.7	30.82	1343.4	3093.7			
21	-7.6	30.82	1424.7	3124.1			
20	-6.9	30.82	1455.1	3142.2			
19	-6.4	30.82	1473.2	3154.8			
18	-6.0	30.83	1485.8	3164.0			
17	-5.7	30.83	1495.0	3170.7			
16	-5.5	30.83	1501.7	3175.6			
15	-5.4	30.83	1506.6	3179.2			
14	-5.3	30.83	1510.2	3181.8			
13	-5.2	30.83	1512.8	3183.7			
12	-5.1	30.83	1514.7	3185.1			
11	-5.1	30.84	1516.1	3186.1			
10	-5.1	30.84	1517.1	3186.9			
9	-5.1	30.84	1517.9	3187.4			
8	-5.0	30.84	1518.4	3187.8			
7	-5.0	30.84	1518.8	3188.1			
6	-5.0	30.84	1519.1	3188.2			
5	-5.0	30.84	1519.2	3188.1			
4	-4.9	30.85	1519.1	3187.4			
3	-4.8	30.85	1518.4	3184.6			
2	-4.4	30.85	1515.6	3165.4			
1	-2.7	30.85	1496.4			1669.0 LIQ	3.225

FEED STREAM 1 TO STAGE 23 = 183.7 KGMOL/HR
FEED STREAM 2 TO STAGE 23 = 457.6 KGMOL/HR
FEED STREAM 3 TO STAGE 23 = 591.6 KGMOL/HR
FEED STREAM 4 TO STAGE 25 = 377.4 KGMOL/HR
FEED STREAM 5 TO STAGE 35 = 1345.0 KGMOL/HR
DECANTED WATER STREAM 12 = 0.0 KGMOL/HR
BOTTOMS STREAM 11 = 1669.0 KGMOL/HR
EXTERNAL REFLUX RATIO= 0.321
INTERNAL REFLUX RATIO= 0.509

	VAPOR				LIQUID			
	WT FLOW	MW	VOL FLOW	DENSITY	WT FLOW	MW	VOL FLOW	DENSITY
STAGE	KG/HR		M3/HR	KG/M3	KG/HR		M3/HR	KG/M3
35	6078.5	4.7	562.17	10.8123	10591.4	25.7	19.63	539.5
34	7710.8	21.8	174.04	44.3055	17295.6	26.4	41.44	417.4
33	14415.0	24.2	290.28	49.6587	18795.0	27.1	45.96	408.9
32	15914.4	25.1	308.26	51.6272	19416.7	27.4	47.77	406.5
31	16536.1	25.5	315.06	52.4862	19688.6	27.6	48.56	405.4
30	16808.0	25.6	318.01	52.8537	19800.2	27.6	48.89	405.0
29	16919.7	25.7	319.26	52.9956	19836.3	27.7	49.02	404.7
28	16955.7	25.8	319.75	53.0276	19832.3	27.7	49.04	404.4
27	16951.7	25.8	319.87	52.9950	19800.3	27.8	48.99	404.1
26	16919.7	25.8	319.79	52.9091	19740.7	27.8	48.90	403.7
25	16860.1	25.9	319.54	52.7631	36807.6	27.9	91.31	403.1
24	23461.3	26.6	428.85	54.7076	37423.0	28.2	93.55	400.0
23	24076.6	27.0	441.06	54.5878	87256.7	29.0	215.58	404.8
22	37274.2	27.7	650.92	57.2631	89929.4	29.1	224.32	400.9
21	39946.9	28.0	686.86	58.1582	91043.1	29.1	227.76	399.7
20	41060.6	28.2	699.92	58.6639	91738.2	29.2	229.84	399.1
19	41755.6	28.3	707.53	59.0156	92232.4	29.2	231.31	398.7
18	42249.8	28.4	712.81	59.2714	92592.5	29.3	232.37	398.5
17	42609.9	28.5	716.63	59.4586	92855.7	29.3	233.15	398.3
16	42873.1	28.5	719.40	59.5951	93047.1	29.3	233.72	398.1
15	43064.5	28.6	721.41	59.6947	93186.1	29.3	234.13	398.0
14	43203.6	28.6	722.86	59.7674	93287.1	29.3	234.43	397.9
13	43304.6	28.6	723.90	59.8205	93360.5	29.3	234.65	397.9
12	43377.9	28.6	724.65	59.8597	93413.9	29.3	234.81	397.8
11	43431.4	28.6	725.20	59.8887	93453.0	29.3	234.93	397.8
10	43470.5	28.7	725.58	59.9106	93482.1	29.3	235.01	397.8
9	43499.5	28.7	725.87	59.9271	93503.6	29.3	235.08	397.8
8	43521.0	28.7	726.07	59.9399	93519.5	29.3	235.13	397.7
7	43537.0	28.7	726.22	59.9500	93531.7	29.3	235.17	397.7
6	43549.1	28.7	726.33	59.9577	93540.5	29.3	235.20	397.7
5	43557.9	28.7	726.40	59.9634	93545.8	29.3	235.24	397.7
4	43563.2	28.7	726.46	59.9660	93544.6	29.3	235.27	397.6
3	43562.1	28.7	726.49	59.9618	93514.9	29.4	235.26	397.5
2	43532.4	28.7	726.48	59.9220	93125.9	29.4	234.31	397.4
1	43143.4	28.8	726.61	59.3757	49982.6	29.9	124.52	401.4

Table 2. Process Flow Scheme, Mass Balance.

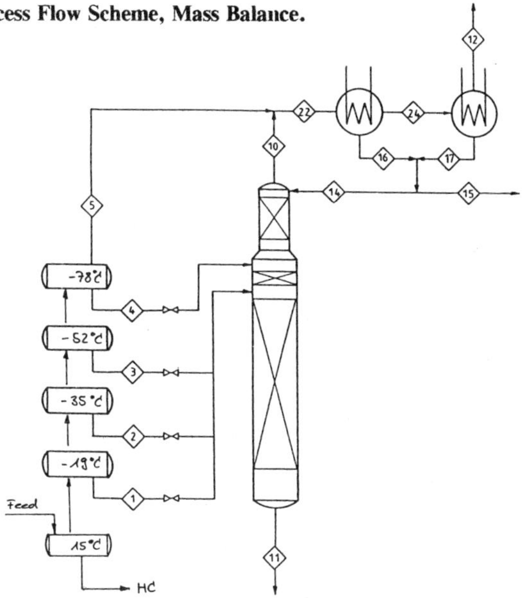

Mass Balance	FEED		STREAM 11		STREAM 10	
	KGMOL/HR	MOL FR.	KGMOL/HR	MOL FR.	KGMOL/HR	MOL FR.
CARBON MONOXIDE	25.49	0.0086	0.00	0.0000	4.57	0.0124
HYDROGEN	1079.25	0.3638	0.00	0.0000	34.49	0.0939
METHANE	158.92	0.0536	0.01	0.0000	128.42	0.3496
ETHANE	621.03	0.2094	617.61	0.3707	14.94	0.0407
ETHYLENE	1011.00	0.3408	984.32	0.5908	184.89	0.5034
PROPANE	2.06	0.0007	2.03	0.0012	0.00	0.0000
PROPYLENE	18.45	0.0062	18.27	0.0110	0.00	0.0000
PROPADIENE	0.57	0.0002	0.56	0.0003	0.00	0.0000
C4+	49.66	0.0167	43.39	0.0259	0.00	0.0000
Total	2966.42	1.0000	1666.20	1.0000	367.31	1.0000

Table 3. Tower Loadings.

DESIGN FOR SULZER MELLAPAK

DATE : 23/11/1989

COLUMN : T-3401 Demethanizer

BED DIMENSIONS :

Bed	Height mm	Diameter mm	Bedtype	Material	Packing Type
1	3780.	1200.	Mellapak	Alu	350.Y/350.Y
2	840.	2282.	Mellapak	Alu	350.Y/350.Y
3	2520.	2282.	Mellapak	Alu	350.Y/350.Y
4	5670.	2282.	Mellapak	Alu	350.Y/350.Y
5	6930.	2282.	Mellapak	Alu	350.Y/350.Y

BED LOADINGS :

Bed	RhoV kg/m3	RhoL kg/m3	V kg/hr	L kg/hr	F Pa^.5	% of flood	l m3/hr/m2	Sigma mN/m
1 ,Top	43.210	495.5	5761.	9384.	.22	26.	16.75	2.5
Bot.	51.460	394.6	12438.	16060.	.43	58,	35.99	2.5
2 ,Top	51.460	430.1	12438.	28166.	.12	20.	16.01	2.0
Bot.	54.560	393.1	19983.	35711.	.18	31.	22.21	2.0
3 ,Top	54.560	417.9	19983.	70783.	.18	40.	41.41	2.0
Bot.	60.980	393.6	39284.	90085.	.34	66.	55.96	2.0
4 ,Top	61.480	393.3	40121.	90921.	.35	69.	56.52	1.8
Bot.	63.000	392.2	42749.	93549.	.37	72.	58.32	1.8
5 ,Top	63.100	392.2	42931.	93732.	.37	72.	58.43	1.8
Bot.	62.950	391.8	43147.	93948.	.37	73.	58.63	1.8

PRESSURE DROP, NUMBER OF STAGES :

Bed	dP top mbar	dP Bottom mbar	dP Avg. mbar	NTS Req.	NTS exp. = measured
1	.1	.4	.3	8.0	8.0
2	.0	.0	.0	2.0	2.0
3	.1	.2	.1	5.0	5.0
4	.6	.8	.7	8.0	8.0
5	1.0	1.1	1.1	10.0	10.0

Table 4. Tower Internals.

distributor-collector-tray for high pressure applications

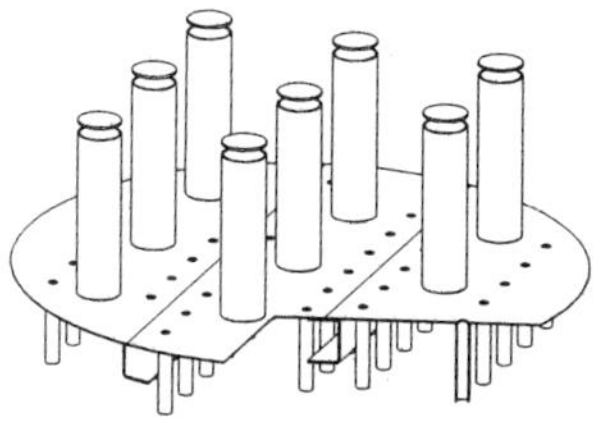

pat. *)

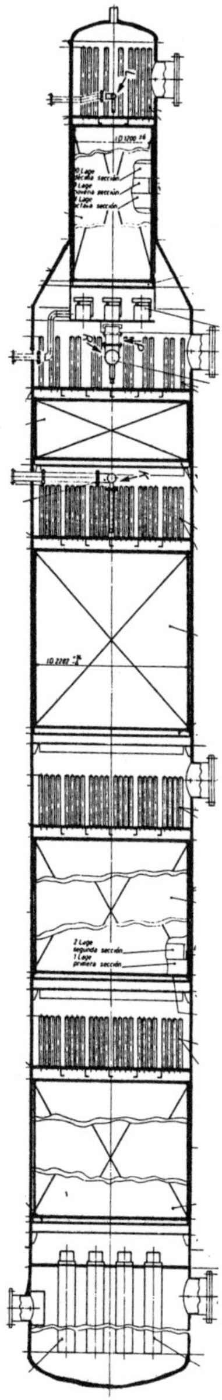

Distributor-collector-tray with ground holes (type G) for high pressure applications. The open area for the gas flow is reduced to 5 to 8 %. The circular chimneys are covered with conventional chimney hats or optionally have lateral holes at the top. To prevent entrainment of the liquid by the vapor flow, the liquid is guided through downconducting tubes to the packing.
To provide radial liquid remixing, collecting and predistributing elements may be added.

These are chimney tray collectors with the function of a liquid distributor at the same time. They are primarily used when the remaining space between two packed sections of a column is strongly limited or for special applications (absorption, refinery) or conditions (high liquid loads, foaming liquids). Like the chimney tray collectors the distributor-collector-trays are usually bolted or clamped and are therefore installable and removable through a manhole.

EFFECTS OF TILT AND MOTION ON A PACKED DEAERATION COLUMN

R.K.Tanner, S.A. Baker and B.Waldie
Department of Chemical and Process Engineering, Heriot-Watt University, Riccarton, Edinburgh EH14 4AS

The mass transfer performance of different packings are compared in water deaeration columns under tilt and motion conditions. The reduction in performance with tilt is greatest for those packings performing well under vertical conditions and is more severe in tall, narrow columns. A method is proposed and tested for scaling-up the effects of tilt in a packed column based on the concept of geometric similarity. With cyclic motion about the vertical, deaeration performance can approach that in vertical operation.

INTRODUCTION

In recent years, there has been significant interest in the effects of tilt and motion on processing equipment due to the increasing use of floating production systems for the recovery of offshore oil and gas. Such systems can allow viable development of otherwise marginal fields and can be relocated to other fields at a later date. Waldie[1] describes how equipment used on semi-submersible and tension leg platforms, converted super tankers and specially designed barges can be exposed to varying degrees of tilt and motion under marine conditions.

Onshore process equipment can also experience these effects. Lockett and Augustyniak explain that tilt can result from misalignment during construction or settlement of foundations, whilst tall towers can sway in high winds[2]. The performance of any process equipment requiring steady fluid interfaces is affected by tilt and motion. On an offshore production platform the primary separators, produced water separators and liquid level measurement and control systems are particularly at risk[1]. Contacting columns are also susceptible as the countercurrent gas and liquid flows are disturbed; uneven liquid levels on column trays cause weeping and loss of downcomer seals, whilst channelling of the liquid and gas streams is likely in packed columns.

This research investigates the effects of tilt and motion on a packed deaeration column, an item of equipment commonly used on offshore production platforms for deoxygenating sea water prior to injection into the oil reservoir for pressure maintenance. Oxygen removal is necessary to avoid corrosion in the downstream components of the injection system. In a gas strip deaerator, the subject of this study, oxygen removal is achieved using a continuous flow of oxygen lean gas (normally produced hydrocarbons) countercurrent to the water flow.

Studies into the effects of tilt and motion on packed columns have mainly been by contractors considering the design of floating plants for processing offshore associated gas. Mitsui Engineering and Shipbuilding Company[3], Linde[4] and Swedyards Development Corporation[5] have all published results. In addition, Sulzer Brothers have tested their structured packing for use in offshore conditions[6]. An early study by Weedman and Dodge[7] considered the effect of tilt on an air distillation plant, commonly used for shipboard oxygen production. Table 1 summarises the experimental details of these studies.

The present work is the first investigation of tilt and motion on the deaeration process. It follows Waldie's preliminary study of deaeration by recycled nitrogen in a simple sieve plate

TABLE 1
Details of previous tilt and motion studies on packed columns[13].

Investigator(s)	Process investigated	Packing type(s)	Packed bed dimensions
Yoshinaga, Izawa, Kotani[3]	CO_2 absorption from air into aqueous Na_2CO_3	1" Pall rings	0.45 m diam/4.5 m tall
Hoerner, Wiessner, Berger[4]	Distillation of methyl cyclohexane/toluene	Regular packing	0.7 m diam/5 m tall
Svensson[5]	Distillation of methanol/water	Norton Intalox	(1) 0.6 m diam/ 4.7 m tall (2) 1.0 m diam/ 2.2 m tall
Pluss, Bomio[6]	Distillation of chlorobenzene/ ethylbenzene and transdecalin/ cis-decalin	Mellapac structured packing	0.45 m diam/up to 6 m tall
Weedman, Dodge[7]	Distillation of liquid air	Various random packings	50 mm diam/up to 1.2 m tall

column[8] where tray efficiencies were reduced by up to 55 per cent at 6 degrees tilt. This research is also the first to compare the performance of different packings under tilt and motion conditions.

EXPERIMENTAL DETAILS

Figure 1 is a diagram of the equipment used. Mains water flowed countercurrent to nitrogen, the stripping medium, in the atmospheric pressure column. Although foaming can affect the performance of a sea water deaerator, it is reasonable to assume that the effect of tilt and motion on mass transfer efficiency will be the same using mains water and sea water.

Two different diameter columns — 0.22 m and 0.40 m — and packed heights of up to 2.45 m were investigated. Both columns were manufactured from high optical clarity acrylic so that liquid flow patterns could be witnessed.

Uniform initial liquid distribution was achieved in both columns using drip pan distributors with more than 1700 distribution points per square metre of column cross-sectional area. The water supply system was designed to provide the very high fluxes used in deaeration offshore. Tests were carried out at 38.0 kg/m s, a typical design loading, and lower loadings of 8.9 and 22.3 kg/m s.

The nitrogen, produced at a constant purity by a membrane generator (Calor model J1), was supplied to the bottom of the column through a pipe distributor as shown in Figure 1. Gas loadings of 0.006, 0.013, 0.026 and 0.054 kg/m s were investigated, covering the likely range used in practice.

Both columns could be tilted in one plane about the vertical. The axis for tilt was located 0.5 m above the bottom of the 0.22 m diameter packed bed and 1.2 m above the bottom of the 0.4 m bed. The angle of tilt was measured with an electronic inclinometer and angles of tilt of up to 8 degrees were investigated. Figure 2 shows the 0.4 m diameter column inclined at 5 degrees to the vertical.

Cyclic motion in one plane about the vertical was also possible on the 0.4 m diameter column. This was achieved using an electrically driven crank mechanism. Motion of up to 8 degrees maximum inclination and with periods between 6 and 30 seconds was investigated. Cyclic motion in one plane was selected as previous research has shown that this creates greater liquid maldistibution than irregular multidirectional motion of the same maximum amplitude[4].

Four different polypropylene packings were tested; three random packings (Norton 16 mm and 25 mm Pall rings and No. 1 Super Intalox saddles) and one structured packing (Sulzer Mellapak 250Y). The Mellapak, supplied in 6 sections, was installed with the manufacturer's circumferential sealing bands in place and with the flow channels of alternate sections perpendicular to each other.

A tracer response technique allowed the residence time to be determined for water taking different routes through the packed bed. This consisted of a system for injecting a pulse of salt solution into the inlet water stream and a series of electrical conductivity cells for measuring the salt concentration in the water — one located above the bed and several at different radial positions directly below the bed.

After steady state conditions had been achieved for a particular trial, flow rates and inlet and outlet oxygen contents and temperatures were recorded for the water and gas streams. Dissolved oxygen measurements were made using a high resolution meter with temperature compensation (Orbisphere model 2606). The sampling system was carefully designed to avoid the ingress of gas bubbles which would affect the measurement. A turbine meter (Barton model 8023) was used to measure the high water flows required.

The oxygen contents were measured on the inlet and outlet lines which meant that the calculated mass transfer performance included that of any end effects above and below the bed. It is assumed that these end effects are unchanged by tilt and motion at given liquid and gas loadings.

DATA ANALYSIS

To allow direct comparison of the mass transfer performance from trials involving different inlet water temperatures and dissolved oxygen contents, the deaeration performance is presented in terms of number of transfer units, N_{OL}. Deaeration is a liquid-film controlled process, and the following equation applies under these circumstances[9]:

$$N_{OL} = \frac{X_1 - X_2}{(X_1 - X_{1e}) - (X_2 - X_{2e})} \quad \mathrm{Log_e} \frac{(X_1 - X_{1e})}{(X_2 - X_{2e})} \tag{1}$$

The equilibrium dissolved oxygen contents in the water streams entering and leaving the column, X_{1e} and X_{2e}, are calculated from the following expression, with $i = 1$ and $i = 2$ respectively:

$$Xi_e = \frac{1.78\,Yi}{(0.073T_w) + 2.546} \tag{2}$$

This expression is applicable to water temperatures between 0 and 25°C and assume that Henry's Law applies for the low dissolved oxygen concentrations involved. It is based on oxygen-in-water solubility data quoted by Perry and Chilton[10].

As the mass transfer coefficient is temperature dependent, the N_{OL} values are corrected to a common reference temperature (16°C) using the following relationship proposed by Sherwood and Holloway[11]:

$$N_{OL} \propto \exp\,(0.023\,T_w) \qquad (3)$$

RESULTS AND DISCUSSION

EFFECTS OF TILT

Figure 3 compares the effect of tilt on the four packings in the 0.4 m diameter column packed to a height of 2.45 m. These results are for a typical deaerator liquid loading of 38.0 kg/m s.

The marked difference in the performance of the packings in vertical operation is discussed by the authors in other publications[12,13]. With the uniform initial liquid distribution provided here, packings with a low radial spreading coefficient (such as Pall rings) perform better than those with a high coefficient (such as Super Intalox saddles and Mellapak).

High radial spread accelerates the development of the equilibrium liquid distribution. The equilibrium distribution is less uniform than the initial distribution due to liquid channelling, in particular the development of a high liquid loading at the column walls. This maldistribution reduces effective liquid-gas contact resulting in reduced mass transfer performance. Figure 3 shows that differences in the performance of the packings reduces as angle of tilt increases. At 8 degrees tilt, the performance of the two sizes of Pall ring and the Super Intalox saddles lie within 10 per cent of each other. The improving performance of Mellapak relative to the other packings can be explained by the high radial spread it creates which carries water into the relatively 'unwetted' region of the bed.

A corollary of this levelling out in performance is that packings that perform best under vertical conditions suffer most with tilt. For instance at 3 degrees tilt there is only 8 per cent deterioration with Mellapak but 30 per cent with 25 mm Pall rings, although the absolute performance of the Pall rings is still higher.

MALDISTRIBUTION IN A TILTED PACKED COLUMN

Figure 4 illustrates the maldistribution in a tilted packed column. Although there is some radial spreading of the liquid, it tends to flow towards the lower inclined wall. This was clearly demonstrated in a simple experiment involving the injection of dye into the water inlet stream.

Once at the wall, this liquid is less likely to return to the bulk of the packing than in the vertical condition. The column wall provides a rapid route through the column, with limited opportunity for further contact with the gas stream. Table 2 shows that increased inclination reduces the residence time in the column for liquid flowing down the wall. The effect of tilt was found to be more pronounced at lower liquid loadings. This is because the proportion of the total liquid stream taking this wall route increases as the liquid loading reduces.

A consequence of liquid migration to the lower wall is the existence of a poorly wetted region of the bed as shown in Figure 4, contributing little to mass transfer. The poorly wetted region also

TABLE 2
Residence times in packing for liquid leaving close to lower inclined wall.

Mode of operation	Residence time
Vertical column	15.1 seconds
3° tilt	12.4 seconds
5° tilt	12.0 seconds

$D = 0.22$ m, $L = 38.0$ kg/m^{2}s,
$H = 2.45$ m, 25 mm Pall rings

offers a preferential route for gas flow, further reducing effective liquid-gas contact. This bypassing of the liquid and gas streams explains the results of a parallel study by the authors which showed that the pressure drop over a packed column reduces with increasing angle of tilt[14].

SCALING-UP EFFECTS OF TILT

Figure 5 shows the effect of tilt on different sizes of packed bed using 16 mm Pall rings. It is clear that the reduction in mass transfer performance increases as packed height increases and column diameter reduces. These results confirm findings in a previous study by Pluss and Bomio[6] and they can be explained by considering Figure 4 again. The severity of the maldistribution increases with the proportion of the packed bed volume that is poorly wetted. By making the simplifying assumption of a vertical boundary between the well wetted and poorly wetted regions, geometrical considerations show that this proportion increases with increased packed height and reduced diameter.

The concept of geometrically similar columns has been proposed by Porter and Jones as a tool for scale-up of packed columns[15]. If the column diameter, packed height and packing size are changed in equal proportion, with loadings and quality of initial liquid distribution remaining unchanged, then liquid and gas distribution patterns are assumed to be unaltered resulting in the same mass transfer performance.

It is proposed that geometric similarity can also be used to scale-up the effects of tilt on a packed column. As liquid radial spread is proportional to packing size[15], by increasing packing size with column dimensions the shape of the boundary between the well wetted and poorly wetted regions is unchanged. Therefore the proportion of the packed bed volume that is poorly wetted remains the same, resulting in the same reduction in mass transfer performance.

Experimental trials were carried out to test this proposal using the two sizes of packed deaerator illustrated in Figure 6. The scale-up of packing size was not in exact proportion to column dimensions but was the best that could be achieved with the available sizes of Pall ring. The results were promising at lower liquid loadings, the reduction in deaeration performance being within 10 per cent agreement in the two columns at $L = 8.9$ kg/m s. However, as liquid loading was increased, the effect of tilt became much more pronounced for the larger column. Figure 7 gives typical results.

Future work will study liquid distribution in tilted packed columns in more detail to allow this scale-up technique to be refined.

EFFECTS OF CYCLIC MOTION

Figure 8 shows the effect of cyclic motion about the vertical on the packings. It shows the deaeration performance for a range of periods of motion for ± 3 degrees maximum deviation from the vertical for a specified liquid and gas loading. Note that this figure gives deaeration performance as a percentage of that under vertical conditions and therefore does not give a direct comparison of the absolute performance of the packings.

It is clear that with cyclic motion, deaeration performance can approach and even exceed that for vertical operation, as is the case with the Super Intalox saddles. The motion effectively redistributes the liquid over the packing. These results confirm previous studies that showed tilt to be the worst operating condition[3,4,5].

It is also worth noting that over the range of periods of 5 to 20 seconds, where Waldie has explained most sea wave energy and resulting induced motion is contained, these packings are insensitive to motion for an amplitiude of ± 3 degrees. At higher periods, it is reasonable to expect the performance to reduce and approach that at a fixed angle of tilt of 3 degrees.

CONCLUSION

The reduction in mass transfer performance of a packed water deaeration column in tilted operation depends upon the packing used. Packings performing best under vertical conditions suffer most with tilt, resulting in a levelling out of relative packing performance at higher angles of tilt. The effect of tilt is most severe for tall, narrow columns.

A consideration of maldistribution in a tilted packed column suggests that geometrically similar columns will be affected to the same degree. Experimentally testing this proposal has produced some promising results. With cyclic motion of ± 3 degrees amplitude about the vertical, deaeration performance can be an improvement over that in vertical operation and is insensitive to the period of motion for the range of periods most often experienced on floating production systems.

ACKNOWLEDGEMENTS

Industrial and SERC funding from MTD Ltd and the loan of a nitrogen generator by Calor Air Separation allowed this research to proceed.

REFERENCES

1. Waldie, B., 1986, *The Chemical Engineer*, 426: 17.
2. Lockett, M.J. and Augustyniak, J.D., 1991, *Trans IChemE*, 69A: 99.
3. Yoshinaga, K., Izawa, Y. and Kotani, T., 1981, *AIChE 9th National Meeting*.
4. Hoerner, B.K., Wiessner, F.G. and Berger, E.A., 1982, *Proc 14th Annual Offshore Technology Conference*, 527.
5. Svensson, H.V., November 1982, *Chem Eng Prog*, 78: 43.
6. Pluss, R.C. and Bomio, P., 1987, *IChemE Symposium Series No. 104*, A259.
7. Weedman, J.A. and Dodge, B.F., 1947, *Ind and Eng Chem*, 39: 732.
8. Waldie, B., 1991, *Proc Conf on Floating Production Systems — Blueprint for the 90s, Monaco*.

9. Coulson, J.M. and Richardson, J.F., 1978, *Chemical Engineering — Volume 2* (Pergamon Press, Oxford), 562.
10. Perry, R.H. and Chilton, C.H., *Chemical Engineers' Handbook*, Fifth Edition (McGraw-Hill), 3-98.
11. Sherwood, T.K. and Holloway, F.A.L., 1940, *Trans AIChE*, 36: 39.
12. Baker, S.A., Tanner, R.K. and Waldie, B., 1992, *IChemE Research Event*, 567.
13. Baker, S.A., Tanner, R.K. and Waldie, B., *ChERD*, Submitted 1992.
14. Baker, S.A., Livingstone, J., Tanner, R.K. and Waldie, B., 1991, *IChemE Research Event*, 285.
15. Porter, K.E. and Jones, M.C., 1987, *IChemE Symposium Series No. 104*, A245.

NOMENCLATURE

D	Column diameter (m)
G	Gas flux (kg/m s)
H	Height of packing (m)
L	Liquid flux (kg/m s)
N_{OL}	Number of Transfer Units
T_w	Temperature of water (°C)
X	Dissolved oxygen content of water (mg/l)
Y	Oxygen content of gas (mol per cent)

SUBSCRIPTS

1(2)	Conditions at top (at bottom) of column
e	Equilibrium conditions

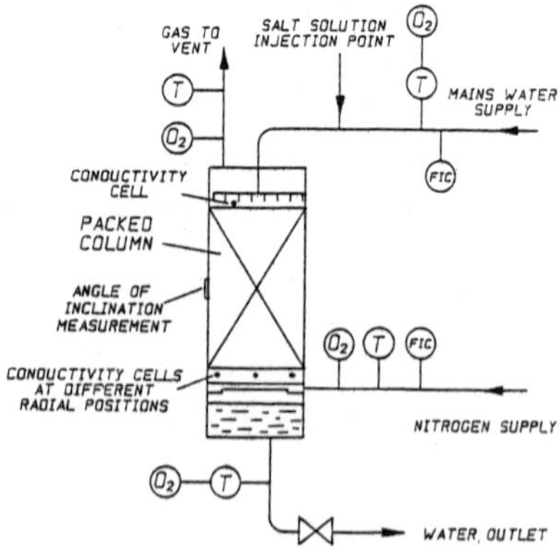

Fig. 1 DIAGRAM OF EQUIPMENT

Fig. 2 0.4m. DIAMETER COLUMN INCLINED AT 5 DEGREES TO VERTICAL

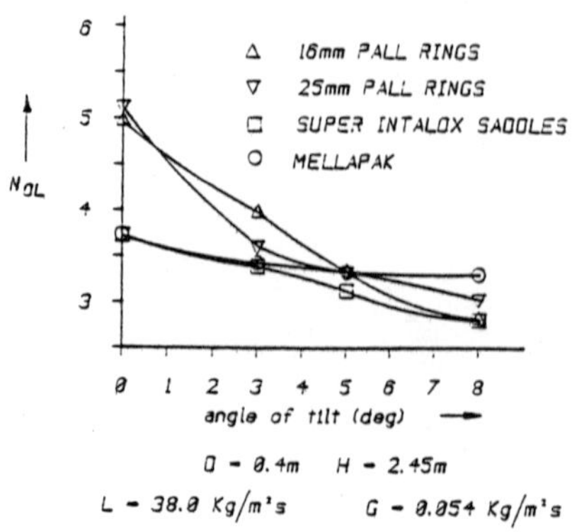

Fig. 3 EFFECT OF TILT ON DIFFERENT PACKINGS

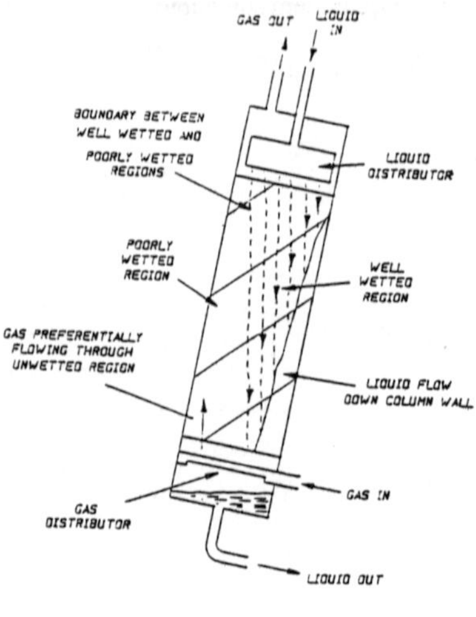

Fig. 4 MALDISTRIBUTION IN A TILTED PACKED COLUMN (13)

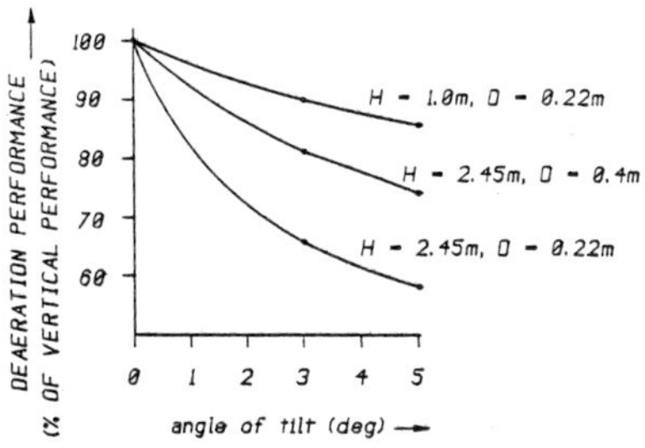

16mm PALL RINGS

L = 38.0 Kg/m²s   G = 0.013 Kg/m²s

Fig.5 EFFECT OF TILT ON DIFFERENT PACKED BED SIZES

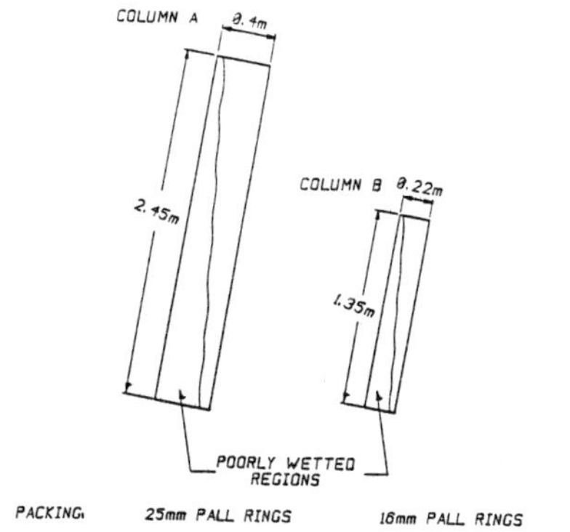

BASIS:

1. EQUAL H/D FOR BOTH COLUMNS
2. SIMILAR SHAPED BOUNDARY BETWEEN WELL WETTED AND POORLY WETTED REGIONS DUE TO SCALING-UP OF PACKING SIZE
3. PROPORTION OF TOTAL BED THAT IS POORLY WETTED IS SIMILAR IN BOTH COLUMNS

Fig.6 PROPOSAL OF EQUAL EFFECT OF TILT FOR GEOMETRICALLY SIMILAR COLUMNS

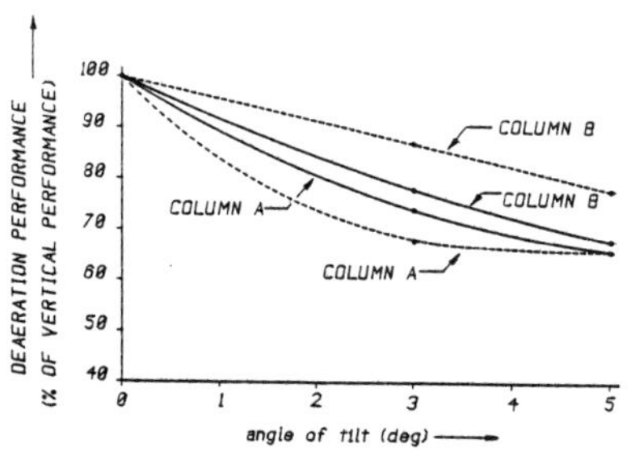

——— L = 8.9 Kg/m² s   G = 0.026 Kg/m² s

------- L = 38.0 Kg/m² s   G = 0.026 Kg/m² s

Fig.7 COMPARISON OF EFFECTS OF TILT ON COLUMNS A AND B

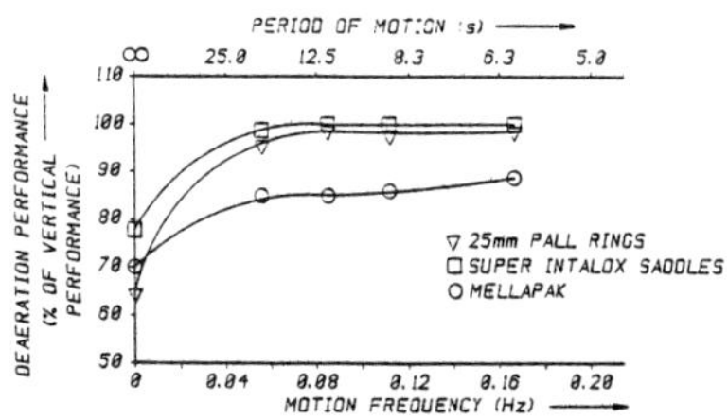

D = 0.4m   H = 2.45m

L = 22.3 Kg/m s   G = 0.013 Kg/m s

3° MAX.DEVIATION FROM VERTICAL

Fig.8 EFFECT OF MOTION ON DIFFERENT PACKINGS (13)

SEPARATION EFFICIENCY OF A STRUCTURED PACKING IN VACUUM DISTILLATION

A. Roselló and A. Martín
Department of Chemical Engineering, University of Seville, Spain

Distillation efficiencies were obtained in a laboratory column with a Sulzer packing by fractionating trans.decaline/cis.decaline and heptanol/octanol systems. Variations of mass-transfer coefficients suggest that the performance of structured packing in vacuum distillation cannot be completely explained by a correlation in which only the vapour side resistance is significant.

INTRODUCTION

Structured packing is gaining acceptance for its interesting performance characteristics, in spite of its high cost. Because of its low pressure drop, it is often considered in vacuum distillation.

In order to obtain the necessary height of a packed column for a given separation, information about mass transfer efficiency must be available. Bravo and co-workers[1] used the conventional model of two resistances to correlate their data with a gauze packing. As the flow in this structured packing resembles the flow in a wetted-wall column, they suggested a relationship for the most important resistance (vapour) following the work of Johnstone and Pigford: $Sh_V Re_V^{0.80} Sc_V^{0.33}$. The liquid side mass transfer coefficient was based on penetration theory (therefore, $k_L \propto (D_L/\theta)^{0.50}$), but Bravo *et al* found a very low liquid phase resistance. In their model, the specific surface a_e is equal to the total although it is now thought that a_e may depend on the liquid flow. On the other hand, the liquid side resistance is usually ignored.

The assumption of negligible liquid resistance is very convenient, because liquid phase properties do not have to be considered in the design. Nevertheless, this one-resistance model has little flexibility, because of the small changes in vapour properties. Hence, the efficiency of structured packings would be largely independent of the pressure or mixture composition or even of the system[5]. In contrast, data for dumped packings show an influence of pressure on HETP[6], and changes of HTU are often found when composition varies in normal pressure distillation.

EXPERIMENTAL METHOD

Separations were carried out at total reflux in a column with a Sulzer EX laboratory packing. Bed height was 24 cm; diameter, 3 cm. Samples were collected at the ends of the bed, and the volumetric flow of liquid was measured at the top of the column. The composition of the mixtures was analyzed by gas chromatography in a HP 5890 chromatograph (columns: UCC-W982 10% on Cromosorb W-AW-DMGS and FFAP 10% on Cromosorb AW; oven temperatures: 140 and 180°C). Mixtures of known concentration by weight were used as reference standards.

Experimental data of HTU were determined for the trans.decaline/cis.decaline system as a function of the molar flow at the following pressures: 13.3, 10.0, 6.7, 3.3, 1.3 and 0.7 kPa. Mean composition was maintained constant. Vapour-liquid equilibrium data were taken from Marsh[7].

Experimental data of HTU in the heptanol-octanol system were obtained at 13.3, 6.7, 4.0 and 1.3 kPa, varying the flow and the system composition. Vapour-liquid equilibrium data were calculated assuming ideal behaviour of both components in the vapour and liquid phases, in agreement with equilibrium data of Rose *et al*[8] for alcohol/alcohol systems.

RELATIONSHIPS

Height of a transfer unit and molar flow were determined for each run. Starting from the definition of HTU, experimental values of volumetric global mass transfer coefficient were obtained:

$$Ka_e = V/\text{HTU}S \tag{1}$$

If K' is the mass transfer coefficient in the usual form of Sherwood number, $K' = K\,\rho_M$, being ρ_M the molar density.

In general

$$\frac{1}{K'} = \frac{1}{k_V'} + \frac{m}{k_L'} \tag{2}$$

$$k' \propto D Re^a Sc^b \tag{3}$$

Assuming the behaviour of vapour phase in packed column is similar to the behaviour in wetted-wall columns, and assuming a liquid side mass transfer based on penetration theory, for a given geometry:

$$k_V \propto \left(\frac{M}{\mu_V}\right)^{-0.2} Sc_V^{-0.667} V^{0.8} \tag{4}$$

$$k_L \propto \left(\frac{M}{\mu_L}\right)^{-0.5} Sc_L^{-0.5} V^{0.5} \tag{5}$$

If the gauze packing is completely wetted, independent of the liquid load[2], $a_e \approx a$. Then:

$$k_V a_e \propto Sc_V^{-0.667} \left(\frac{M}{\mu_V}\right)^{-0.2} V^{0.8} \tag{6}$$

$$k_L a_e \propto Sc_L^{-0.5} \left(\frac{M}{\mu_L}\right)^{-0.5} V^{0.5} \tag{7}$$

at total reflux, $V = L$

It would appear that fitting the model to a set of experimental results or determining whether the liquid phase resistance is negligible or not would be simple. However, in practice, these simple decisions are not possible, given the typical scattering of distillation data.

RESULTS

Values of (Ka_e) versus molar flux for the trans.decaline/cis.decaline system are plotted in Figure 1.

Relationships between (Ka_e) and V were established at each pressure. Straight lines of different slopes were obtained:

P (kPa)	13.3	10.0	6.7	3.3	1.3	0.7
Slope	0.93	0.73	0.73	0.76	0.69	0.71
LRC	0.9877	0.9791	0.9658	0.9907	0.9820	0.9591

If the liquid phase resistance is negligible

$$ka_e \approx k_V a_e \propto Sc_V^{-0.667} \left(\frac{M}{\mu_V}\right)^{-0.2} V^{0.8} \tag{8}$$

Hence, $Ka_e \propto V^{0.8}$ at a given pressure and constant composition. The average slope for our results is 0.76, close to the suggested value in Bravo *et al*'s model.

When pressure varies, at each flow:

$$Ka_e \propto Sc_V^{-0.667} \mu_V^{0.2} \tag{9}$$

This product varies little. In the trans.decaline/cis.decaline mixture, with molar fraction 0.80:

P (kPa)	13.3	6.7	1.3	0.7
$Sc_V^{-0.667} V^{0.2}$ (viscosity in μP)	6.38	6.54	6.41	6.66

The theoretical mass-transfer coefficient is a little higher when pressure drops, but it can be assumed to be constant. The model will fit if empirical (Ka_e) is constant too. Experimentation shows that lower (Ka_e) values are found at lower pressures (1.3 and 0.7 kPa), but they cannot be statistically separated. Therefore, the Bravo, Rocha and Fair model including negligible liquid resistance may be considered an acceptable approximation to our results with the trans.decaline/cis.decaline mixture.

Nevertheless, other conditions related to variations in composition and efficiencies of different systems have to be fulfilled. When molar flow is a constant:

$$Ka_e \approx k_V a_e \propto \left(\frac{M}{\mu_V}\right)^{-0.2} Sc_V^{-0.667} = f(P, x) \tag{10}$$

A comparison between efficiencies in the trans.decaline/cis.decaline system and the heptanol/octanol system can be made for a given concentration of each mixture. Taking $x = 0.80$ for trans/cis.decaline and $x = 0.50$ for heptanol/octanol:

P (kPa)	13.3	6.7	1.3
Factor in decaline mixture	2.382	2.442	2.393
Factor in alcoholic mixture	1.386	1.375	1.370

The model predicts volumetric mass-transfer coefficients of trans/cis.decaline 1.75 times the coefficients of heptanol/octanol.

On the other hand, differences in the factor tend to be small for a given mixture. For instance, at two different molar fractions of heptanol/octanol:

P(kPa)	13.3	6.7	4.0	1.3
***x*, 0.20**	1.386	1.431	1.442	1.426
***x*, 0.40**	1.386	1.375	1.376	1.370

The model predicts slightly higher mass-transfer coefficients at $x = 0.20$ than at $x = 0.50$, but differences are very small. Hence, if only vapour resistance is taken into account, (Ka_e) has to be virtually independent of composition in the heptanol/octanol system.

Experimental variations of (Ka_e) as a function of V at three average compositions and the different pressures are shown in Figures 2a, 2b and 2c.

Of course, it is not possible to define an accurate relationship between (Ka_e) and P, but these results provide information about other significant aspects:

- (Ka_e) is dependent on composition in the heptanol/octanol system: (Ka_e) at $x = 0.519 >$ (Ka_e) at $x = 0.600 >$ (Ka_e) at $x = 0.205$ (see Figure 3).
- (Ka_e) of trans.decaline/cis.decaline are approximately equal to $(Ka_e)_{0.50}$ in the heptanol/octanol system. A comparison between the latter coefficients and their deduced values starting from those of the trans.decaline/cis.decaline system are presented in Figure 4.
- There is a change in the function $Ka_e = f(V^n)$. The power is now 0.65.

CONCLUSIONS

Experimental data of the trans.decaline/cis.decaline system at different pressures and flows correlate approximately with the Bravo, Rocha and Fair model. But when data of heptanol/octanol (a mixture with a sharper variation in viscosity) are considered, the model does not fit as well and an important liquid resistance contribution must be included.

NOMENCLATURE

a	Specific area of packing
a_e	Effective specific area
D	Diffusion coefficient
HETP	Height equivalent to a theoretical plate
HTU	Height of a transfer unit
K, K'	Global mass-transfer coefficients
k, k'	Individual mass-transfer coefficients

L	Liquid molar flow
M	Molecular mass
m	Slope of equilibrium line
P	Pressure
Re	Reynolds number
S	Section of column
Sc	Schmidt number
Sh	Sherwood number
V	Vapour molar flow
x	Molar fraction

Greek letters

θ	Exposure time
μ	Dynamic viscosity
ρ_M	Molar density

Subscripts

L	Liquid phase
V	Vapour phase

LITERATURE

1. Bravo, J.L., Rocha, J.A. and Fair, J.R., 1985, *Hydrocarbon Processing* 64 (1): 91–95.
2. Spiegel, L. and Meier, W., 1987, *IChemE Symp Series 104*: A203–A215.
3. Fair, J.R. and Bravo, J.L., 1990, *Chem Eng Prog* 86 (1): 19–29.
4. Shah, G.C., 1991, *Chem Eng Prog* 87 (4): 49–56.
5. Rütti, A. and Spielmann, T., 1988, *Laboratory and Pilot Scale Distillation, Sulzer Brothers Ltd, Brochure 22.28.06.40.*
6. Kister, H.Z. and Doig, I.D., 1982, *Can J Chem Eng* 60: 155–158.
7. Marsh, K.N., 1987, *Recommended Reference Materials for the Realization of Physico-Chemical Properties*, Blackwell.
8. Rose, A., Papahronis, B.T. and Williams, E.T., 1958, *J Chem Eng Data* (3): 216–219.

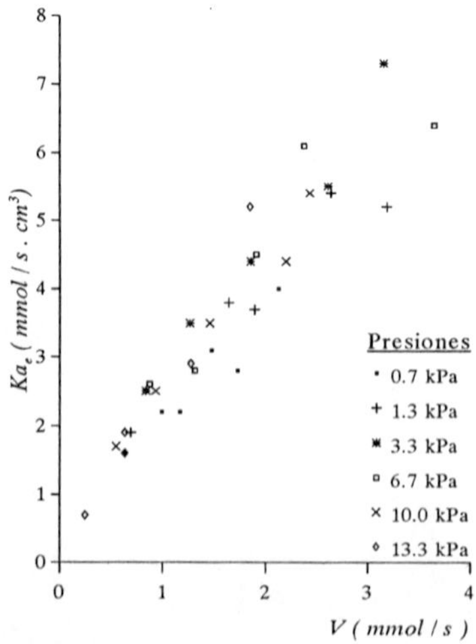

Figure 1.- Global mass-transfer coefficient Mixture: Trans.Decaline/Cis.Decaline

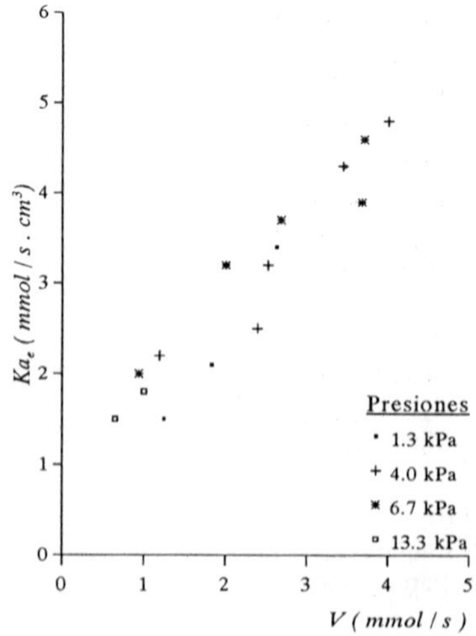

Figure 2a .- Global mass-transfer coefficient in heptanol / octanol at $\overline{x} = 0.205$

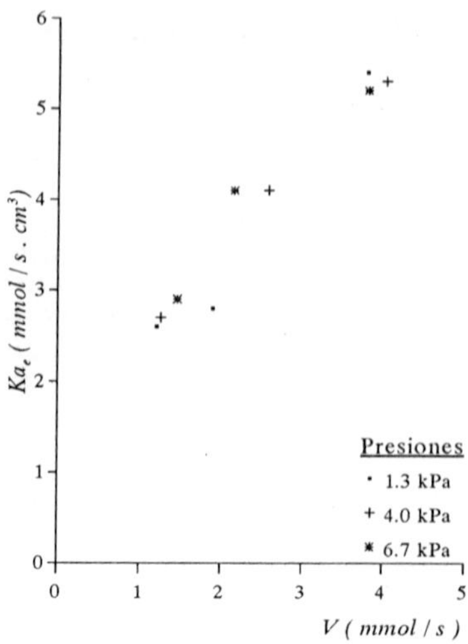

Figure 2b .- Global mass-transfer coefficient in heptanol / octanol at $\overline{x} = 0.600$

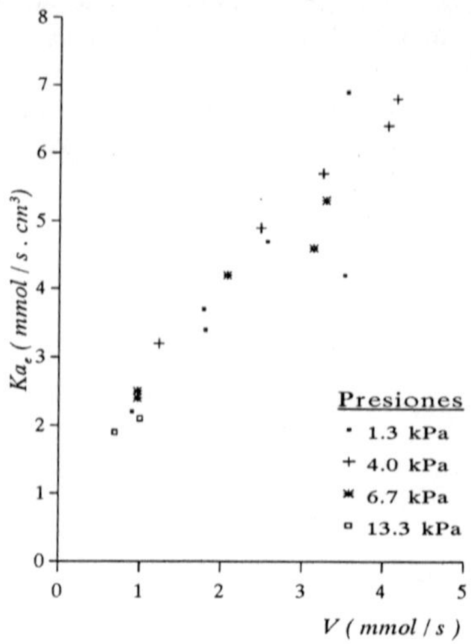

Figure 2c .- Global mass-transfer coefficient in heptanol / octanol at $\overline{x} = 0.519$

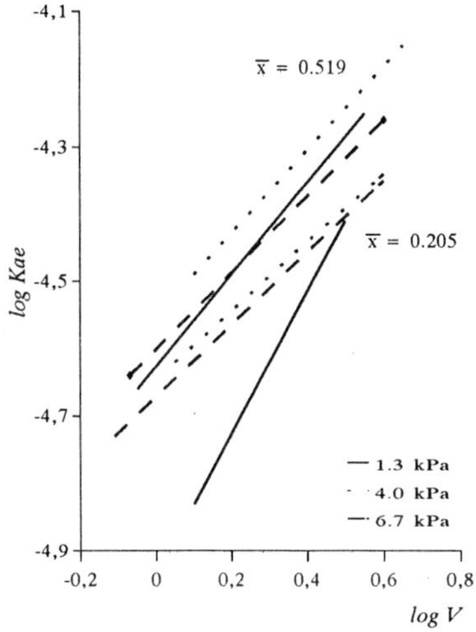

Figure 3 .- Mass-transfer coefficients in heptanol / octanol at $\overline{x}$ = 0.205 and $\overline{x}$ = 0.519

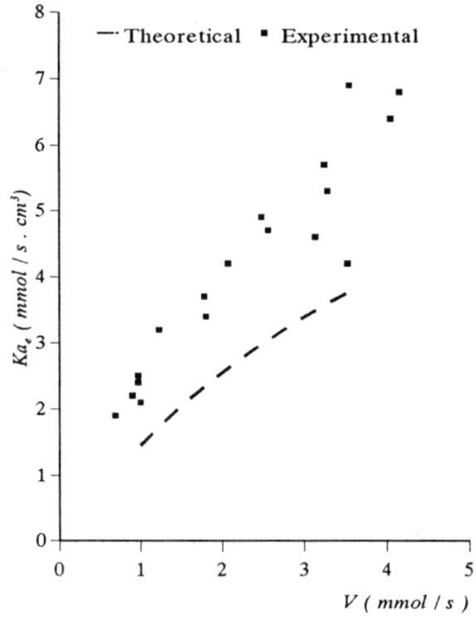

Figure 4.- Mass-transfer coefficients in heptanol / octanol at $\overline{x}$ = 0.519

ADVANTAGE IN CORRELATING PACKED COLUMN PERFORMANCE

R. Billet and M. Schultes
Ruhr University Bochum, Department of Thermal Separation Processes, Universitätsstraße 150, D-4630 Bochum, Germany

This paper concerns the determination of liquid holdup, pressure drop and mass transfer in gas/liquid two-phase countercurrent columns filled with random or structured packing. The equations resulting from an established physical relationships are checked against experimental values on a large number of different column packings and gas/liquid systems. The experimental and calculated results agree well with only slight deviations. This also applies to the range above the loading point for two-phase countercurrent flow up to the flooding point.

INTRODUCTION

Packed columns are being widely adopted for numerous processes in the chemical industry and in environmental technology. In comparison with plate towers they are advantages for the separation of heat sensitive mixtures because of the lower liquid residence time and the lower bottom temperature.

An accurate knowledge of the flow characteristics of both phases is a fundamental requirement for the optimum design of absorption, desorption and rectification columns from the aspects of fluid dynamics and process engineering.

For this reason, it is intended here to present a model that is valid for conventional and modern packing in two-phase countercurrent columns and allows reliable results to be obtained for numerous different two-phase systems.

PHYSICAL MODEL

The assumption in describing the model is that the effective free space in the column is equivalent to a multiplicity of vertical flow channels. If the forces of inertia are neglected within the film equilibrium will be established between the force of gravity and the frictional force within the film, Equation (1), with the shear forces resulting from countercurrent flow of gas stream, Equation (2)[1].

$$\frac{d\left(\eta_L \cdot \frac{d\bar{u}_{L,S}}{ds}\right)}{ds} = -\rho_L \cdot g \tag{1}$$

$$\tau_V = -\frac{1}{2} \cdot \psi_L \cdot \rho_V \cdot \bar{u}_V^2 \tag{2}$$

PREDICTING LIQUID HOLDUP

Equating the forces acting on the film and integration gives Equation (3) for the liquid holdup up to the loading point, if it is taken into consideration, that in practice at normal liquid loads the surfaces of the packing are not completely wetted. In this case, the thickness of the film can be described by $s_O = h_L/a_h$ in terms of the liquid holdup and the hydraulic area.

$$h_L = \left(12 \cdot \frac{\eta_L \cdot a^2 \cdot u_L}{\rho_L \cdot g}\right)^{1/3} \cdot \left(\frac{a_h}{a}\right)^{2/3} \tag{3}$$

Empirical figures for a large number of different systems revealed that the ratio a_h/a could best be given by Equations (4) and (5) for Reynolds numbers below and above $Re_L = 5$.

$$Re_L < 5: \frac{a_h}{a} = C_h \cdot Re_L^{0.15} \cdot Fr_L^{0.1} = C_h \cdot \left(\frac{u_L \cdot \rho_L}{a \cdot \eta_L}\right)^{0.15} \cdot \left(\frac{u_L^2 \cdot a}{g}\right)^{0.1} \tag{4}$$

$$Re_L \geq 5: \frac{a_h}{a} = C_h \cdot 0.85 \cdot Re_L^{0.25} \cdot Fr_L^{0.1} = C_h \cdot 0.85 \cdot \left(\frac{u_L \cdot \rho_L}{a \cdot \eta_L}\right)^{0.25} \cdot \left(\frac{u_L^2 \cdot a}{g}\right)^{0.1} \tag{5}$$

The constant C_h in Equations (4) and (5) is listed in Table 1. The mean relative deviation between the liquid holdup calculated from Equation (3) and the experimental result is only 6.7%.

The liquid holdup below the loading point is independent of the gas velocity. Afterwards, it rises owing to the shear stress in the countercurrent stream of gas — slightly at first and then to a greater extent until the flood point is reached. These relationships can be described by Equation (6) empirically.

$$h_L = h_{L,u_V<u_{V,S}} + (h_{L,Fl} - h_{L,u_V<u_{V,S}}) \cdot \left(\frac{u_V}{u_{V,Fl}}\right)^{13} \tag{6}$$

The average liquid holdup at the flood point given by Equation (7) is about 2.2 times higher than the value given by Equation (3). The term in brackets in Equation (7) is a connection for the viscosity and density of the system (subscript L) referenced to the corresponding values for water at 20°C (subscript W). The liquid holdup between the loading and flood point can be predicted by means of Equation (6) to within a mean relative deviation of 10.8 %.

$$h_{L,Fl} = 2.2 \cdot h_{L,Eqn.(3)} \cdot \left(\frac{\eta_L \cdot \rho_W}{\eta_W \cdot \rho_L}\right)^{0.05} \tag{7}$$

PREDICTING PRESSURE DROP IN UNWETTED COLUMNS

From the shear force/pressure equilibrium, the pressure drop $\Delta p_O/H$ in a stream of gas flowing through an untrickled bed of height H given by Equation (8) can be derived from Equation (2) with ψ_O as the resistance coefficient for the untrickled bed[2].

$$\frac{\Delta p_0}{H} = \psi_0 \cdot \frac{a}{\varepsilon^3} \cdot \frac{F_V^2}{2} \cdot \frac{1}{K} \tag{8}$$

$$\frac{1}{K} = 1 + \frac{2}{3} \cdot \frac{1}{1-\varepsilon} \cdot \frac{d_P}{d_S} \tag{9}$$

The wall factor K, given by Equation (9), takes into account the difference of the local void fraction at the wall of the column from the theoretical value ε[6]. The resistance coefficient ψ_O is determined from the empirical Equation (10). ψ_O embraces the effect on the gas stream exerted by the vapour Reynolds number Re_V. This number is defined by Equation (11). The constant C_P is specific for any given type of packing. C_p values are listed in Table 1.

$$\psi_0 = C_P \left(\frac{64}{Re_V} + \frac{1.8}{Re_V^{0.08}} \right) \tag{10}$$

$$Re_V = \frac{u_V \cdot d_P}{(1-\varepsilon) \cdot \nu_V} \cdot K \quad \text{with} \quad d_P = 6 \cdot \frac{1-\varepsilon}{a} \tag{11}$$

PREDICTING PRESSURE DROP IN WETTED COLUMNS

If the packing is wetted by the liquid, the column volume available for gas flow is reduced by an amount corresponding to the liquid fraction. Hence, the effective void fraction $\varepsilon_e = \varepsilon - h_L$ can be substituted for ε in Equation (8). Equation (12) is then obtained by taking into consideration the change of the packing surface as a result of the wetting, by the resistance coefficient for the two-phase flow ψ_L[2]. If the packing is theoretically assumed to be wholly wetted, the thickness of the film can be described by $s_O = h_L'/a$ and the theoretical liquid holdup up to the loading point can be calculated by Equation (13)[1].

$$\frac{\Delta p}{H} = \psi_L' \cdot \frac{a}{(\varepsilon - h_L')^3} \cdot \frac{F_V^2}{2} \cdot \frac{1}{K} \tag{12}$$

$$h_L' = \left(12 \cdot \frac{\eta_L \cdot a^2 \cdot u_L}{\rho_L \cdot g} \right)^{1/3} \tag{13}$$

Evaluation of the experimental data gives rise to Equation (14) for the resistance coefficient ψ_L, which depends on the liquid Reynolds number Re_L. Values for the constant C_p in Equation (15) are listed in Table 1.

$$\psi_L' = C_P \cdot \left(\frac{64}{Re_V} + \frac{1.8}{Re_V^{0.08}}\right) \cdot \left(\frac{\varepsilon - h_L'}{\varepsilon}\right)^{1.5} \cdot \exp\left(\frac{Re_L}{200}\right) \cdot \left(\frac{h'_{L,Eqn.(6)}}{h'_{L,Eqn.(13)}}\right)^{0.3} \tag{14}$$

$$Re_L = \frac{u_L \cdot \rho_L}{a \cdot \eta_L} \tag{15}$$

Prediction of the theoretical liquid holdup h_L' from Equation (13) is restricted to the range extending up to the loading point. Equation (6) can be applied for the liquid holdup in the entire loading range with h_L' as the theoretical holdup described by Equation (13) and $h'_{L,Fl}$ as the theoretical liquid holdup at the flood point, given by Equation (16)[1].

$$h'^{3}_{L,Fl} \cdot (3 \cdot h'_{L,Fl} - \varepsilon) = 6 \cdot \varepsilon \cdot \frac{\nu_L \cdot a^2}{g} \cdot u_{L,Fl} \tag{16}$$

Comparison of pressure drops calculated both from Equation (8) and from Equation (12) with the values determined by experiments up to the loading point shows an average relative deviation of 9.1%.

Instationary gas and liquid flows above the loading point makes it difficult, especially in the vicinity of the flood point, to determine precisely the pressure drop by experiment. If the comparison were to be extended up to 90% of the flood point, the mean relative error would rise to 10.8%.

PREDICTING MASS TRANSFER FOR ABSORPTION AND DESORPTION

If in a packed column mass transfer occurs by instationary diffusion, Equation (17), according to Higbie applies with D as the diffusion coefficient of the transferred component and as the duration necessary for renewal of the interfacial area[3,4,5].

$$\beta = \frac{2}{\sqrt{\pi}} \cdot \sqrt{D \cdot \frac{1}{\tau}} \tag{17}$$

$$\beta_L \cdot a_{Ph} = C_L \cdot \left(\frac{g}{\nu_L}\right)^{1/6} \cdot \left(\frac{D_L}{d_h}\right)^{1/2} \cdot a^{2/3} \cdot u_L^{1/3} \cdot \left(\frac{a_{Ph}}{a}\right) \tag{18}$$

If τ is determined by the liquid holdup, the length of contact and the phase loading, Equation (17) gives rise to Equation (18) for the volumetric mass transfer coefficient $\beta_L\, a_{Ph}$. C_L is a constant for the packing and a_{Ph} is the interfacial area for mass transfer.

If it is assumed that mass transfer in the gas phase also follows the law of instationary diffusion, Equation (17) gives rise to Equation (19) for the volumetric mass transfer coefficient $\beta_V a_{Ph}$ in which the exponents m = ¾ and n = ⅓ of the gas Reynolds number $u_V/(a\,\nu_V)$ and the Schmidt number ν_V/D_V allow the best correlation of the test results[4,5]. C_V is a constant which has to be determined experimentally.

$$\beta_V \cdot a_{Ph} = C_V \cdot \frac{1}{(\varepsilon - h_L')^{1/2}} \cdot \frac{a^{3/2}}{d_h^{1/2}} D_V \cdot \left(\frac{u_V}{a \cdot \nu_V}\right)^m \cdot \left(\frac{\nu_V}{D_V}\right)^n \left(\frac{a_{Ph}}{a}\right) \tag{19}$$

A dimensional analysis of the influencing parameters showed that the volumetric mass transfer coefficients could be determined most accurately if the ratio a_{Ph}/a were to be given by Equation (20)[4,5].

$$\frac{a_{Ph}}{a} = 1.5 \cdot (a \cdot d_h)^{-0.5} \cdot \left(\frac{u_L \cdot d_h}{\nu_L}\right)^{-0.2} \cdot \left(\frac{u_L^2 \cdot \rho_L \cdot d_h}{\sigma_L}\right)^{0.75} \cdot \left(\frac{u_L^2}{g \cdot d_h}\right)^{-0.45} \tag{20}$$

Constants C_V and C_L are specific for any given type of packing, and are listed in Table 1. The average relative deviations of the calculated mass transfer coefficients from the experimentally determined values are 8.3% for the liquid side and 12.4% for the gas side.

PREDICTING MASS TRANSFER FOR RECTIFICATION

During rectification the physical properties of the liquid- and vapour phase are changing along the height of the column. If the surface tension of the components of the mixture to be separated differs, surface tension gradients occur, which influence the interfacial area. Systems with increasing surface tension along the liquid flow are called positive systems, whereas for negative systems surface tension decreases. If the surface tension is nearly constant during the process the system is called neutral.

The Marangoni number, Equation (21), describes the differential change of surface tension along the column with the composition in the liquid phase $d\sigma_L/dx$ caused by a concentration gradient from the bulk of the liquid to the surface Δx. Since the liquid phase equilibrium concentration at the surface cannot be measured, Δx has to be calculated from the mass transfer resistance distribution, and the overall concentration difference[5].

$$Ma_L = \frac{d\sigma_L}{dx} \cdot \frac{\Delta x}{D_L \cdot \eta_L \cdot a} \tag{21}$$

$$\left(\frac{a_{Ph}}{a}\right)_{neg.Sys.} = \left(\frac{a_{Ph}}{a}\right)_{Eqn.(20)} (1 - 2.4 \cdot 10^{-4} \cdot |Ma_L|^{0.5}) \tag{22}$$

Test results carried out in rectification columns led to Equation (22) for describing the ratio a_{Ph}/a for negative systems while the afore-mentioned Equation (20) is still valid for positive resp. neutral systems.

For rectification columns the calculated height of an overall mass transfer unit on the gas side HTU_{OV} can be compared with the experimental results, if the constants C_V and C_L of Table 1 are used. With an average relative deviation of 14.1% the cited equations are again more accurate than previous methods.

CONCLUSIONS

The equations presented in this paper permit the holdup, pressure drop and mass transfer in absorption-, desorption- and rectification columns to be predicted in the entire loading range. Their evaluation requires physical and packing geometry data as well as a constant, characteristic for the shape and structure of the packing, cf. Table 1. Table 2 indicates figures describing the capacity range, test facilities and physical properties for all investigations.

REFERENCES

1. Billet, R, 1987, *IChemE Symp Ser No 104*, A171–A182.
2. Billet, R. and Schultes, M., 1991, *Chem Eng Technol*, 14: 89–95.
3. Billet, R., 1983, *Festschrift der Fakultät Maschinenbau, Ruhr-Universität Bochum*, 24–31.
4. Billet, R., 1990, *FAT Sci Technol*, 92 (9): 361–370.
5. Schultes, M., 1990, *Dr. Thesis; Einfluß der Phasengrenzfläche auf den Stoffübergang in Füllkörperkolonnen*, Fortschritt-Berichte VDI, Reihe 3, No. 230, 4000 Düsseldorf, VDI-Verlag.

NOMENCLATURE

a	surface area of packing, m^2/m^3
a_h	hydraulic area of packing, m^2/m^3
a_{Ph}	interfacial area of packing, m^2/m^3
C_h, C_L, C_P, C_V	constants
D	diffusion coefficient, m^2/s
d	diameter, m
$d_h = 4\,\varepsilon/a$	hydraulic diameter, m
Fr	Froude number
F_V	vapour or gas capacity factor, $m^{-1/2}kg^{1/2}s^{-1}$
g	acceleration due to gravity, m/s^2
H	height, m
HTU_{OV}	overall mass transfer unit on the gas side, m
h_L, h_L'	liquid holdup, m^3/m^3
K	wall factor
Ma	Marangoni number
Re	Reynolds number
s	film thickness, m
Δp	pressure drop, Pa

u_L	specific liquid velocity, m^3/m^2s
u_V	vapour or gas velocity, m/s
$\bar{u}$	effective velocity, m/s
x	mole fraction, mol/mol
β	mass transfer coefficient, m/s
ε	void fraction, m^3/m^3
η	dynamic viscosity, kg/(ms)
ν	kinematic viscosity, m^2/s
ρ	density, kg/m^3
σ	surface tension, kg/s^2
τ	duration of contact, s
τ_V	shear stress in gas stream, N/m^2
ψ_O, ψ_L, ψ_L'	resistance factor

SUBSCRIPTS

Fl	flood point
L	liquid
O	surface
P	particle
S	loading point
S	column
V	vapour or gas

Table 1: Characteristic data and constants for dumped and regular packings

DUMPED AND REGULAR PACKINGS		SIZE mm	N $1/m^3$	a m^2/m^3	ε m^3/m^3	C_h	C_P	C_L	C_V
Pall ring	Metal	50	6242	112.6	0.951	0.784	0.763	1.192	0.410
		38	15772	149.6	0.952		1.003	1.227	0.341
		25	53900	223.5	0.954	0.719	0.957	1.440	0.336
	Plastic	50	6765	111.1	0.919	0.593	0.698	1.239	0.368
		35	17000	151.1	0.906	0.718	0.927	0.856	0.380
		25	52300	225.0	0.887	0.528	0.865	0.905	0.446
	Ceramic	50	6215	116.5	0.783	1.335	0.662	1.227	0.415
Ralu ring	Plastic	50	5770	95.2	0.938		0.468	1.520	0.303
NOR PAC ring	Plastic	50	7330	86.8	0.947	0.651	0.350	1.080	0.322
		35	17450	141.8	0.944	0.587	0.371	0.756	0.425
		25	50000	202.0	0.953	0.601	0.397	0.883	0.366
Hiflow ring	Metal	50	5000	92.3	0.977	0.876	0.421	1.168	0.408
		25	40790	202.9	0.962	0.799	0.689	1.641	0.402
	Plastic	50	6815	117.1	0.925	1.038	0.327	1.478	0.345
		25	46100	194.5	0.918		0.741	1.577	0.390
	Ceramic	50	5120	89.7	0.809		0.538	1.377	0.379
		20	121314	286.2	0.758	1.167	0.628	1.744	0.465
TOP-Pak ring	Alu	50	6871	105.5	0.956	0.881	0.604	1.326	0.389
Raschig ring	Ceramic	25	47700	190.0	0.680	0.577	1.329	1.361	0.412
VSP ring	Metal	50 - 2	7841	104.6	0.980	1.135	0.773	1.222	0.420
		25 - 1	33434	199.6	0.975	1.369	0.782	1.376	0.405
Envi Pac ring	Plastic	80 - 3	2000	60.0	0.955	0.641	0.358	1.603	0.257
		60 - 2	6800	98.4	0.961	0.794	0.338	1.522	0.296
		32 - 1	53000	138.9	0.936	1.039	0.549	1.517	0.459
DIN-PAK	Plastic	70	9763	110.7	0.938	0.991	0.378	1.527	0.326
		47	28168	131.2	0.923	1.173	0.514	1.690	0.354
Ralu pack	Metal	YC-250		250.0	0.945		0.191	1.334	0.385
Impulse packing	Ceramic	100		91.4	0.838	1.900	0.417	1.317	0.327
Montz packing	Metal	B1-200		200.0	0.979	0.547	0.355	0.971	0.390
		B1-300		300.0	0.930	0.482	0.295	1.165	0.422
Euroform	Plastic	PN-110		110.0	0.936	0.511	0.250	0.973	0.167

Table 2: Capacity range and test facilities

		liquid holdup	pressure drop	mass transfer
Gas capacity factor	F_V $[m^{-1/2} \cdot kg^{1/2} \cdot s^{-1}]$	0.10 + 2.78	0.21 + 5.09	0.003 + 2.77
Liquid load	u_L $[m^3/m^2h]$	1.33 + 82.8	0.61 + 60.1	0.256 + 118
Liquid density	ρ_L $[kg/m^3]$	800 + 1810	738 + 1115	758 + 1237
Kinematic viscosity of liquid	ν_L $[m^2/s]$	0.74 + 142	0.30 + 99.0	0.30 + 1.66
Surface tension of liquid	σ_L $[kg/s^2]$	20.8 + 86.3	—	17.2 + 74.0
Diffusion coefficient in liquid	D_L $[m^2/s]$	—	—	1.04 + 6.50
Gas density	ρ_V $[kg/m^3]$	—	0.06 + 3.28	0.07 + 4.93
Kinematic viscosity of gas	ν_V $[m^2/s]$	—	2.9 + 106	2.20 + 126
Diffusion coefficient in gas	D_V $[m^2/s]$	—	—	3.70 + 87.4
Investigated system		20	24	45

PREDICTING THE LIQUID PHASE MASS TRANSFER COEFFICIENT — k_L — FOR THE SULZER STRUCTURED PACKING MELLAPAK

M. Henriques de Brito[1], U. von Stockar* and P. Bomio[2]
*Institute of Chemical Engineering, Swiss Federal Institute of Technlogy, CH 1015, Lausanne, Switzerland.
Current addresses:
1 Sulzer do Brasil SA, Caixa Postal 6859, CEP 01051 São Paulo, Brazil.
2 Sulzer Brothers Limited, CH 8401 Winterthur, Switzerland.

A model for predicting the liquid phase mass transfer coefficient, k_L, for the SULZER structured packing MELLAPAK 250.Y and 500.Y is proposed. In the development of this model, experimental data was considered together with the penetration theory. The value of k_L increases with the liquid and gas flow and decreases with the geometric area of the packing. Higher k_L values are associated with irregular packings of equal geometric area.

INTRODUCTION

For the design of the height of industrial separation columns for distillation and absorption processes, the *HTU-NTU* method is used. The liquid phase height of transfer units, HTU_L, depends on the specific liquid load, on the effective transfer area and on the liquid phase mass transfer coefficient, k_L. This coefficient expresses the resistance to mass transfer by diffusion in the liquid boundary layer at the gas-liquid interface. No methods seem available for measuring the liquid phase mass transfer coefficient of packings. Thus, k_L cannot be directly determined experimentally, but only in the form of a product given as k_La. In the literature, no measurements of k_La and effective areas of structured packings from which k_L has been calculated seem to be reported. In the present work, k_L is determined for MELLAPAK types 250.Y and 500.Y, using experimental measurements of k_La and of the specific mass transfer area. We decided to carry out measurements with 250.Y, due to the broad utilization of the packing. Experiments with 500.Y were also carried out, to investigate the influence of the specific packing geometric area on the k_L. Performance characteristics of the packing MELLAPAK are described by Spiegel and Meier (1987).

k_L FOR 250.Y and 500.Y OBTAINED FROM EXPERIMENTAL DATA

The k_L can be obtained by dividing experimental values of the product k_La by effective area values for mass transfer:

$$k_L = \frac{k_La}{a} \tag{1}$$

Experimental values for k_La and for the mass transfer area for 250.Y and 500.Y were measured under industrial-scale operating conditions in a pilot plant column with a diameter of 295 mm, a packing height of 420 mm and with a liquid distributor having a high drip point density of 527 points/m^2, thus eliminating possible entrance effects. The pilot plant and associated analytical equipment is described elsewhere (Henriques de Brito (1991)). The working temperature and pressure were the same for all experiments, thus eliminating their influence.

The volumetric liquid phase mass transfer coefficient, $k_L a$, was obtained by desorption of O_2 from demineralized water into air. These measurements are reported by Henriques de Brito (1991). The values are considered to have been measured within a precision of less than 6%.

The effective mass transfer area was measured using the fast pseudo first order reaction of CO_2 with NaOH. These measurements are reported by Henriques de Brito (1991). The error in the evaluation of mass transfer area is 9 and 8% for 250.Y and 500.Y respectively.

Although different gas-liquid systems were used to measure values of $k_L a$ and effective mass transfer area, it is assumed in this work that the system did not influence the data in any way. The effective area during desorption and chemical absorption experiments would be expected to be all equal with structured packings. The flow in defined channels avoid the existence of stagnation zones of mass transfer. Consequently, the whole gas-liquid interfacial area created would be expected to be identical to effective areas for chemical and physical absorption.

For the $k_L a$ and effective area measurements, for both MELLAPAK 250.Y and 500.Y, it was verified that $k_L a$ changes little with respect to the F-Factor, presenting higher values only near flooding conditions. This is in agreement with the fact that there should be little effect of the F-Factor below the first break point in the pressure drop versus gas rate curve. Often this covers most of the efficient operating range of packed beds. So it is likely that the k_L values in Table 1, obtained with equation (1) at F-Factors of 0.8, will be applicable in most industrial cases. Table 2 gives the coefficient and the exponent of an empirical correlation for the k_L data.

In the evaluation of k_L of 250.Y and 500.Y, the errors are less than 15% and 14% respectively, considering the errors associated with $k_L a$ and effective mass transfer area measurements. These errors are conservative estimation.

MODEL FOR PREDICTING k_L

A model for k_L with a sound theoretical basis is, however, desirable. According to the penetration theory proposed by Higbie (1935),

$$k_L = 2\sqrt{\frac{D}{\Pi t}} \tag{2}$$

TABLE 1
Values for k_L

B [m³/m²h]	kL * 10+4 [m/s] 250.Y	500.Y
21.29	0.85	0.47
42.6	1.03	0.61
51.2	1.09	0.66
60.8	1.14	0.70

TABLE 2
Constants for the correlation: $k_L = C_1 B^{C_2}$

MELLAPAK	$C_1 * 10^{+5}$	C_2
250.Y	3.62	0.28
500.Y	1.42	0.39

For the contact time, t, Billet (1983) proposed a model, which, adapted for the structured packing MELLAPAK, yields:

$$t = \frac{h_L L}{\frac{B}{\cos 45^\circ} \frac{1}{3600}} \frac{1}{100} \tag{3}$$

The contact time has an adjustable parameter which is the contact length, L. This parameter is a characteristic flow dimension associated to the packing. The liquid hold up was measured by Suess and Spiegel (1992) and is principally a function of the specific liquid load, with F-Factor being of negligible importance far from flooding conditions. Hence, the data can be correlated in the form of:

$$h_L = C_3 B^{C_4} \tag{4}$$

From equations (2), (3) and (4), an expression for k_L is obtained:

$$k_L = 2\sqrt{\frac{D}{\Pi L C_3 \, 36 \cos 45^\circ}} \, B^{(1-C_4)/2} \tag{5}$$

This semi-empirical model for predicting k_L requires knowledge of the liquid hold up and of the contact length, L. Average values of the liquid hold up for given specific liquid loads were correlated by least squares regression to obtain expression (4). Table 3 gives the values for the coefficients and exponents of correlation (4). The contact length, L, was determined by comparing k_L values given in Table 1 with the predicted k_L given by equation (5). As both values should be the same, the evaluation of the adjustable parameter (contact length) is possible. The diffusivity of O_2 in water is 2.4×10^{-9} m^2/s (Incropera and DeWitt (1981)). Table 4 shows the contact length for different specific liquid loads. The average contact lengths are 61 and 102 mm for 250.Y and 500.Y respectively. From equation (5) and Table 2, it can be verified that the contact length depends on the specific liquid load with the power of 0.043 and −0.177 for 250.Y and 500.Y respectively.

The contact length does, however, vary significantly with the packing size. Packing 500.Y, with a smaller corrugation angle measured in an horizontal plane, retain the liquid film longer than 250.Y. The larger corrugation angle of 250.Y is associated with a greater wetted area and

TABLE 3
Values for correlation (4)

MELLAPAK	C_3	C_4
250.Y	1.76	0.397
500.Y	3.04	0.397

TABLE 4
The contact length

B	$L*10^{+3}$ [m] 250.Y	500.Y
21.29	59.6	114.2
42.6	61.4	101.1
51.2	61.9	97.8
60.8	62.4	94.9

consequently greater viscous losses of momentum and earlier detachment due to gravity. The corrugation angle of 250.Y and 500.Y was measured and was approximately 82° and 63° respectively. The contact length would also be expected to be a function of F-Factor and physical properties.

A simple experiment ckecked whether the values were reasonable. Water was injected at the top of the packing and the length of the liquid film was observed. Figure 1 shows the results. The measured values for the contact length of approximately 62 mm and 102 mm for 250.Y and 500.Y respectively, compare well with the values in Table 4.

From Table 4 and Figure 1, it is then reasonable to assume constant contact length values in the range of the experimental range of liquid rates, since at worst (case with 500.Y at a specific liquid load of 21.29 m^3/m^2h) this will incur an 11 % error. This error is within experimental uncertainty in the determination of k_L. From equation (5), using for the contact length the values of 61 mm and 102 mm for MELLAPAK 250.Y and 500.Y respectively, the following expressions are obtained:

for MELLAPACK 250.Y: $$k_L = 3.34 \times 10^{-5} B^{0.302} \qquad (6)$$

for MELLAPACK 500.Y: $$k_L = 1.97 \times 10^{-5} B^{0.302} \qquad (7)$$

COMPARISON BETWEEN THE MODEL AND EXPERIMENTAL DATA

Figure 2 shows the comparison between k_L data given in Table 1 and values from correlations (6) and (7). For low F-Factors, far from 100 % capacity, the k_L for MELLAPAK 250.Y is about 69% higher than for MELLAPAK 500.Y, as hold up and contact length of 500.Y are greater, the contact time increases (equation (3)); and the value of k_L is then, according to the penetration theory, smaller (equation (2)).

Increased F-Factors would be expected to increase k_L by reducing the contact time, ie by increasing the liquid renewal rate due to turbulence. The increase of k_L with F-Factor may be more pronounced with 500.Y, since the kLa and the effective mass transfer area for 500.Y are strongly dependent on the F-Factor (Henriques de Brito (1991)). High F-Factors are responsible for spreading the liquid over the 500.Y packing and for creating turbulence at the gas-liquid interface. As a result, near flooding, the k_L of 500.Y would be expected to be higher than for 250.Y.

COMPARISON WITH kL VALUES FOR IRREGULAR PACKINGS

Effective areas and k_La of irregular packings were measured by Mohunta *et al* (1969a) and Mohunta *et al* (1969b) respectively. For both parameters, correlations are proposed, which seem accurate not alone for their own data, but for other data for different packings and systems in the literature also. Using their equations for k_La and effective area and considering the liquid density, viscosity and diffusivity of oxygen in water to be respectively 998 kg/m^3, 1×10^{-3} kg/ms and 0.24×10^{-8} m^2/s, yields:

$$k_La = 2.356 \times 10^{-3}\, a_p^{-1/12}\, B^{0.75} \qquad (8)$$

and

$$a = 1.414\, a_p^{2/3}\, B^{1/3} \qquad (9)$$

With equations (1), (8) and (9), it is found that

$$k_L = 2.065 \times 10^{-3}\, a_p^{-3/4}\, B^{0.417} \qquad (10)$$

Equation (10) with values of geometric area of 250 m^2/m^3 and 500 m^2/m^3 is compared to equation (6) and (7), for MELLAPAK 250.Y and 500.Y respectively, in Figure 3. Considering that equations (8) and (9) have each a potential error of ±20% associated with them, the relative error in the evaluation of the k_L with equation (10) is ±40%.

Equation (10) predicts k_L values from 28 to 51% higher for irregular packings of specific geometric area equivalent to MELLAPAK. It is possible that k_L of irregular packings is higher, since the contact length may be smaller, similar to the size of the packing, and the contact time may be therefore reduced. If irregular packings present higher k_L values, the ratio of effective area to the geometric area is, however, smaller than for MELLAPAK, with the result that the value of $k_L a$ is lower compared to that of structured packings (Henriques de Brito (1991)).

The exponent on the liquid load in equation (10) is 0.417. The work of Dharwadkar and Sawant (1985), in which $k_L a$ and mass transfer area were measured, without evaluation of k_L, suggests that k_L varies with the specific liquid load in the power range of 0.177–0.373 for the packings which were studied. The k_L correlations from Shulman *et al* (1955), Semmelbauer (1967) and Zech and Mersman (1979) suggest exponents of 0.45, 0.59 and 0.5 respectively. This exponent may be theoretically estimated modelling the velocity profile of a completely developed laminar film flowing down an inclined plate with constant wetted area (Henriques de Brito (1991)). With such a model, k_L is proportional to the liquid flow to the power of 1/3. Correlations (6) and (7) propose an exponent of 0.302, which is in accordance with the theoretical estimate and is within the range suggested by Dharwadkar and Sawant (1985).

These authors state additionally that 'the value of the k_L decreases with increase in the packing size for the same type of packing'. However, this statement is in disagreement with their own data for Pall rings, since 50 mm metal and polypropylene Pall rings have higher k_L values than the corresponding packings of 38 mm at a superficial liquid flow of 6×10^{-3} m/s· What may happen is that k_L decreases with an increase in the specific geometric area, as suggested by equation (10). The specific geometric area of irregular packings reduces with an increase of the packing size.

The contact length, which defines when a liquid renewal occurs, would be expected to increase with increased size of an irregular packing, thus reducing k_L. However, the liquid hold up may well decrease, thereby creating conditions which favour liquid renewal, which would increase k_L. Depending upon the packing and the system, one or the other effect may be predominant.

For Bolles and Fair (1982) 'the individual mass transfer coefficient should be more heavily dependent on the system itself than on the geometric characteristics of the packing'. Indeed, systems with different physical properties have an influence on both the contact length and hold up, and thus on the k_L. Nevertheless, the geometry of the packing is also an extremely important parameter in the determination of the flow over the packing and of the liquid hold up. It is therefore unsound to assume a negligible influence of the packing geometry on k_L.

The k_L values of MELLAPAK decrease with an increase of the specific geometric area, as for irregular packings, since the contact length becomes bigger. The k_L values of 250.Y are 69 % higher than for 500.Y. Equation (10) predicts that when the specific geometric area is divided by two, k_L becomes 68% higher.

CONCLUSIONS

The use of the penetration theory in predicting values for k_L of MELLAPAK 250.Y and 500.Y furnishes a model with a theoretical basis. There is a good agreement with data obtained experimentally. The model is valid especially far from flooding conditions. This operating range is frequently encountered in industrial packed columns.

Although there is some variation of contact length with the specific liquid load, it is useful to assume a constant, average value as a characteristic dimension of the packing. This fact enabled the use of the penetration theory for building a model for k_L. The contact length of 500.Y is greater than for 250.Y. The contact length would be expected to depend also on the F-Factor and physical properties.

An increase of the specific geometric area of a MELLAPAK type causes a decrease of k_L. For low F-Factors, far from 100% capacity, the k_L for 250.Y is about 69 % higher than for 500.Y. Irregular packings, with the same specific geometric area as MELLAPAK, may present higher k_L values. This is possibly due to the fact that the contact length is smaller, and the liquid renewal rate of irregular packings therefore higher. A possible higher liquid hold up of irregular packings may reduce the effect of a higher liquid renewal rate.

NOMENCLATURE

a	[m²/m³]	specific mass transfer area
B	[m³/m²h]	specific liquid load
C		coefficient of empirical model
D	[m²/s]	diffusivity
F-Factor		
h_L	[%]	liquid hold up
HTU	[m]	height of a transfer unit
k_L	[m/s]	liquid phase mass transfer coefficient
$k_L a$	[1/s]	volumetric mass transfer coefficient
L	[m]	contact length
t	[s]	contact time

Subscript

L	liquid
p	packing
1, 2, ...	indexes

REFERENCES

1. Billet, R., 1983, *Festschrift der Fakultät für Maschinenbau*, Ruhr-Universität Bochum, 24–31.
2. Bolles, W.L. and Fair, J.R., 1982, *Chemical Engineering* 89 (July 12): 109–116.
3. Bravo, J.L. and Fair, J.R., 1982, *Ind Eng Chem Process Des Dev* 21: 162–170.
4. Dharwadkar, S.W. and Sawant, S.B., 1985, *The Chemical Engineering J*. 31: 15–21.
5. Henriques de Brito, M., 1991, Lausanne: PhD Thesis EPFL (984), Swiss Federal Institute of Technology.
6. Higbie, R., 1935, *Trans Am Instn Chem Eng* 31: 365–389.

7. Incropera, F.P. and DeWitt, D.P., 1981, New York: John Wiley & Sons, 785.
8. Mohunta, D.M., Vaidyanathan, A.S. and Laddha, G.S., 1969a, *Indian Chemical Engineer* April: 39–42.
9. Mohunta, D.M., Vaidyanathan, A.S. and Laddha, G.S., 1969b, *Indian Chemical Engineer* July: 73–79.
10. Semmelbauer, R. (1967). *Chem Eng Sci* 22: 1237–1255.
11. Shulman, H.L., Ullrich, C.F., Proulx, A.Z. and Zimmerman, J.O., 1955, *AIChE J.* 1(2): 253–258.
13. Spiegel, L. andMeier, W., 1987, *IChemE Symposium Series 104*: A203–A215.
14. Suess, P. and Spiegel, L., 1992, *Chem Eng and Processing*, to be published.
15. Zech, J.B. and Mersmann, A., 1979, *IChemE Symposium Series 56*: 2.5/39–47.

a) 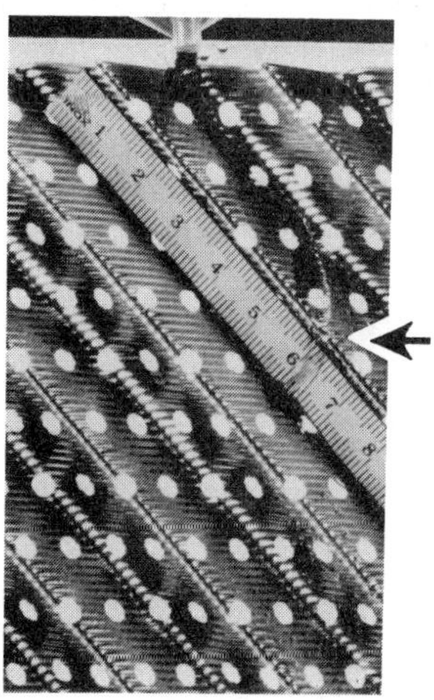 b) 

Figure 1 Liquid film over the packing.
a) Mellapak 250.Y b) Mellapak 500.Y

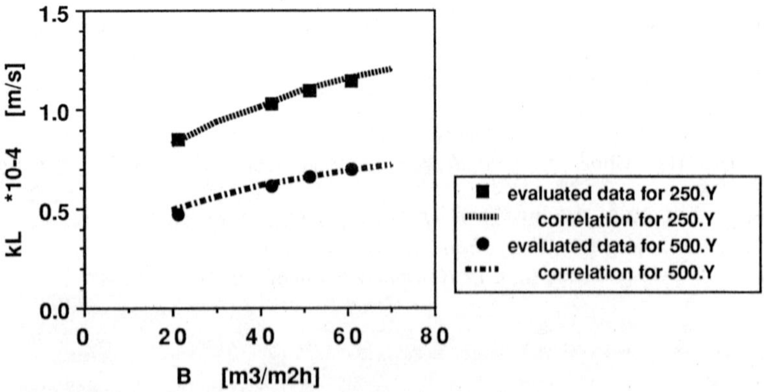

Figure 2 Model for predicting k_L compared with experimental data.

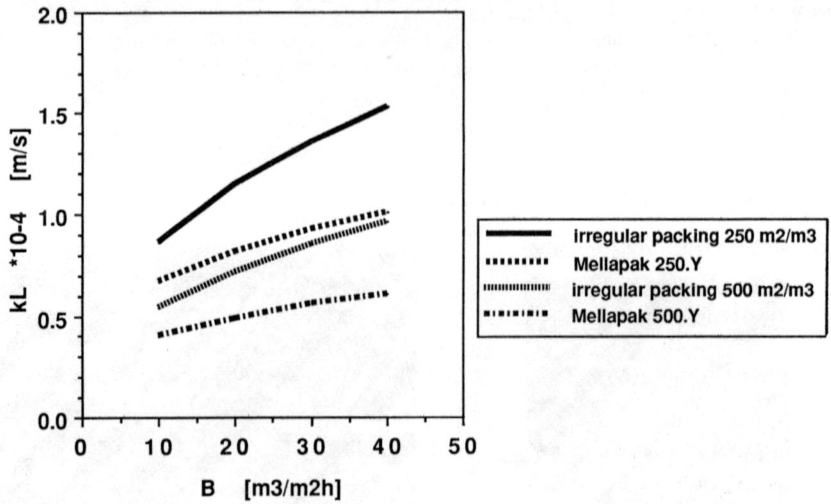

Figure 3 Comparison of k_L values.

WASTE GAS SCRUBBING IN PACKED TOWERS

Werner Geipel and Karl-Heinz Maier
Rauschert Verfahrenstechnik, 8645 Steinwiesen, Germany

Layout of packed beds and design of internals for a packed bed flue gas scrubber are described. A comparison of dimensions and parameters to a conventional equipped tower is presented.

Demands for greater efficiency in terms of the cleaning of industrial waste gases continue to increase. For example in the Federal Republic of Germany with regard to municipal waste incineration facilities, emissions from hydrochloric acid cannot exceed 10 mg/m^3 while sulphur dioxide emissions must remain less than 50 mg/m^3. In some instances, sulphur dioxide emissions cannot exceed 20 mg/m^3. These demands lead to a search for cost effective, highly efficient waste gas cleaning technologies. Packed towers can provide a highly effective means of cleaning waste gases.

Packed towers utilizing random packings have been successfully operated in the chemical processing industry for a wide variety of applications for many years. The first random packing, the Raschig Ring, was used over 80 years ago and this type of packing is still used today. In general, packed towers provide the following advantages:

- Very good mass transfer capabilities. The mass transfer efficiency can be varied by varying the height of the packed bed.
- Low energy usage due to the fact that the operating differential pressure over the packed bed is low.
- Flexibility in terms of the substances that can be removed from a waste gas stream. This is due to the fact that random packings and packed tower internals can be manufactured from a variety of heat and corrosion-resistant materials.

In general, some reasons for being hesitant to use a packed tower for the cleaning of waste gases are:

- Packed towers cannot efficiently handle waste gas streams that are also laden with solid particulate. Fouling or pluggage of the void space in the packed bed occurs increasing overall pressure drop and reducing cleaning efficiency.
- Large diameter packed towers are difficult to operate on a consistent basis.
- The packed bed has to receive a minimum, continuous liquid load at all times when the packed tower is in operation or the packed bed will not be sufficiently wetted thus reducing the packed tower's cleaning efficiency.

Contrary to all these reasons listed for not using a packed tower, it has been found, through many actual applications, that if certain operating parameters are maintained, the packed tower can be an excellent means of cleaning waste gas streams. This is demonstrated by the following example of a packed tower used for a flue gas scrubbing application.

The generalized layout of a flue gas cleaning plant utilizing a packed tower is the following. Cooled flue gas is first passed through a quench system so that saturation temperature is reached. Then the flue gas flows through a two circuit packed tower. In the first circuit (acid circuit), hydrochloric acid and fluorides are removed through absorption by water. In the second circuit (caustic circuit), sulphur dioxide is removed through absorption with a sodium hydroxide solution. After these two circuits, the gas stream passes a droplet separator and is reheated before exiting through a stack. Random packings can be used in the quench system as well as both the acid and caustic circuits of the packed tower.

Many common principles for the sizing of a packed tower exist. The diameter of the packed bed is dependent upon the volume of the waste gas that is required to be handled. Today's high performance random packings allow for large hydraulic loadings without the problems of high differential pressure. The packed bed height is calculated using principles of mass transfer. The dimensioning of the packed bed is based on the equation by Colburn.

The height of a transfer unit is best determined empirically. Diagram 1 shows results from an experimental scrubber unit where high performance random packings have been used for the absorption of sulphur dioxide by a sodium hydroxide solution. Based on the experimental results of the chemical absorption of SO_2 by caustic HTU-values for HCl/H_2O had been calculated by approved formulas (Billet, Onda). Practical experience shows, however, that designing large scrubbers by these parameters is not sufficient. Measurements at several large scale units (up to 9,4 m diameter) indicate by far higher HTU-values. Moreover the HTU-value of HCl/water is always significantly higher than for SO_2/NaOH. We think that the difference between pilot and full scale unit for SO_2 is based on scale up effects, while the difference on HCl is based on the mass transfer equilibrium.

With regard to a packed tower scrubber's ability to function efficiently when solid particulate is contained in the gas stream, there are different opinions. Based on our experience, under normal operation conditions, there is no loss of efficiency in a packed tower scrubber handling a solid particulate laden gas stream (example: fly ash or mineral dusts) if the packed bed receives a continual, sufficient liquid load and the packings used have an open structure. Fouling of packings or crystallization build-up on packings only occurs. Areas exist that receive no liquid load ('dead areas') and the packing does not receive a sufficient liquid irrigation.

The results of tests clearly indicate that for coarser particles good separation efficiency exists. For aerosols or very small particulates this however has not proven to be the case. In terms of these tests, no fouling of the packings occurred due to the particulate being suspended in the gas stream. The particulate was removed from the gas stream by the scrubbing liquid and removed from the system.

These results shown before can be used for large scale units if a homogeneous gas and liquid distribution can be achieved in a homogeneous packed bed of random packings. When designing a large scale scrubber it is essential that the following be considered:

- proper selection of random packing type and material
- proper design of the liquid distributor
- proper design of other internals

Poorly designed internals or internals that are not properly installed can hinder the proper functioning of a packed tower scrubber. Their design and proper installation is more critical to an effective scrubber application than the actual type of random packing used.

The material most commonly used for random packings in flue gas scrubbers is polypropylene. It is relatively inexpensive, resistant to chemical attack by a large variety of substances and commonly used scrubbing liquids. The maximum continuous operating temperature of polypropylene is 90°C. Short term temperature excursions up to 110°C can be handled. Disadvantages of polypropylene are its low tensile strength which has to be considered by the strength calculations of the internals and high coefficient of thermal expansion. That means that internals for packed tower diameters greater than two meters require additional support. Support bars constructed of either metal or FRP can be used. Typically, the internals themselves are also constructed from polypropylene. It is important when designing a packed bed scrubber to keep the packed bed height and the height of the internals as low as possible to keep the overall scrubber vessel height low, thus reducing overall costs.

To ensure that a packed tower scrubber operates efficiently, most attention must be paid to the proper design of the liquid distributor. It is essential that the liquid distributor can evenly distribute liquid on a continuous basis over the entire cross section of the packed bed. A review of all types of liquid distributors reveals that the trough-type distributor with holes in the bottom of the troughs provides the most efficient distribution of liquid to the packed bed. Its advantage is that as long as a minimum liquid heat is maintained in the troughs, variations of the rate of liquid supplied to the distributor and undulations on the surface of liquid maintained in the trough will not result in unequal or intermittent distribution of liquid to the packed bed. A minimum water level of 40 mm should be maintained in the troughs.

It is important that the overflow troughs are carefully installed and levelled to ensure an equal distribution of liquid over the packed bed cross-section. Especially in terms of large diameter scrubbers, it is essential to avoid inproper levelling or deflections of the troughs. If a distribution trough must be longer than approx. one meter, it should be segmented and liquid should be supplied to it by a pre-distribution system. This system must be under constant positive pressure to ensure that each segment of the troughs receives a sufficient volume of liquid. A back washing system prevents build-ups of material in the closed liquid pipe system.

Criteria which should be regarded designing a liquid distributor are listed in Table 2. Converting those criteria to other internals (support and collecting trays) will give proper specifications.

The proper design of the collecting tray between the acid and caustic circuit is also critical. This collecting tray must prevent leakage of liquid into the lower circuit and must also be designed so that the airflow passing between the circuits has a slow velocity to prevent droplet entrainment. A common type of collecting tray is a chimney-type tray.

For large packed towers an alternative form of collecting tray as depicted in Diagram 3 is more effective. This open construction does not require flange connections and gaskets which may eventually leak. With this design liquid leakage rates between the two circuits are between 0.01 and 1.0% of the total liquid volume being handled. Normally separate internals are required for the collection of liquid and the support of the packed bed. This requires a larger overall vessel height and two support rings. The collecting tray construction shown in Diagram 3 avoids this as it also acts as a support for the packed bed. The construction allows for the gas stream to pass through an opening that is 45% of the total packed bed cross section.

To build and to operate large scrubbers following requirements have to be met:

- save scale up of packing performance (reliable HTU-values)
- low energy costs to handle large gas volumes (low pressure drop)
- stable operation (no fouling)
- low investment (reduction of the scrubber volume)

Table 3 compares some dimensions of two existing scrubbers both of which are designed to handle 800,000 m^3/h of waste gas. One of the scrubbers uses conventional packings like Pall Rings or saddles while the other scrubber uses high performance packings like Hiflow Rings. It is apparent that by using high performance packings and well designed internals packed bed diameters can be reduced and gas loading volumes can be increased significantly. The reduction in the liquid load reduces the size of the pump required. The total height of the scrubber is effected by the number and the height of internals required. By using combined support trays, liquid collecting units and two step liquid distributors, the overall scrubber height can be significantly reduced.

By using high performance packings and well designed internals the overall size of a packed tower scrubber can be significantly reduced resulting in lower operating costs. A packed bed scrubber can and should be regarded as an economical, efficient waste gas scrubbing device.

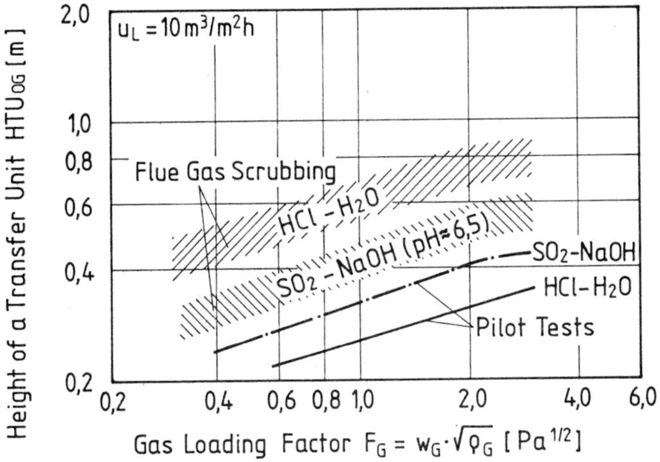

Diagram 1: HTU_{OG} of Hiflow Rings

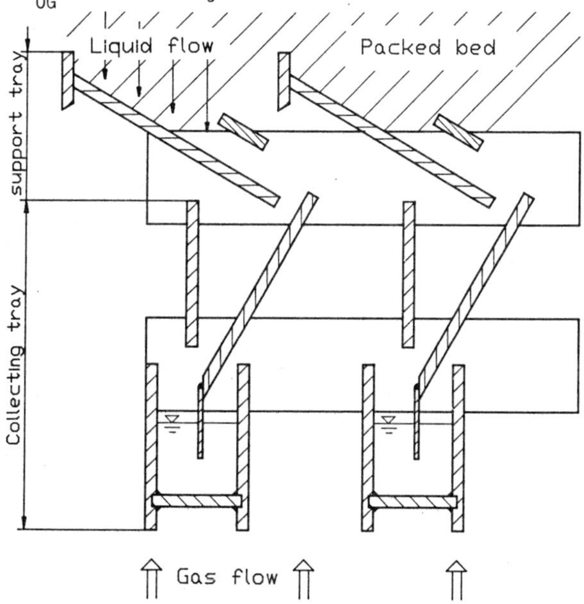

Diagram 2: Collecting and Support Tray

Chemical, thermal and mechanical resistance of the material Strength calculation based on lifetime and max design temperature		
Liquid load	design +/- 30 % operation range	
Uniformity of liquid distribution better than +/- 15 %		
Min liquid head in troughs		40 mm
Number of drip points		> 100 1/m²
Horizontal adjustment better than		0.5 mm/m
Complete cleaning must be possible		

Table 1: Criteria for Liquid Distributors

Flue Gas Scrubbing 800 000 m3/h	Conventional Packings (Pall Rings)	High Performance Packings and Internals
Packed Tower		
Diameter	12,5 m	9,4 m
Total Height	50,0 m	36,0 m
Packed Height	2 trays + 3 x 1,5 m	4 x 1,4 m + 2 x 1,5 m
Volume	6 200 m3	2 500 m3
F-Factor	2,0 $Pa^{1/2}$	3,2 $Pa^{1/2}$
Liquid Load	10 - 15 m3/m2h	3 - 10 m3/m2h
Pressure Drop	20 - (40) mbar	< 20 mbar
Energy		
Liquid Pumps	400 kW (approx.)	50 kW (1)
Blower	630 - (1 260) kW	630 kW

(1) no pump tank

Table 2: Dimensions and Operating Parameters of Waste Gas Scrubbers

MEASUREMENT OF LARGE SCALE GAS AND LIQUID MALDISTRIBUTION IN STRUCTURED PACKINGS

Z. Olujic, F. Stoter and J. de Graauw
Delft University of Technology, Laboratory for Process Equipment, Leeghwaterstraat 44, 2628 CA Delft, The Netherlands

The results are reported of an academic effort to find answers to questions about the origin, the extent, and the nature of the maldistribution of gas and liquid in large diameter beds consisting of corrugated sheet structured packings. The measurements were made using air and water at atmospheric conditions in a rectangular packed bed simulator representing an axial slice of a 3 m ID column.

INTRODUCTION

Large scale liquid maldistribution which proved to be a major cause of malperformance of large packed columns can result from inadequate initial liquid distribution or from 'structural deviations' within the packed bed[1–5]. The quality of initial liquid distribution depends directly on design and operation of distributor. The importance of a good working distributor and the most recent developments and trends in distributor design are discussed elsewhere[6].

A common structural deviation is the wall flow. The problem with pilot scale experimentation is that in columns with diameters up to 0.5 m even the liquid supplied centrally as a point source will be brought to the wall within the length (height) of one packing element (usually 0.2 m). With an efficient wall wiper band around each packing element the wall flow is avoided and, normally, three to four layers of packing are needed to establish a stable more or less even liquid distribution[5,7,10–13].

This seems not always to be the case in large diameter columns. Most probably because of another type of 'structural deviation' which becomes more pronounced with increasing diameter. Namely, the packing is normally installed through manholes, which means in discrete segments, including unevitable gaps between them. The number and sizes of segments depend on column diameter, manhole dimensions, the layout, etc. The imperfections in assembling packing layers may cause discontinuities in liquid flows and in extreme situations could be a source of a large scale maldistribution ie the malperformance of column.

To find answers to this and other industrial scale problems, liquid and gas flow distribution measurements were made using a rectangular packed bed simulator.

EXPERIMENTAL

The experimental equipment is shown in Figure 1. It consists of the water circuit, air circuit, and the rectangular packed column. The water is pumped from a storage tank and fed to the column via 15 pipes each equipped with a valve and a magnetic flowmeter and feeding a narrow trough distributor with 10 drip pipes (an equivalent to 100 drip points per square meter) provided with a number of vertically placed holes. In this way a wide range of liquid loads is covered and the lay out of drip points is fixed. After having passed the packed bed, water is collected in 15 separate compartments (0.2 m × 0.5 m). The liquid flow rate in each compartment was determined simply by measuring the time interval needed to fill the compartment up to a certain level, after closing the outlet valve. Outlets of the compartments are connected via a central return pipe with the storage tank. Installed centrifugal pump enables volumetric flowrates equivalent to liquid loads per compartment up to 60 m^3/m^2h.

The capacity of the fan (7 m^3/s) enables superficial air velocities high enough to flood common structured packings at lower liquid loads. The air is supplied via 15 pipes into compartments below the packed bed. Each of the pipes has a valve and a flowmeter. As in the case of liquid, initial profiles can be arranged in an arbitrary manner. After leaving the packing air enters the compartments where gas velocity is measured with an anemometer. These as well as bottom compartments are also equipped with pressure taps to measure the pressure drop along the column so that pressure profiles across the inlet and outlet of the packed bed can be established.

The rectangular column is made mainly from plexiglass, with a total height which allows packed beds up to 4.4 m height. The width and the depth of the bed are 3 and 0.5 m, respectively, with a total cross section area of 1.5 m^2. The rectangular packed bed has one dimension large enough and represents actually an axial slice of a packed column with an internal diameter of 3 m (see Figure 2).

The packing tested was a corrugated metal sheet packing with a surface area of 250 m^2/m^3 (Montz-pak B1–250). The height of packing elements (segments) is 0.195 m, and the length 1 or 1.5 m. In contrast to small scale tests (diameters usually < 0.5 m) where a layer of packing consists of one element of packing the layers of the packed bed in our column consist of two or three packing elements (segments), including smaller or larger gaps amidst adjacent elements. The structure of the rectangular bed is illustrated in Figure 2.

A number of different initial gas and liquid profiles has been used to study the redistribution properties of the structured packing. The uniform liquid profile means an equal flow rate from each of 15 distributors. In the case of gas phase it was not possible to get a completely uniform initial profile because of the inadequacy for fine tuning of butterfly valves which were used to keep the pressure drop in the gas circuit at a minimum. As a representative of a severe gas and liquid maldistribution, a profile corresponding to blanking half distributor was used, and for point source type studies single-compartment profiles were used, moving from one to other side of the column. A symetrical reduction in the number of drip points was arranged by blocking (plugging) each second distributor. Further in the text the characteristic inlet profiles are shown in symbolic form (one arrow denotes one compartment) as illustrated below (for gas phase) and/or above (for liquid phase) of each of outlet velocity profiles shown in figures illustrating results of this study.

RESULTS AND DISCUSSION

A preliminary study[8] carried out with full packed height (4.2 m) and fans which were not able to produce gas velocities close to flooding point was limited to the hydraulic behaviour of packing below loading range. The obtained outlet profiles of gas and liquid proved to have a typical, reproducible form, independent of the presence and the load of the other phase. Point source tests have shown that gas phase spreads easily driven by local pressure gradients reaching an even distribution within short packed height. In contrast to gas the lateral spreading of liquid from a point source proved to be limited to an unexpectedly narrow range. A thorough prewetting of packing improved this situation only slightly. The most striking observation was the evidence of the total incapability of the bed to correct any kind of severe disturbance in initial maldistribution.

Figure 3 suggests that in the case of half distributor inactive practically one half of the bed will remain dry. The gas flow will not improve this situation neither be influenced by this situation. Such initial disturbances proved to be completely restorable in small pilot scale columns[7]. The experimental results discussed below are also concentrated around gas and liquid distribution and include tracer spreading tests and effects of the initial distribution, height of the packed bed and the loads of gas and liquid. In the latest case the emphasis is on the most intriguing issue ie on loading range behaviour of phases.

GAS DISTRIBUTION

As shown in Figure 3 in the case of gas flow with an uniform initial distribution it can be expected that pressure and consequently velocity will exhibit a plug flow profile. This is probably not far from the truth. Figure 4 shows gas profiles leaving a 4.2 m bed as function of both gas and liquid load. As already mentioned, even with a point source inlet the gas velocity profile will reach uniformity within short packed height. But this large spreading ability, illustrated in Figure 5a, is not a measure for lateral mixing in normal situation, with more or less uniform gas flow profile. Tracer spreading experiments have been carried out to find out the answer to this question. Figure 5b shows the results for full bed height obtained with centrally introduced tracer gas, illustrating the effect of gas load. Comparison of the obtained profiles representing percentages of tracer gas measured in all compartments suggests that radial gas dispersion increases with decreasing superficial gas velocity.

LIQUID DISTRIBUTION

In contrast to gas the lateral spreading of the liquid from a point source seems to be identical to tracer spreading from the point source (see Figure 6). The obtained profiles suggest a limited lateral spreadability of liquid. However, the distribution seems to be at least stable, and a full outlet profile obtained by summation of profiles resulting from single compartment feeds proved to be identical to that obtained with an uniform initial liquid distribution.

According to the form of profiles shown in Figure 6b it seems that there is no liquid load effect, at least within the range investigated which corresponds to operating range of common vacuum applications. This can also be conjectured from Figure 7 where outlet liquid profiles are shown resulting from symmetrically introduced maldistributions in initial liquid distribution, with and without including the wall zone (side compartments). Practically identical profiles have been obtained without presence of gas flow, indicating absence of any gas flow effect at low superficial gas velocities.

On the other hand, this figure can be used to illustrate the practical relevance of such observations, ie to demonstrate the sensitivity of large diameter beds to an initial underirrigation of wall zone. At the bottom of the 4.2 m bed there is a difference of about 30% in the liquid flow rates on the wall compartments. The distance of feed pipes from the side walls was 0.05 and 0.25 m, respectively. With an initial liquid distribution beginning 0.8 m from the wall liquid will practically not reach the wall after passing a bed of 4.2 height. This is certainly so with the packing tested, which spreads less than other types of corrugated sheet packings, according to double plate spreading tests[9].

The profiles of liquid distribution shown in above figures are related to a bed height of 4.2 m. To obtain the informations about profiles along the packed height the measurements have been repeated with a bed height of 1 m. The outlet profiles are shown for an uniform initial distribution in Figure 8. The effect of length ie a deterioration in the profile is obvious.

Generally, the obtained outlet liquid profiles exhibit a typical shape, which is practicaly independent of liquid load. The shape of the profiles is most probably a product of the structure of our bed (see Figure 2) and related discontinuties in the liquid flow caused by gaps between elements (blocks) of packing in each layer of the packed bed. Especially at very high liquid loads the gaps may cause an abrupt break in channel flow in the packing, and liquid falling down to the next layer enters the channels in excess well beyond what is expected. This means considerably higher local liquid loads as well as an underirrigation at surrounding locations, resulting consequently in local liquid to gas ratios much higher or much lower than expected. The best way to avoid gaps is to use element lengths equal to distances from wall to wall, and in the case of very large diameters the longest feasible elements and install them carefully.

CONCLUDING REMARKS

In view of results obtained, industrial scale hydraulic experiments provided informations on effects of structural bed deviations which cannot be obtained from pilot-scale experiments. Generally, gas spreads more and the liquid less than expected with respect to the channel geometry of the packed bed. The presence of gas phase did not affect the distribution of liquid and vice versa, within normal operating range. Also, it cannot be expected that gas phase will help to smooth out the disturbances in initial liquid distribution. On the other hand, at low to moderate liquid load the gas distribution itself will not be disturbed by the presence of a severe liquid maldistribution. It should be noted that a similar initial liquid maldistribution affected gas flow adversely in a 0.5 m i.d. column[13] and that at very high liquid loads the maldistributions present in liquid phase may induce maldistribution in gas phase by choking the channels with liquid[3].

With respect to initial maldistribution, gas and liquid exhibit completely different behaviour, as well as small and large diameter beds. Striking is that large diameter beds seems to be practically incapable to correct an uneven initial liquid distribution. Even with an uniform initial distribution, the liquid maldistribution can develop in the bulk of packing to some extent as a consequence of structural deviations, such as gaps between adjacent elements in each layer of packing of bed itself. Careful installation of both distributor and packing is a major prerequisite for achieving expected performance of large diameter columns containing structured packings[14].

ACKNOWLEDGEMENTS

The construction of our large diameter column simulator was initially supported by SHELL and DSM. Packing and distributors were provided by J. Montz. We are also obliged to graduate students who performed experimental work.

REFERENCES

1. Olujic, Z. and de Graauw, J., 1989, Appearance of maldistribution in distillation columns equipped with high performance packings, *Chem Biochem Eng Q*, 3: 181–196.
2. Shah, G.C., 1991, Effectively troubleshoot structured-packing distillation systems, *Chem Eng Progr*, 87: 49–56.

3. Kurtz, D.P., McNulty, K.J. and Morgan, R.D., 1991, Stretch the capacity of high pressure distillation columns, *Chem Eng Progr*, 87 (3), 43–49.
4. Meier, W., 1990/1991, New separation columns for old, *Chemical Processing Technology International*, 47–51.
5. Hoek, P.J., Wesselingh, J.A. and Zuiderweg, F.J., 1986, Small scale and large scale maldistribution in packed column, *Chem Eng Res Des*, 64: 431–449.
6. Boniface, A., 1990, Distributors, *The Chemical Engineer*, No. 469, 19–24.
7. Olujic, Z. and de Graauw, J., 1990, Experimental studies on the interaction between the initial liquid distribution and the performance of structured packings, *Separation Science and Technology*, 25: 1723–1735.
8. Olujic, Z., Stoter, F. and de Graauw, J., 1991, *Gas Separation & Purification*, Gas distribution in large-diameter packed columns (in press).
9. Stoter, F., Olujic, Z. and de Graauw, J., 1990, Measurement and modelling of liquid distribution in structured packings, Paper No. 199a, *AIChE Annual Meeting, Chicago, USA, November 1990.*
10. Stikkelman, R.M., de Graauw, J., Olujic, Z., Teeuw, H. and Wesselingh, J.A., 1989, A study of gas and liquid distributions in structured packings, *Chem Eng Technology*, 13: 445–449.
11. Meier, W., Hunkelar, R. and Stoecker, D., 1979, Performance of the new regular tower packing 'Mellapak', *IChemE Symposium Series, No 56*, 3.3/1–17.
12. Potthoff, R. and Stichlmair, J., 1991, Maldistribution in Packungskolonnen, *Chem-Ing-Techn*, 63: 72–73.
13. Kouri, R.J. and Sohlo, J.J., 1987, Liquid and gas flow patterns in random and structured packings, *IChemE Symposium Series, No 104*, B193–211.
14. Spiegel, L., 1990, Applications of structured packings in the process industries, Paper No. 199h, *AIChE Annual Meeting, Chicago, USA, November 1990.*

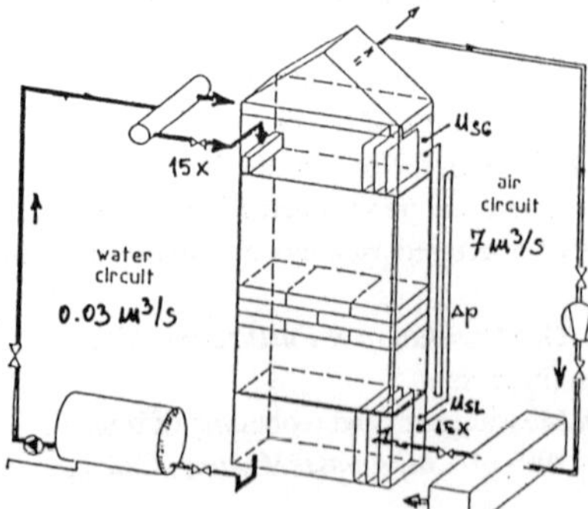

Fig. 1 Schematic diagram of large diameter packed column simulator

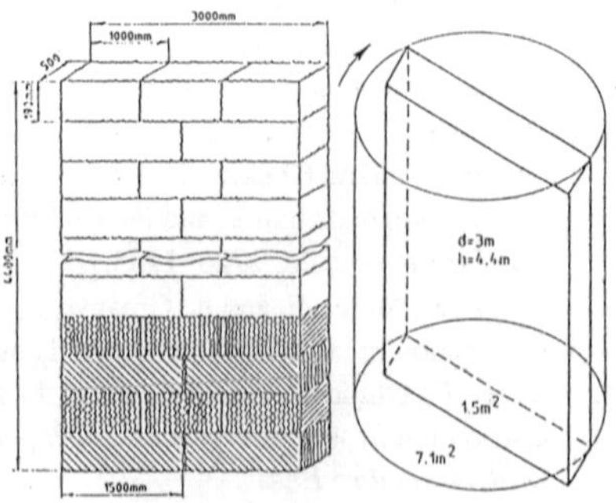

Fig. 2 Structure and dimensions of rectangular packed bed, representing an axial slice of a 3 m i.d. column

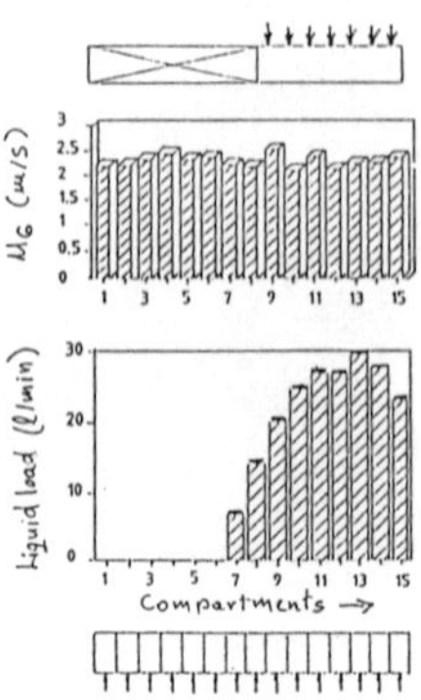

Fig. 3 Profiles of gas and liquid leaving, respectively, the top and the bottom of a 4.2 m packed bed, originating from an uniform initial profile of gas and an initial liquid profile covering the right half of the bed only

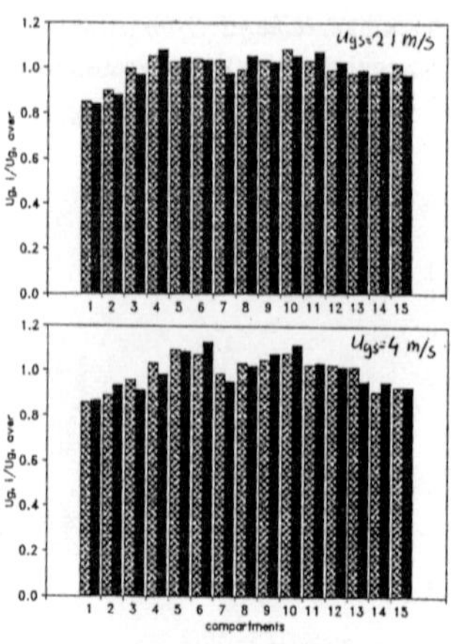

Fig. 4 Gas profiles leaving the top of a 4.2 m bed as a function of gas and liquid loads, resulting from an uniform initial profile

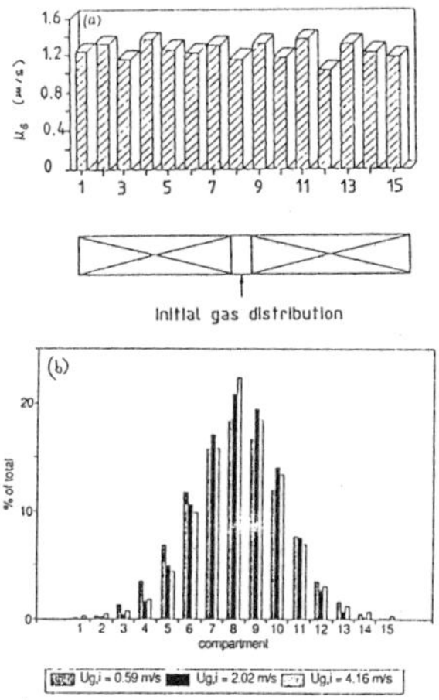

Fig. 5 Outlet profiles of (a) gas flow and (b) a tracer gas resulting from a point source introduced in compartment eight below a 4.2 bed

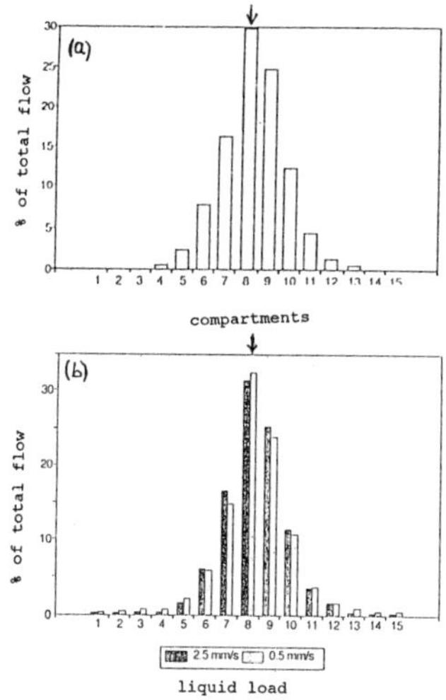

Fig. 6 Outlet profiles of (a) liquid flow and (b) tracer liquid resulting form a point source introduced in compartment eight at the top of a 4.2 m bed

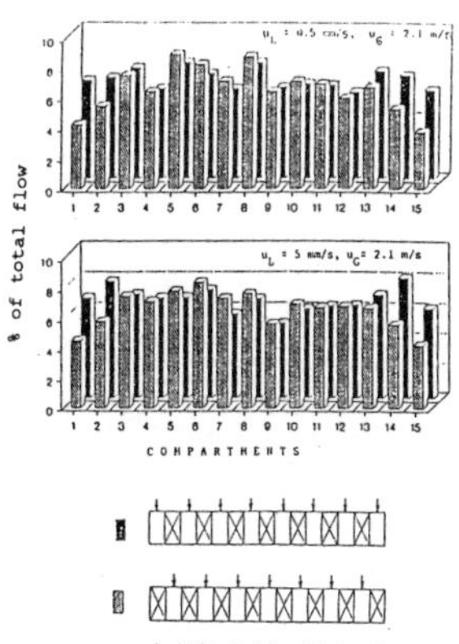

Fig. 7 Liquid profiles leaving a 4.2 m bed, resulting from inlet profiles respresenting symetrically introduced large scale maldistribution, with and without including the wall zone

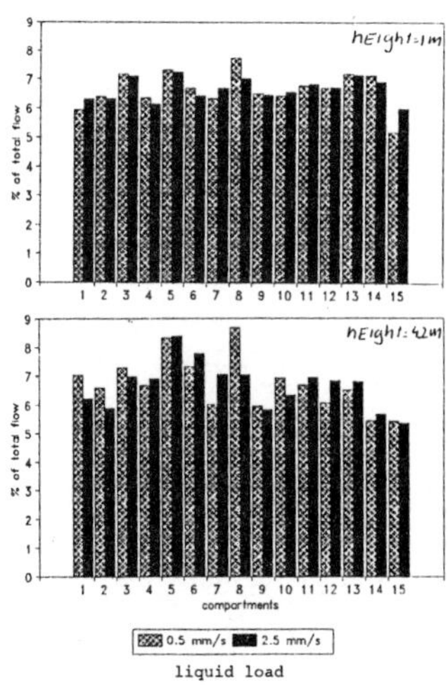

Fig. 8 Effect of packed height on liquid distribution for a number of liquid loads

THE EFFECT OF INTEGRATING CRYOGENIC DISTILLATION WITH MEMBRANES ON ENERGY REQUIREMENTS FOR NITROGEN AND ARGON PRODUCTION

R. M. Thorogood
Air Products and Chemicals, Inc, Allentown, PA, USA

The availability of efficient air separation membranes has prompted the evaluation of hybrid processes in combination with a single column cryogenic nitrogen generator. Various cycles are described for the production of nitrogen and co-production of argon without oxygen. It is found that the energy requirement for nitrogen production from a single column process can be reduced by the use of waste recycle schemes; however, a hybrid using current membranes is no more effective than improved cryogenic schemes. But high performance membranes in a recycle hybrid offer the possibility of efficient co-production of argon with nitrogen.

Keywords: Nitrogen, argon, membranes, cryogenic distillation, hybrid process.

INTRODUCTION

Cryogenic distillation is the principal method for the separation of air into its components. However, membrane and kinetic adsorption technologies have now displaced distillation for smaller capacity nitrogen plants where high purity nitrogen is not required. Thus, the opportunity has arisen to combine separation processes with the intent of reducing power or capital requirements. This paper will discuss some processes which combine a single distillation column with a membrane separator for the production of high purity nitrogen with and without co-production of argon.

Most argon is now obtained as a by-product from distillation of air to produce high purity oxygen. A very small amount is recovered from ammonia purge gas. Over the past ten years argon and nitrogen consumption have increased rapidly whilst oxygen has declined[1]. This trend has stimulated interest in obtaining argon as a co-product with nitrogen where oxygen is not required.

SINGLE COLUMN DISTILLATION FOR NITROGEN PRODUCTION

The most frequently used process for production of high purity nitrogen has a single high pressure distillation column. This process is applied commercially for production in the range 10 to 300 t/d of nitrogen and could thus provide a useful supplementary source of argon if an economic process can be devised. The single column process in its simplest form is shown in Figure 1. In this process, air is compressed and purified to remove water, carbon dioxide, and some hydrocarbons. It is then cooled to near its condensation temperature. The air is fed to the bottom of a distillation column with provision for typically 40 to 70 theoretical stages. The overhead vapour is pure nitrogen, part of which is heat exchanged with the incoming air to give the product nitrogen. The remainder is condensed to provide reflux. The bottom liquid stream approaches equilibrium with the feed air and is enriched with oxygen. This stream is then reduced in pressure and revaporized against the condensing nitrogen reflux. The dewpoint temperature of the oxygen rich stream and the reboiler-condenser temperature difference determine the pressure required to give condensation of the nitrogen flux. The vaporized oxygen rich stream is partially reheated against feed air and then expanded through a turbine to provide refrigeration for the process. Finally, the stream is reheated to ambient temperature against feed air and discharged to atmosphere.

Energy for the nitrogen separation and process refrigeration is provided entirely by the incoming air pressure. It has been determined[2] that this process can operate at a minimum feed pressure of about 5 bar (abs) to maintain a cold box refrigeration balance at zero liquid nitrogen production. At higher pressures, both the flow and the pressure at the expander inlet increase and it becomes possible to produce liquid nitrogen which can be stored as backup for plant outages. However, if liquid is not required at product pressures exceeding 4.5 bar (abs), it is generally necessary to waste refrigeration by providing a Joule-Thompson bypass expansion in parallel with the turbine. The majority of industrial nitrogen requirements correspond to this situation. Conversion of this energy loss to useful separation work has been the subject of several process studies and inventions over the past few years[2]. One means to convert the excess expander flow is to utilize a membrane to selectively reject oxygen from the process and recycle the nitrogen enriched stream.

DISTILLATION/MEMBRANE HYBRID FOR NITROGEN PRODUCTION

A process to utilize a membrane in a hybrid scheme with distillation for nitrogen production has been described by Agrawal and Auvil[3]. This scheme is shown in Figure 2. The process operates by utilizing the excess pressurized flow which is not required in the expansion turbine to meet plant refrigeration requirements. This gas is rewarmed to ambient temperature against feed air and is fed to the single stage membrane, where it is separated into nitrogen rich and oxygen rich streams. The oxygen rich permeate stream is vented to atmosphere. The residual nitrogen rich stream, with a composition close to air, is recycled at pressure to an intermediate stage in the air compressor.

Current polymeric membrane materials offer oxygen/nitrogen selectivities from approximately 4 to 8.5. Using a specific material with a high selectivity of 8.2 in order to favour a low power consumption, the performance of the hybrid has been calculated[2] for comparison with the conventional waste expander process. The results are compared in Table 1 for processes with a standardized cold box design as detailed in Table 2. The membrane cycle scheme offers a specific power reduction for nitrogen production at 7.9 bar abs of about 14% compared to the standard waste expander process. The power reduction can be achieved with only small modifications to the cryogenic equipment. However, the assumed membrane performance in this process calculation is beyond that of current commercial systems and alternative processes which do not require a membrane have been devised. These use enhanced separation schemes in the cold box to provided pressurized recycle streams. Surprisingly, these have been found to save more energy than the membrane hybrid scheme[1, 2]. This has been shown to be due to the higher separation efficiency of distillation compared to a membrane, especially in the high purity range[4].

One feature of the hybrid membrane recycle process which is not immediately apparent from specific energy comparisons is the high recovery of nitrogen from fresh feed air. This is achieved by selectively rejecting oxygen from the process. Since oxygen permeates faster than argon, a secondary effect of the hybrid process is to recycle argon to the distillation process. Advantage may be taken of this to separately recover argon with the hybrid process.

CO-PRODUCTION OF ARGON WITH NITROGEN

The co-production of argon with oxygen in the standard double column cycle for air separation depends upon the accumulation of argon in the low pressure column. This is a consequence of the intermediate relative volatility between nitrogen and oxygen coupled with the stripping and rectifying actions of the oxygen reboil and nitrogen reflux in the upper column. By this means, argon concentration can be increased to 10 to 15% in the middle of the column. A side stream may then be tapped from the column and further distilled to provide a high concentration argon product stream (95+% argon). The residual stream depleted of argon is returned to the primary distillation column. Such a scheme is illustrated in Figure 3.

To achieve argon recovery from a single column requires a similar strategy of argon accumulation. The nitrogen reflux is sufficient to achieve high purity and thereby to prevent argon loss in the nitrogen product, but the absence of an oxygen reboiler and the large O_2/N_2 liquid waste flow from the bottom of the distillation column cause the ready elution of argon from the system. Thus, it is necessary to recover the argon along with nitrogen from the enriched oxygen waste stream and recycle it to the column. It is also necessary to structure the distillation separation in such a manner that the recycled argon is retained in the column. The feasibility of doing this has been explored in two different process cycles and for differing membrane performance5.

TWO STAGE MEMBRANE RECYCLE OF ARGON

Polymeric membranes exhibit a permeability for argon which is intermediate between oxygen and nitrogen. Consequently, the selectivity for argon is quite small and only a modest enrichment of argon purity can be achieved in a single stage. With this in mind, a scheme was devised using two membrane stages for recovery of argon (and nitrogen) from the oxygen rich waste stream of the single column nitrogen generator cycle. This process is shown in Figure 4.

Air is compressed and purified to remove water, carbon dioxide, and most hydrocarbons. The air is combined with the oxygen rich waste stream from the distillation system. This stream also contains residual argon. The combined stream is fed to the first membrane stage, where it is separated into a relatively oxygen and argon rich low pressure permeate stream and a larger nitrogen rich pressurized reject stream. The latter stream is cooled to near condensation and becomes the primary feed to the bottom of the main distillation column.

The low pressure permeate is then recompressed and fed to a second membrane stage through which an enriched oxygen low pressure permeate stream is expelled from the process. The residual high pressure product stream is enriched in argon and nitrogen. This stream becomes the secondary feed to the main distillation column. The composition and pressure of this feed are such that it is partially condensed at the boiling temperature of the product crude argon stream from the secondary distillation column. Thus, it may be used to reboil the secondary column and is thus condensed before becoming the secondary feed to the primary column. This feed is supplied to the column a few trays above the main feed. The enriched argon concentration is thus matched to the accumulated argon concentration in the distillation column, whereas the argon content of the waste stream is determined by the lower argon purity of the primary feed.

The main distillation column can thus retain and accumulate argon contained in the two feed streams. The accumulated argon is tapped from the column as a liquid stream and fed to the second argon enrichment column where it is concentrated to an argon content sufficient for shipment to a separate argon purification plant.

This process concept was evaluated for its energy efficiency by assuming membrane selectivity typical for a UV treated poly (trimethyl silylpropyne) membrane of 5.5 for oxygen and 2.9 for argon relative to nitrogen. The result is shown in Table 3, Column 1 and may be compared to the performance of the single column nitrogen generator without argon recovery (Table 1). The energy penalty is rather substantial due to the recompression of the dual recycle streams. Thus, such a scheme would only be attractive if a severe argon shortage from other sources would develop.

SINGLE STAGE MEMBRANE RECYCLE FOR ARGON AT HIGH SELECTIVITY

The high energy requirement determined for the argon recycle process at the modest selectivity of current polymeric membranes leads to the question of process performance if a high selectivity membrane becomes available. To evaluate the potential of such a hybrid scheme, a process was devised to utilize a membrane with a very high hypothetical oxygen selectivity of 100 to nitrogen. Argon selectivity was chosen with the same permeability ratio to oxygen as for the current membrane at 52.8.

With such a high selectivity, only a single membrane stage is needed to recover argon. This single stage process is shown in Figure 5. In this process the membrane receives the oxygen rich waste stream directly from the reboiler (after it has been reheated to ambient temperature). The membrane pressure is optimized for power minimization by compression before and/or after the membrane. The oxygen rich permeate is vented to atmosphere whilst the nitrogen/argon rich stream is recycled to the main distillation column a few trays above the bottom.

Compressed air feed is provided directly to the bottom of the column. The bottom liquid waste stream is thus equilibrated with the air feed which minimizes argon escape. Also in this process, the refrigeration cycle is incorporated as part of the air feed. A part of the air is condensed against the crude argon stream at the bottom of the secondary argon enrichment column. This provides reboil for the secondary column. A further part of the air is used to provide refrigeration for the process by turbine expansion and is then recombined with the feed air to the column.

This process was evaluated for various conditions of the oxygen enriched recycle stream. Results for three different membrane feed pressures are shown in Table 3, Columns 2 through 4. All three cases show a very encouraging result that the power requirement for the process is less than that of the conventional waste expander process for nitrogen production without argon. It is also closely comparable to improved recycle processes for nitrogen production. Thus, argon would be produced without energy penalty for separation (as a gas). The process calculations were made without regard for the number of compressors required. Column 4 of Table 3 represents a preferred case using only one recycle compressor and providing a minimum membrane area requirement due to the high feed pressure. In all of the process comparisons made, identical performance parameters have been chosen for equipment as detailed in Table 2.

CONCLUSIONS

The addition of a membrane separation system to the oxygen rich waste stream of a single column nitrogen generator has been evaluated. The membrane further enriches the oxygen content of the waste stream before it is discarded to atmosphere. The retained nitrogen is recycled to the distillation system and greatly increases the recovery of nitrogen product from the air feed. The incorporation of the membrane improves process efficiency at product pressures above 4.5 bar abs. However, alternative distillation schemes can provide yet higher efficiency for nitrogen production.

The membrane does, however, provide a means for argon recovery and recycle. Given that a very high selectivity membrane can be achieved, it is shown that argon can be co-produced with nitrogen at very low energy cost with a recycle hybrid process.

ACKNOWLEDGEMENTS

The author would like to acknowledge the key contributions of R. Weimer and D.H.S. Ying in the development of hybrid cycles for argon recovery and of R. Agrawal and T. Roden in the development and investigation of processes for nitrogen recovery.

REFERENCES

1. Thorogood, R.M., 1991, Developments in Air Separation, *Gas Separation and Purification*, 5, 83–94.
2. Agrawal, R. and Thorogood, R.M., Production of Medium Pressure Nitrogen by Cryogenic Air Separation, *Gas Separation and Purification* (In Press).
3. Agrawal, R. and Auvil, S.R., 1986, Process for the Generation of Gaseous and/or Liquid Nitrogen, US Patent 4, 595, 405; June 17.
4. Thorogood, R.M., 1990, Ideal Energy Requirements for Nitrogen Separation by Single Column Distillation, Single Stage Membrane, and a Recycle Hybrid, *Gas Separation Technology - Process Technology Proceedings, 8*, Editors E.F. Vansant and R. Dewolfs, Elsevier, 647–654.
5. Thorogood, R.M., Tsai, W., Weimer, R.F. and Ying, D.H.S., 1991, Hybrid Membrane Cryogenic Generation of Argon Concurrently with Nitrogen, US Patent 5, 049, 174; September 17.

TABLE 1 - COMPARISON OF POWER FOR NITROGEN PRODUCTION FROM WASTE EXPANDER CYCLE WITH A MEMBRANE DISTILLATION HYBRID

Process		Waste Expander Cycle (Figure 1)	HybridCycle (Figure 2)
Feed Composition to Cold Box (% O_2)		21	21
Membrane Feed:			
	Composition (% O_2)	-	35.6
	Pressure (bar abs)	-	3.7
Nitrogen Product:			
	Composition (ppm O_2)	5	5
	Pressure (bar abs)	7.9	7.9
Power Requirement (kwh Nm^{-3} N_2) (at conditions of Table 2)		0.256	0.218

TABLE 2 - STANDARDIZED CONDITIONS FOR NITROGEN PROCESS PERFORMANCE CALCULATIONS

Distillation:	
Column Theoretical Stages	50
Heat Exchange:	
No. of Heat Transfer Units in Main Heat Exchanger	60-70
Temperature Difference at Top of Reboiler/Condenser (°C)	2.4
Membrane:	
Single Stage Oxygen Selectivity	8.2
Compressor/Expander:	
Air Feed Temperature (°C)/Relative Humidity (%)	2.11/50
Isothermal Efficiency (%)	70
Motor Efficiency (%)	95
Suction Pressure (bar abs)	1
Expander Isentropic Efficiency (%)	85

TABLE 3 - PROCESS CYCLES FOR THE COPRODUCTION OF ARGON WITH NITROGEN[2]

Process	Dual Membrane (Figure 4)	Single Membrane (Figure 5)		
N_2 Product[1]	72.4	71.4	74.3	75.9
Ar Product[1]	0.36	0.37	0.46	0.56
Ar Purity[3]	83.2	71.8	78.6	83.1
Recycle:				
Flow[1]	171	70	78	85
O_2 Concentration[3]	22.4	14.3	12.8	11.8
Membrane Feed Pressure[4]	8.7	3.9	5.5	8.3
Membrane Selectivity				
O_2/N_2	5.5	100	100	100
O_2/Ar	2.9	52.8	52.8	52.8
Specific Power (kwh Nm^{-3} N_2)	.322	.22	.216	.22

[1] moles/100 moles dry air feed
[2] N_2 product purity 5 ppm O_2 at 7.0 bar abs
[3] mol %
[4] bar abs

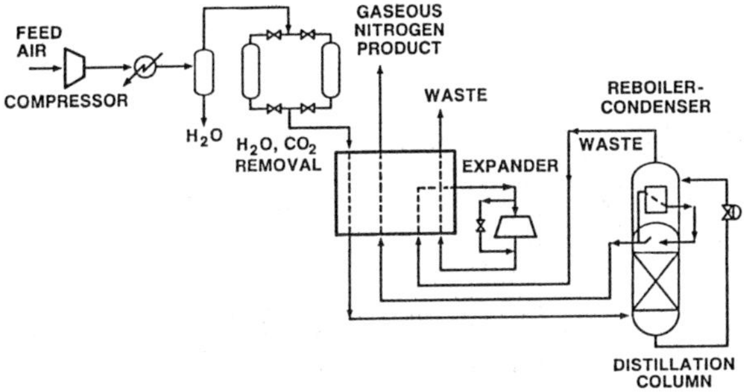

FIGURE 1 - NITROGEN PRODUCTION PROCESS WITH SINGLE DISTILLATION COLUMN AND WASTE STREAM EXPANDER

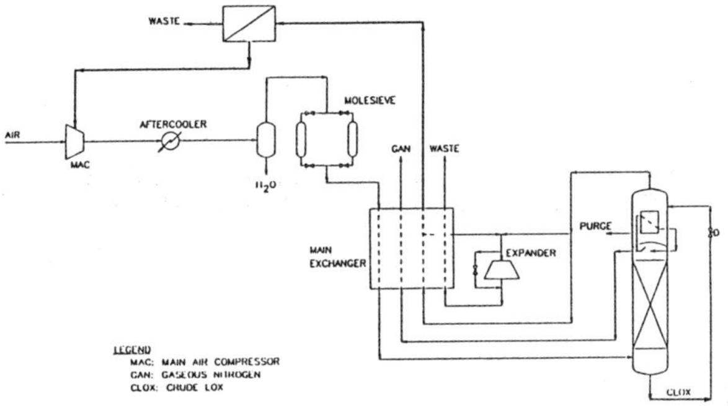

FIGURE 2 - NITROGEN PROCESS WITH WASTE STREAM MEMBRANE SEPARATION AND RECYCLE

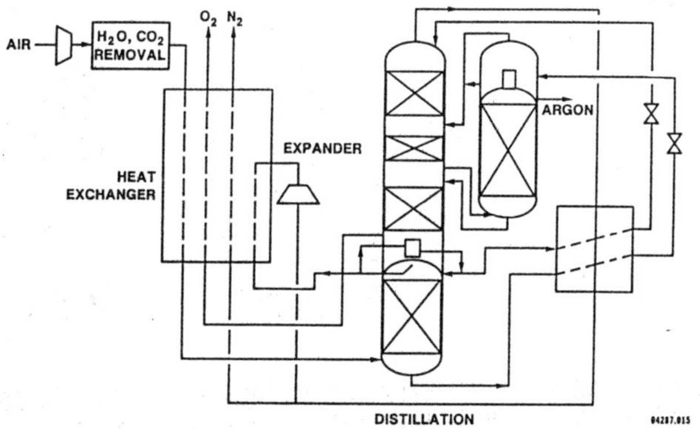

FIGURE 3 - CRYOGENIC SEPARATION OF OXYGEN, NITROGEN AND ARGON BY THE TRADITIONAL THREE COLUMN PROCESS

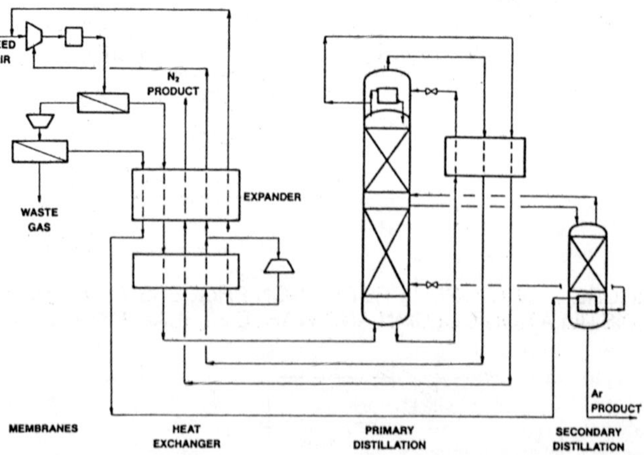

FIGURE 4 - DUAL MEMBRANE PROCESS FOR COPRODUCTION OF ARGON WITH NITROGEN

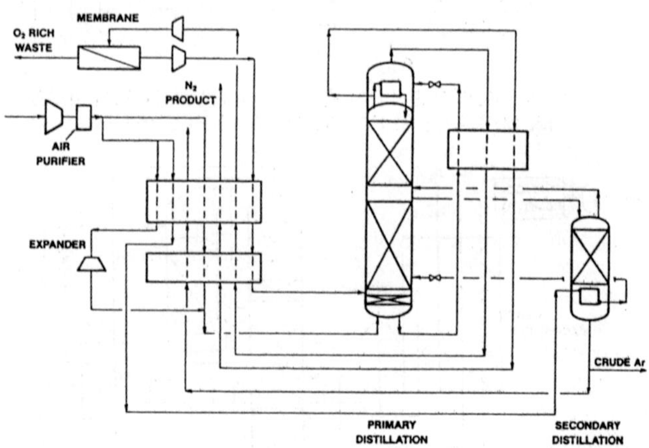

FIGURE 5 - HIGH SELECTIVITY SINGLE MEMBRANE PROCESS FOR COPRODUCTION OF ARGON WITH NITROGEN

NUMERICAL ANALYSIS OF SEPARATION EFFICIENCY IN A CENTRIFUGAL MOLECULAR STILL

H. Ishikawa*, M. Inuzuka*, H. Mori**, S. Hiraoka** and I. Yamada**
* Nippon Sharyo Ltd, Nagoya 456, Japan
** Nagoya Institute of Technology, Nagoya 466, Japan

A mathematical model for molecular distillation with reflux has been developed to calculate flow rates and compositions in a centrifugal molecular still. Numerical calculations have been carried out using an ideal binary mixture for three reflux modes. In the molecular distillation, we hardly expect the effect of a vapour-liquid contact due to a reflux operation as in conventional multistage distillations because of extremely low vapour density. The results of the numerical calculations, however, show the enhancement of separation efficiency by adding the reflux operation.

Besides, the behaviour of the objective function which is defined by summation of yields for the distillate and the residue is also investigated. When the total amount of evaporation is specified, there exists an extremal value of the function and you can use this for finding a optimum operating condition.

INTRODUCTION

Molecular distillation is a separation technique for liquid mixtures using the difference of evaporation rates of components. Since it is operated under vacuum at lower temperatures compared with normal boiling points, the molecular distillation is widely employed for separating low-volatile or thermally unstable substances. On the other hand, because of its sophisticated mechanism, it is usually operated as a simple batch distillation without reflux, which necessarily causes bad separation efficiency for low relative volatility systems. This is thought the major deficiency of the molecular distillation.

In distillation operation, reflux is usually employed to improve separation efficiency. In the molecular distillation, however, the effect of the reflux, coming from a vapour-liquid contact such as occurs in a conventional distillation tower, cannot be expected, since the density of the vapour is extremely low. The reflux in the molecular distillation can only be achieved by returning the condensed vapour from the condensing surface to the liquid on the evaporation surface. In this reflux mechanism, the enthalpy of the vapour cannot be used to evaporate the liquid being contacted and the additional energy is required for re-evaporating a part of the liquid. Several reflux schemes have been proposed so far[5,6]; however, their effects have not been studied experimentally nor mathematically.

In the remainder of this paper, we develop a mathematical model for a centrifugal molecular distillation with reflux and try to make clear the effect of the reflux on a separation efficiency from numerical simulation. An optimization problem on the yield with respect to operating conditions under a fixed total evaporation rate is also presented.

MATHEMATICAL MODEL

The reflux mechanism of a centrifugal molecular still adopted in this paper is shown in Figure 1[6]. A condensing surface is divided into several sections and a condensate of any section is returned to an evaporation surface of inner section where a vapour-liquid contact is occurring. The condensate at the center section is withdrawn as a distillate. The liquid leaving the evaporation surface is collected as a residue.

An assumption is used to develop the mathematical model; a liquid composition change on the evaporation surface occurs mainly due to the evaporation and the component distribution develops only in the radial direction. Thus there is no composition profile in the liquid film normal to the evaporation surface. This assumption is valid under the condition of Schmidt number Sc less than 2000 and a relative volatility less than 10^2. A mass balance of component i around a small element at r on stage j is given as follows where r denotes the coordinate in the radial direction.

$$\frac{dL_j(i)}{dr} = 2\pi r\, G(i) \tag{1}$$

An evaporation flux $G(i)$ is given by the product of a liquid surface composition $x_s(i)$ and a theoretical evaporation flux $G^\circ(i)$ which is derived from the classical kinetics of gases[3,4]. This was confirmed experimentally using a pot still[1]. $x_s(i)$ is given by $L_j(i)$ from the assumption stated above. Thus, $G(i)$ is given as follows.

$$G(i) = x_{sj}(i)G^\circ(i) = \frac{L_j(i)}{\sum_i L_j(i)} G^\circ(i) \tag{2}$$

where

$$G^\circ(i) = P(i)(2\pi RTM(i))^{-1/2} \tag{3}$$

To derive boundary conditions for Equation (1), consider a mass balance around stage j (see Figure 2). A molar flow rate $L'_j(i)$ of component i into stage j is given by summation of a molar flow rate $L_{j-1}(i)$ from stage j–1, a molar feed rate $F_j(i)$ into stage j and a molar reflux rate $D_{j+1}(i)$ from stage j+1.

$$L'_j(i) = L_{j-1} + F_j(i) + D_{j+1}(i) \tag{4}$$

The boundary conditions for Equation (1) are given as follows.

$$L_j(i) = L'_j(i) \quad \text{at} \quad r = r_j \tag{5}$$

When the relative volatility is constant in any stage, Equation (1) is integrated by using Equations (2) and (5). This yields the well-known Rayleigh's relationship of a simple batch distillation.

$$\ln\left\{\frac{L_j(i)}{L'_j(i)}\right\} = \alpha_{ik} \ln\left\{\frac{L_j(k)}{L'_j(k)}\right\} \tag{6}$$

where

$$\alpha_{ik} = \frac{G^{\circ}(i)}{G^{\circ}(k)} \tag{7}$$

Equation (6) shows the relationship between the flow rate of component i into stage j and that of component k. An evaporation rate $D_j(i)$ of component i from stage j is given as follows.

$$D_j(i) = L'_j(i) - L_j(i) \tag{8}$$

Total flow rates and compositions are calculated by the following equations.

$$D_j = \sum_i D_j(i) \quad , \quad L_j = \sum_i L_j(i)$$

$$y_j(i) = \frac{D_j(i)}{D_j}, \quad x_j(i) = \frac{L_j(i)}{L_j} \tag{9}$$

If feed rates $F_j(i)$ and evaporation rates D_j are specified, $L_j(i)$ and $D_j(i)$ are calculated by Equations (4), (6) and (7). The flow rate and composition of the distillate is obtained as D_1, $y_1(i)$ and those of the residue as L_n, $x_n(i)$. Here, the sum of the distillate and residue must be equal to the total feed.

$$\sum_j F_j(i) = F(i) = D_1(i) + L_n(i) \tag{10}$$

The outline of the procedure for calculating $D_j(i)$ and $L_j(i)$ with specified $F_j(i)$ and D_j is described below. Assume α_{ik} are constant.

(1) Specify $F_j(i)$ and D_j. At first, $L_j(i)$ set to 0.
(2) Calculate $L'_j(i)$ by Equation (4).
(3) Calculate $L_j(i)$ by Equations (6) and (7).
(4) Calculate $D_j(i)$ by Equation (8).
(5) Judge whether Equation (10) is satisfied. If Equation (10) is satisfied then stop the calculation; if Equation (10) is not satisfied then go back to Step (2).

CALCULATION RESULTS AND DISCUSSION

Simulations are all carried out using binary systems with constant relative volatility. Thus suffix i for compositions is omitted and the compositions usually mean those of the more volatile component and the flow rates usually show the summation of the component flow rates.

Figure 3 shows the relationship between the distillate composition y and the distillate rate (D/F) with a reflux rate as a parameter for a single-stage reflux operation. When the distillate rate is small, the reflux rate largely affects the distillate composition. But when the distillate rate is large, the reflux has no influence on the distillate composition.

Figure 4 shows the calculation results for the two-stage reflux operation where the condensate D_2 from the second stage returns to the first. The similar behaviour described in Figure 3 is also seen in this figure. The effect of the reflux on the distillate composition is more pronounced in the two-stage reflux operation than in the single-stage reflux operation. On the other hand, the distillate composition with the feed to the second stage is richer than that of the first stage. In this operation, the operating condition is restricted because the reflux rate from the second stage must be larger than the distillate rate from the first stage.

In Figure 5, for the three-stage reflux operation the feed rate to each stage is fixed and the distillate composition contour lines are plotted against the combination of the reflux rates (D_2/F, D_3/F) of the second and third stages. The operation range of the reflux rates is different for the feed stage selected and the vertical or horizontal dotted line denotes the boundary of the feasible operation range. All of the simulation results indicate that large reflux rates generally give higher distillate compositions. Again the operation with the feed to the third stage shows the best separation efficiency. If the total reflux rate (summation of D_2/F and D_3/F) is specified, the operation condition is expressed by the line with slope = −1, eg in Figure 5(b). The contact point of the operation line with the contour line gives the best operation point which yields the highest distillate composition. There is no such point for the operation with the feed to the first stage. If the total reflux rate is specified, this operation can be controlled only by the reflux of the second stage. On the other hand, in the operation with the feed to the second or third stage, both the refluxes of the second and third stages must be adjusted to obtain the optimal operating condition. Up to now, we discuss the separation efficiency only from the distillate composition. From the industrial point of view, how much of the component is recovered must be considered in discussion as well. Let us define the objective function of yield as follows.

$$\varphi = \left(\frac{D_1(1)}{F(1)} + \frac{L_n(2)}{F(2)} \right) / 2 \tag{11}$$

In Figure 6, the objective function φ is plotted against the distillate rate and reflux rate for two-stage reflux operation. A family of contour lines of φ forms a convex profile downward and they are parallel where the reflux rate is large. When the total evaporation rate (summation of D_1/F and D_2/F) is specified, the operation line becomes the line with slope = −1. We can find the best point which gives the largest value of the objective function in the same manner as described above. However, there is no point of contact where the total evaporation rate is small, since the curvature of the contour lines becomes almost negligible.

Experimental verification of the reflux operation is carried out using a two-stage reflux molecular still with a evaporation disk of 0.2 m diameter. The system of EHA(di-2-ethylhexyl adipate) — EHP(di-2-ethylhexyl phthalate) is used in the experiment. The observed distillate compositions are compared with the calculated results in Figure 7. The observed data are located above the line calculated without reflux. This confirms the effect of the reflux which is predicted by the numerical calculations. The deviation from the line calculated with reflux may be caused by the contribution of the condensation on the surface other than the condensing surface.

CONCLUSION

A mathematical model of the molecular distillation with reflux has been developed to evaluate the separation efficiency from numerical analysis. The extensive simulation has been carried out for the ideal binary mixture in the single-stage, two-stage, and three-stage reflux operation.

When we consider the separation efficiency to be a distillate composition, it is found that the separation efficiency increases with the increase of the reflux rate at the small distillate rate. For multiple-stage reflux operations, it is better to introduce the feed to the last stage in order to obtain better separation efficiency and we can find the best operation point which gives the highest distillate composition at a fixed total evaporation rate.

From the industrial point of view, the behaviour of the objective function defined in Equation (11) is investigated. For a fixed total evaporation rate, the optimum operation point can be found. The effect of the reflux is also confirmed by comparing with the experiment.

NOMENCLATURE

D	evaporation rate (mol s^{-1})
F	molar feed rate (mol s^{-1})
G	evaporation flux (mol m^{-2} s^{-1})
G°	theoretical evaporation flux (mol m^{-2} s^{-1})
L	molar rate of liquid (mol s^{-1})
M	molecular weight (kg mol^{-1})
P	saturated pressure (Pa)
r	distance from center of evaporation disk (m)
R	gas constant (J mol^{-1} K^{-1})
T	temperature (K)
x	mole fraction of liquid
y	mole fraction of condensed vapour

SUBSCRIPTS

F	feed
f	feed stage
i	component number
j	stage number
k	component number
n	last stage number
s	liquid surface

GREEK SYMBOLS

α	relative volatility
π	3.14159
φ	objective function defined by Equation (11)

LITERATURE CITED

1. Inuzuka, M., *et al*, 1987, *Int Chem Eng*, 7 (1): 100.
2. Inuzuka, M., *et al*, 1989, *J Chem Eng Japan*, 22 (3): 291.
3. Kundsen, M., 1915, *Ann Phisik*, 47: 697.
4. Langmuir, I., 1913, *Phys Rev*, 2: 329.
5. Ridgway, W., 1963, *Molecular Stills* (Chapman & Hall), 258.
6. Yamada, I., *et al*, *Japan Pat. No. 932367.*

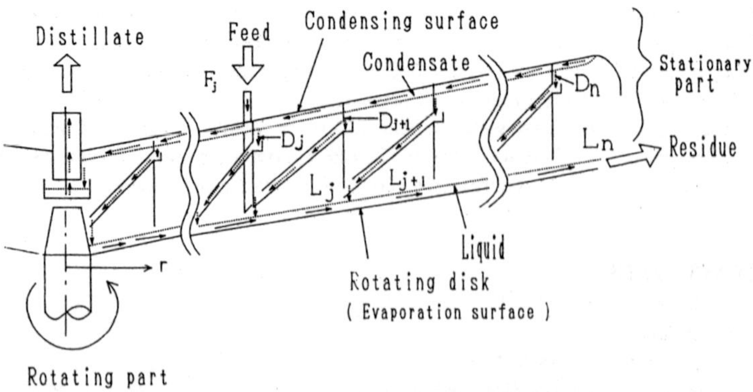

Fig.1 Reflux Scheme of Centrifugal Molecular Still.

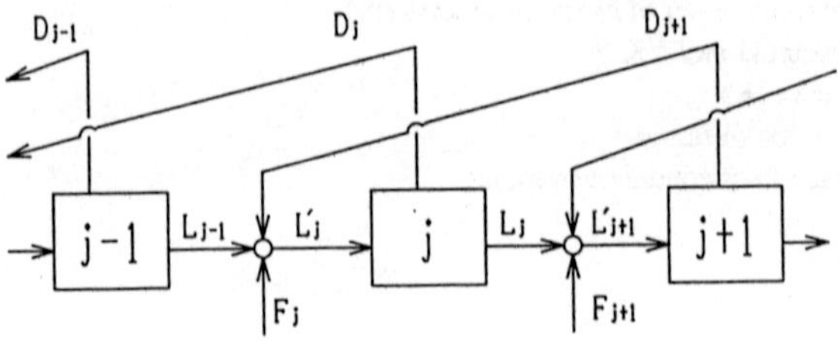

Fig.2 Flow Diagram around Stage j for Numerical Simulation.

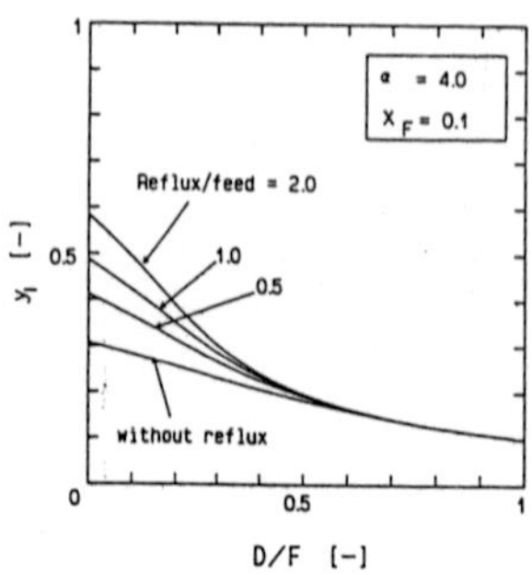

Fig.3 y_1 vs. D for Single-stage Reflux.

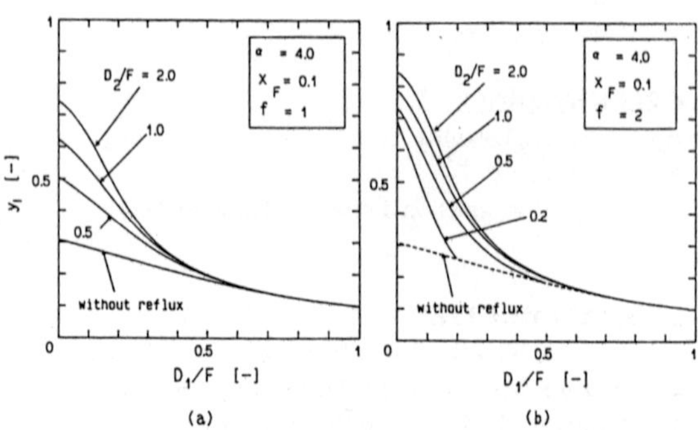

Fig.4 y_1 vs. D_1 for Two-stage Reflux. feed stage : (a) f=1; (b) f=2.

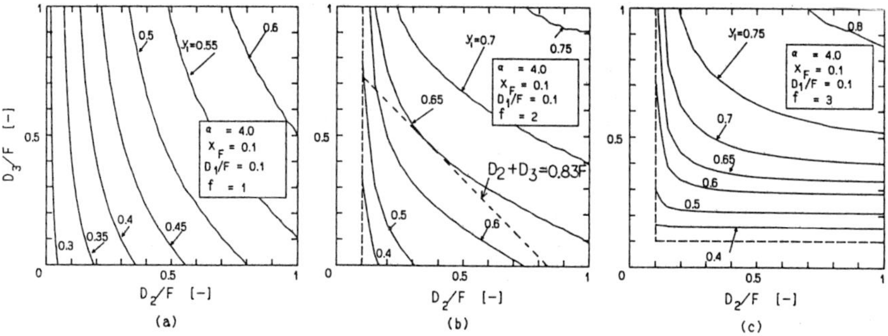

Fig.5 y_1 vs. Reflux Rate for Three-stage Reflux.
feed stage : (a) f=1; (b) f=2; (c) f=3.

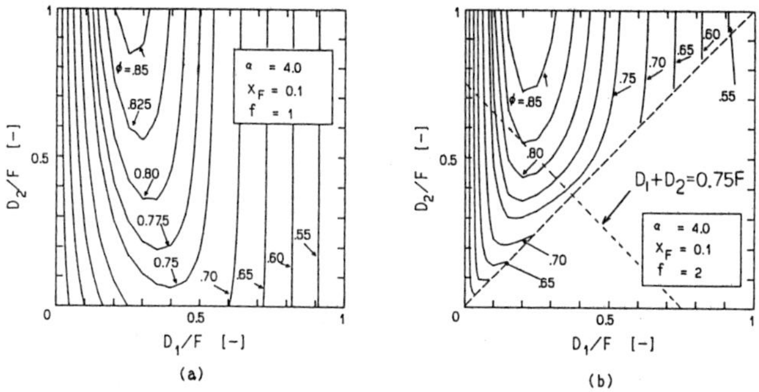

Fig.6 Behaviour of ϕ for Two-stage Reflux.
feed stage : (a) f=1; (b) f=2.

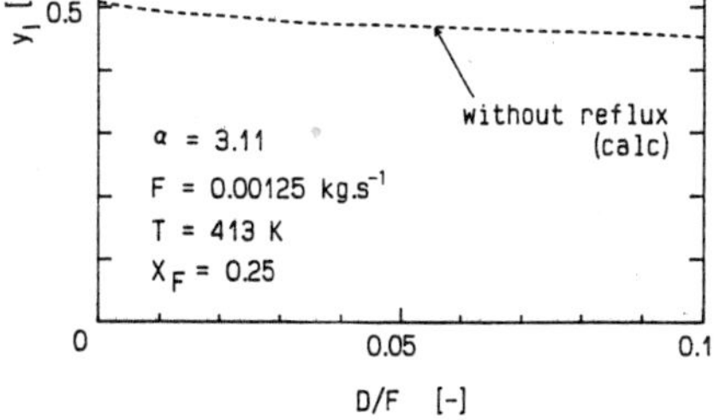

Fig.7 Observed and Calculated y_1 for Two-stage Reflux.

SEPARATION OF AROMATIC HYDROCARBONS BY EXTRACTIVE DISTILLATION WITH DIMETHYLFORMAMIDE AND N-METHYLPYRROLIDONE AS SOLVENTS

C. Ruiz, R. Esteban, A. Vega, F. Diez and J. Coca
Department of Chemical Engineering, University of Oviedo, 33071 Oviedo, Spain

Recovery of aromatic hydrocarbons from a synthetic mixture of toluene, benzene, hexane, cylohexane and heptane was carried out by extractive distillation. Experimental data are reported for dimethylformamide and N-methylpyrrolidone as extractive solvents, using a continuous laboratory-scale bubble-tray column. The influence of operating conditions on solvent selectivity is also reported.

INTRODUCTION

Aromatic hydrocarbons, used as feedstocks for the petrochemical industry, are recovered from aromatic/paraffinic/naphtenic refinery streams, such as reformates or pyrolisis gasoline, by processes such as solvent extraction and extractive distillation.

Components with close boiling points are separated by extractive distillation by adding a high-boiling point compound (extractive solvent) near the upper part of the distillation column. The solvent affects the relative volatility of the compounds, improving the separation. In the case of aromatic/nonaromatic mixtures, the extractive solvent interacts with the aromatic components, which are removed with the solvent as the bottom product of the column.

A large number of extractive distillation solvents have been used for the separation of aromatics (sulfones[1,2,3], morpholine[4,5], mixture of phtalic anhydride, maleic anhydride and adiponitrile[6], phenol[7]), but only a few have been found industrial application. N-methylpyrrolidone[8] (NMP) and dimethylformamide[9] (DMF) are some of the most suitable and widely used solvents for the separation of aromatics from hydrocarbon mixtures.

The purpose of this investigation was to study the performance of both solvents for the separation of aromatics from a paraffinic/naphtenic/aromatic synthetic mixture, using a laboratory-scale bubble-tray column.

EXPERIMENTAL

The separation of aliphatic and aromatic fractions was carried out by extractive distillation in a 21 tray 25 mm internal diameter glass column, each tray consisting of a single circular cap. The vacuum jacketed and thermally isolated column had ports available in 10 trays for varying the feed entry, temperature measurement or sampling. The extractive solvent and hydrocarbon mixture were pumped by metering pumps, preheated and fed into trays 3 and 15, respectively (taking the upper plate as number one in the column). Reflux was controlled by an electrovalve, and liquid from the stillpot was withdrawn continuously. The level was kept constant by means of a siphon. All the experiments were carried out with a fixed reflux ratio of 6. The composition and normal boiling point of the hydrocarbon mixture feed are given in Table I.

Extractive distillation experiments were carried out with NMP and DMF as extractive solvents, and a solvent/hydrocarbon feed mass ratio of between 1 and 2. Experiments were carried out fixing mass flow rates and temperature of the column feeds, and reflux ratio, and using different

TABLE I

Composition and boiling point of the components fed to the distillation column.

Component	Wt. %	B.P. (°C)
Hexane	11.0	69.0
Benzene	80.0	80.1
Cyclohexane	3.5	80.7
Heptane	3.5	98.4
Toluene	2.0	110.6

reboiler heating rates for each run. Effective heat input into the reboiler could not be controlled as it depends not only on the electric power input but also on environmental conditions and the liquid level in the reboiler.

Samples of the feed, overhead and bottoms were analyzed by GC with a 50 m fused silica capillary column (liquid phase OV–101) and a FID detector, using cyclohexene as internal standard and helium as carrier gas.

RESULTS AND DISCUSSION

Column efficiency was estimated by batch distillation with methylcyclohexane/ toluene as a test mixture at total reflux. An overall efficiency of 67.6% was obtained.

The results are presented in Figures 1 and 2, where the feed fraction of each compound recovered as bottoms is plotted vs. the benzene fraction recovered. When both DMF and NMP are the extractive distillation solvents, toluene is quantitatively recovered with the column bottoms (no toluene was detected in the distillate), and very little hexane, typically less than 1 wt % of the total amount of hexane in the feed. The relative amount of hexane in the bottoms product is so small that neither the solvent/hydrocarbon ratio in the feed, nor the type of solvent have a clear influence on the bottoms product composition, as shown in Figure 1. This behaviour is explained by the polar, non-volatile character of toluene, and the non-polar, volatile character of hexane.

As shown also in Figure 1, the selectivity of both DMF and NMP for the heptane/benzene separation is not high, as a very substantial fraction of heptane was entrained by the solvent. As could be expected, a better selectivity is obtained as the solvent/hydrocarbon feed ratio increases, and under the operating conditions studied, DMF yields a slightly better separation than NMP.

Results for cyclohexane, Figure 2, are too scattered for firm conclusions to be drawn, but there is a trend of high selective separation for low benzene recovery. When benzene recovery is high, ie hydrocarbon content in the solvent, selectivity decreases and cyclohexane recovered with the bottom product increases sharply. Hence, selectivity increases as the solvent/hydrocarbon feed ratio increases.

CONCLUSIONS

Selectivity for the separation of aromatics/non-aromatics hydrocarbon mixtures by extractive distillation with DMF and NMP increases when the solvent/hydrocarbon feed ratio is increased. The effectiveness of toluene and hexane separation is very high. The separation of cyclohexane/benzene, components with very close boiling points, shows a high selectivity for low benzene recoveries, but when the hydrocarbon content of the solvent increases, selectivity decreases. The separation benzene/heptane is not so efficient because, although heptane is a non-polar compound, its boiling point is almost 20°C higher than that of benzene. In this case, DMF seems to be slightly more selective than NMP.

These results have implications for industrial extractive distillation. If toluene-free benzene is to be produced, toluene must be separated either before or after the extractive distillation, since no toluene can be separated from benzene by extractive distillation. Hexane is easily separated from benzene, so the hexane content in the feed of the extractive distillation column is not critical. Cyclohexane and mainly heptane, tend to remain in the column bottoms as non-aromatic impurities, so any increase in their concentrations in the feed will represent a more difficult separation, being necessary to increase the number of trays of the column and/or the solvent/feed ratio.

ACKNOWLEDGEMENTS

This work was carried out with the financial support of Repsol Petróleo S.A. (Spain).

REFERENCES

1. Fu-Ming, L., 1986, *Ind Eng Chem Process Des Dev*, 25: 949.
2. Fu-Ming, L. and Coombs, D.M., 1987, *Ind Eng Chem Res*, 26: 564.
3. Fu-Ming, L. and Coombs, D.M., 1988, *Ind Eng Chem Res*, 27: 118.
4. Preusser, G., Richter, K. and Schulze, M., October 14, 1971, *Ger Pat 2,014,262*.
5. Lackner, K. and Emmrich, G., October 14, 1971, *Hydr Proc*, 67.
6. Berg, Ll., 1983, *AIChE J*, 29 (6): 961.
7. Black, C. (Shell Oil), November 22, 1960, *US Pat 2,961,383*.
8. Müller, E. and Eisenlohr, K.H., November 1973, *Ger Pat 2,026,693*.
9. Mikitenko, P. and Asselineau, L., November 1973, *Fr Pat 2,176,488*.

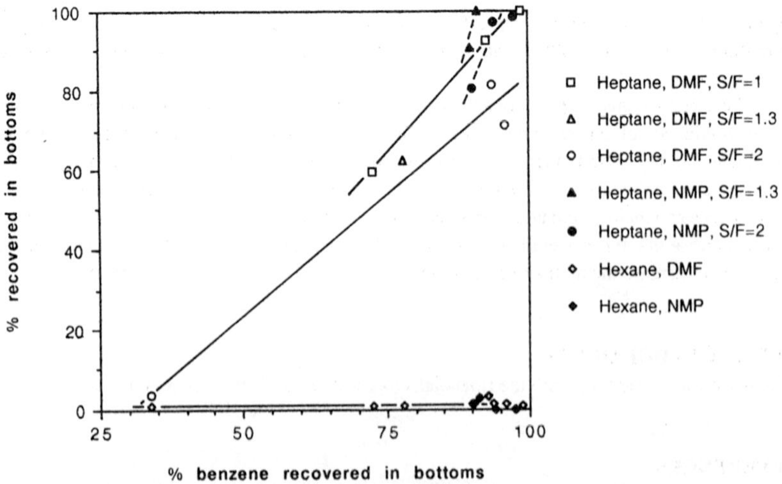

Fig. 1. Percentage of hexane and heptane fed to the column recovered in the column bottoms, as a function of the percentage of benzene recovered in this stream.
(S/F = extractive solvent/hydrocarbon feed ratio)

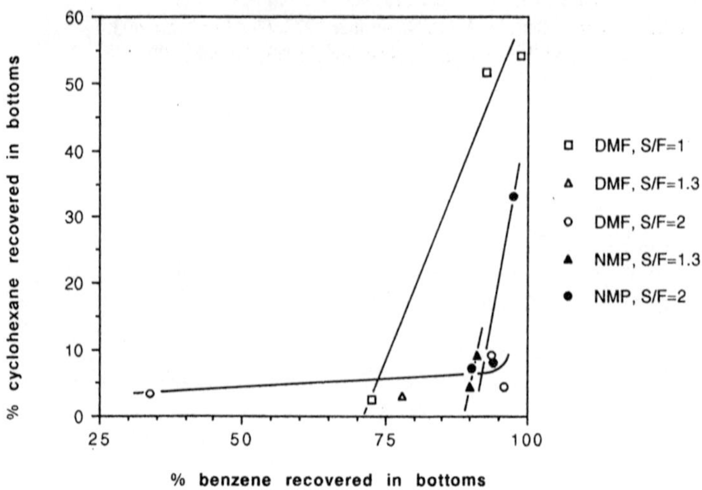

Fig. 2. Percentage of cyclohexane fed to the column recovered in the column bottoms, as a function of the percentage of benzene recovered in this stream.
(S/F = extractive solvent/hydrocarbon feed ratio)

DIRECT CONTACT CONDENSABLE HEAT TRANSFER PERFORMANCE OF TRAYS

S. Banik, K. Majumder and T.K. Sarkar
R&D Division, Engineers India Ltd, New Delhi, India

A comprehensive study has been carried out to evaluate the direct contact condensable heat transfer performance of sieve trays (12.7 mm hole diameter) using steam/air/water in a 1.2 m dia column. It has been observed that flow regime of operation has a strong influence on the heat transfer performance. Overall heat transfer coefficient is found to depend on the air and water flow rates. Heat transfer performance of bubble cap tray, 40 mm Pall ring (CS) and Sheet Metal Structured Packing has also been evaluated.

INTRODUCTION

Direct contact heat transfer process is widely used in refinery operation. In this process hot vapour is contacted with cooled circulating reflux over fractionating devices like trays or packings resulting in transfer of heat and mass. When only heat is transferred, as in the case of ethylene quench column, the process is simpler to design. Large volume of experimental data is also available on direct contact non-condensable heat transfer performance of trays and packings[1]. However, for condensable heat transfer hardly any information is available. Fair[2] has analysed the direct contact heat transfer process and suggested that either heat transfer coefficient or mass transfer analogy could be used for design of heat transfer section. For designing heat transfer section of a heavy hydrocarbon fractionator column, Neeld and O'Bara[3] have developed a correlation for Esso Jet trays. The correlation uses terminal conditions for LMTD and average loads and properties of vapour and liquid over the heat transfer section. Therefore, the correlation has limited applicability.

The present paper describes the results of a systematic study with sieve trays using steam/air/water system. Results of study with Pall ring and Structured Packings are also reported.

EXPERIMENTAL WORK

In Figure 1 flow diagram of the test facilities is shown. The flow diagram shows various equipments and also flow, temperature and pressure measurement points. Table 1 shows sieve tray details.

Saturated steam is generated in the reboiler and mixed with air from the blowers. The mixture is contacted with recirculated cooled water. Some of the steam condenses and rest is vented out with air. Air rate is accurately measured using a set of orifice meter and steam rate is estimated from the rate of high pressure steam condensate from the reboiler. Both inlet and outlet humidity of air and steam-air mixture is monitored using Panametric Hygrometer (model 2100). Temperature profile along the column is measured at the intertray space and at the exit downcomer of each tray. In Table 2 flow rates of various streams for each experimental run is shown. Overall heat balance was done for each run. In most of the runs heat balance was off by 5%. Rate of steam condensation was estimated using liquid side heat balance and also from vapour phase mass transfer.

RESULTS

The temperature profile along the column is shown in Figure 2. Three distinct features representing steam condensation either at bottom, top section or all along the column could be observed.

Effect of water and steam flow rates on steam condensation rate is shown in Figure 3. It could be observed that with lower steam flow rate the effect of water rate is less pronounced. However, at higher steam rate, effect of water rate is appreciable. It could be observed that at low liquid flow rate of 30m^3/hr rate of condensation decreases with increase in steam rate. However, for higher water rate of 60 m^3/hr the effect is proportional. Similarly effect of air flow rate on steam condensation rate is shown in Figure 4 for water flowrate of 30, 40 and 60 m^3/hr. At lower liquid rate slightly higher condensation rate is observed. At 40 m^3/hr of water rate, rate of steam condensation varies from tray to tray. At higher water rate, effect of air rate is negligible. However, at a steam flowrate of 1000 kg/hr this behaviour is observed for all flow rates of water.

DISCUSSIONS OF RESULTS

The direct contact condensable heat transfer process can be described by the following relationships

$$Q = U_o A \Delta T \tag{1}$$

$$W \propto (p_s - p_w^*) \tag{2}$$

Where U_o is over all heat transfer coefficient and A is the characteristic surface area, W is the rate of condensation of steam, p_s is partial pressure of steam and p^*_w is the vapour pressure of water at the average tray temperature. The ΔT in equation (1) is defined as $T_G - T_{AV}$ where T_G is the gas phase average temperature at tray inlet and T_{AV} is average of liquid inlet and outlet temperatures.

Experimental results indicate that under certain condition heat transfer is controlled by equation (1) and in some cases by equation (2) representing transport or equilibrium limited process respectively. For each run, equilibrium driving force expressed as the difference of partial pressure of steam, p_s and vapour pressure of water at T_{AV}, p^*_w was estimated for each tray. It was observed that only for some runs rate of steam condensation was proportional to equilibrium driving force. These runs were identified as equilibrium limited runs. The remaining runs were identified as transport limited runs. For example, in Figure 3 it could be observed that for liquid rate of 30 m^3/hr, rate of steam condensation on trays 4, 5 and 6 is lower at higher steam flow rate (signifying higher equilibrium driving force) compared to lower steam flowrate. In Figure 6, rate of steam condensation is shown as function of equilibrium driving force ($p_s - p^*_w$). Equilibrium limited runs are represented by lines, lower ends of which correspond to top tray and upper ends correspond to bottom tray. The transport limited runs are shown by symbols and each one corresponds to a particular tray, as marked for run no. 1

Figure 5 shows the flow regime of operation of trays. It could be observed that most of the experimental runs correspond to operation in the emulsion flow regime. In this regime liquid is in continuous phase and the vapour remains dispersed as small bubbles. The net surface area is expected to be high. It could be observed that, in general, runs corresponding to flow parameter 0.15 or more are equilibrium limited and runs corresponding to flow parameters 0.1 or less are transport limited. However, runs corresponding to flow parameters 0.1 to 0.15 could be either transport or equilibrium limited. Also at flow parameter more than 0.2, equilibrium limited heat transfer process is observed even at conditions close to flooding. In addition to the tray, it is expected that the downcomer will also contribute to heat transfer, specially in the emulsion regime of operation. Volumetric heat transfer coefficient has been estimated and plotted in Figure 8. It could be observed that it is a function of liquid load for a given hole F-factor.

Heat transfer performance of the bubble cap tray (2 inch caps) was also evaluated for transport limited runs and shown in Figure 7. Bubble cap tray seems to be performing better than sieve tray.

PACKED TOWER PERFORMANCE

Heat transfer performance of 40 mm Pall ring (CS) was also evaluated using the above facilities. Riser Orifice distributor was used. In Figure 9 average rate of steam condensation for 600 mm bed height is shown. No clear trend could be observed. It is difficult to find out exactly the volume of packing being utilised for condensation of steam as only the end conditions could be measured accurately. Also the contribution from the distributor could not be estimated accurately. Therefore, further experimentation is planned using gamma scanning technique.

STRUCTURED PACKING

Limited experiments were carried out with in house developed sheet metal structured packing (250 m^2/m^3 surface area). Preliminary results indicate that at low steam rate, rate of condensation is similar to 40 mm CS Pallring, however, at higher steam rate, it is about 50% higher than Pallring (Figure 9).

CONCLUSION

Process of direct contact heat transfer can be either transport limited or equilibrium limited. Flow regime of operation has strong influence on the heat transfer performance of sieve trays. Heat transfer performance of bubble cap tray appears to be superior to sieve trays.

ACKNOWLEDGEMENTS

The authors are grateful to Engineers India Limited for releasing this work for presentation.

REFERENCES

1. Ognisty, T.P., paper presented at *AIChE Spring National Meeting, Orlando, Florida, March, 1990.*
2. Fair, J.R., Invited Lecture, *Process Heat Transfer Conference, Tulsa, Oklahoma, August 15–18, 1971.*
3. Neeld, R.K. and O'Bara, J.T., 1970, *Chem Eng Prog* 66, (7).

TABLE 1
Sieve tray details

No. of trays	9 operating trays
Material	SS 304
Downcomer area	13 %
Hole size	12.7 mm
Hole area	8 %
Weir height	50 mm

TABLE 2
Range of variables

Run no.	Air rate kg/hr	Water rate m/hr	Steam rate ton/hr	Type of run*
1	8000	40	2.55	TL
2	8000	30	2.55	TL
3	8000	60	2.55	EL
4	6000	60	2.55	EL
5	6000	40	2.55	TL
6	6000	30	2.55	TL
7	6000	30	1.00	EL
8	6000	40	1.00	EL
9	6000	60	1.00	EL
10	8000	60	1.00	EL
11	8000	40	1.00	EL
12	8000	30	1.00	EL
13	6000	30	1.90	TL
14	6000	40	1.90	EL
15	4000	30	2.55	TL
16	6000	15	1.00	TL

*TL Transport limited runs EL Equilibrium limited runs

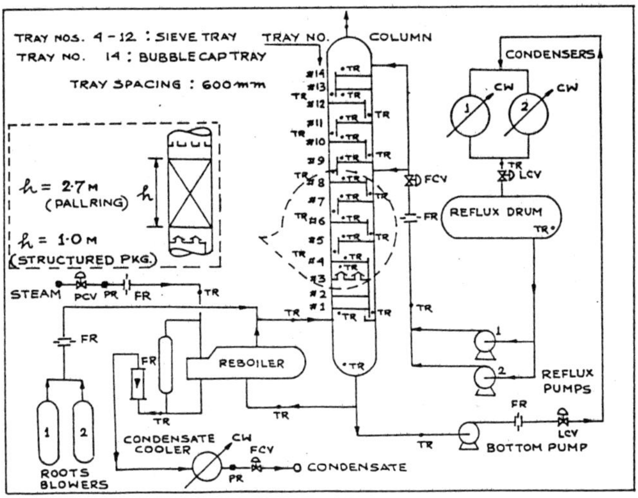

FIGURE 1 : EXPERIMENTAL SETUP.

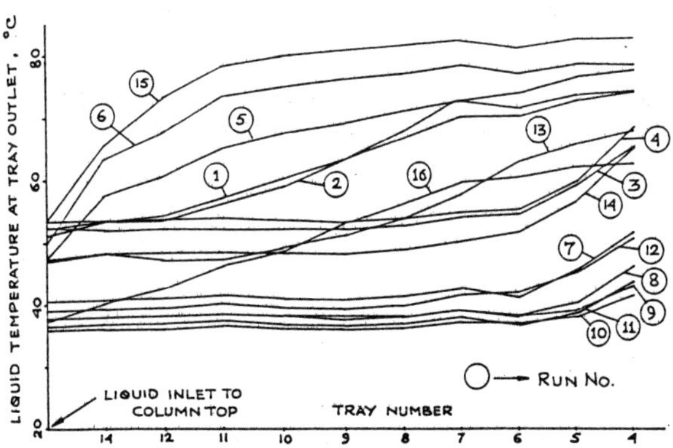

FIGURE 2 : TRAY TEMPERATURE PROFILE.

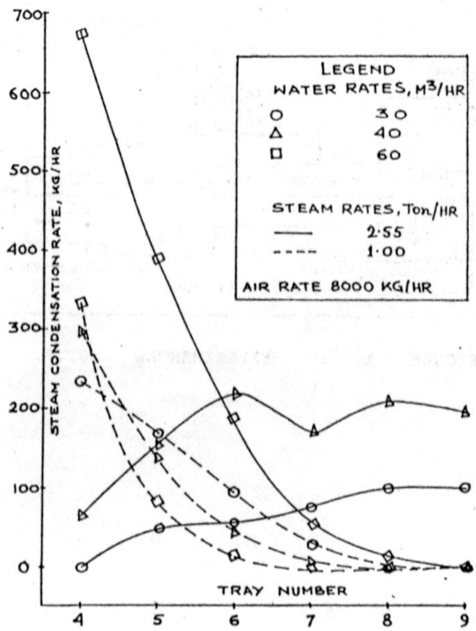

FIGURE 3 : EFFECT OF WATER RATE ON STEAM CONDENSATION ON SIEVE TRAY.

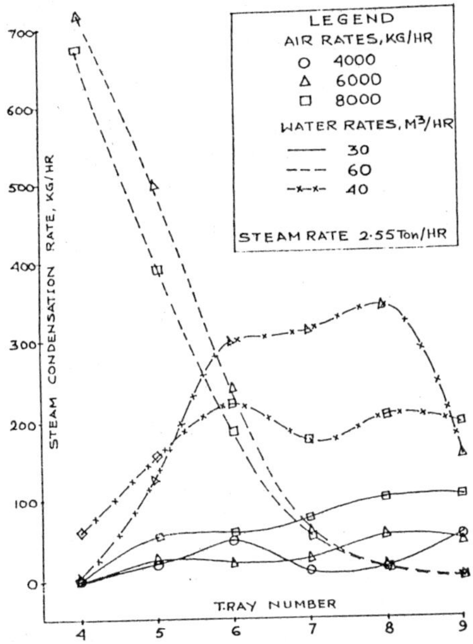

FIGURE 4 : EFFECT OF AIR RATE ON STEAM CONDENSATION ON SIEVE TRAY

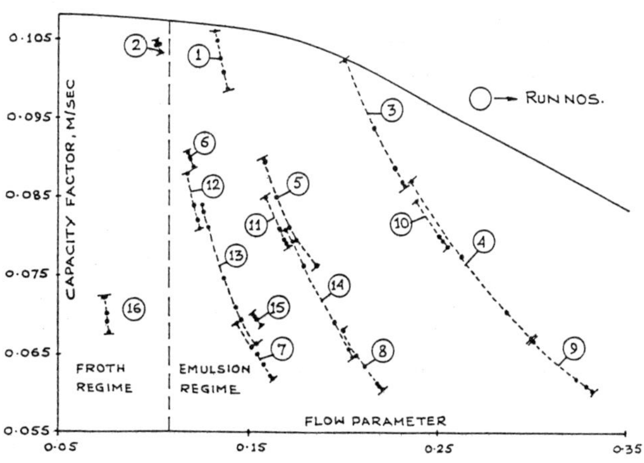

FIGURE 5 : CAPACITY FACTOR (BASED ON BUBBLING AREA) VS FLOW· PARAMETER.

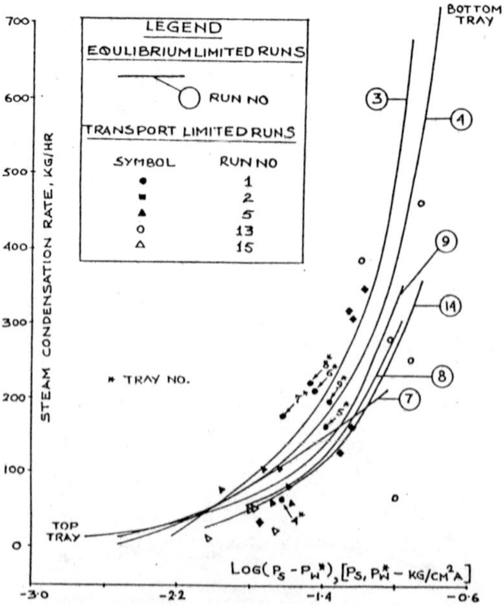

FIGURE 6 : STEAM CONDENSATION RATE VS $\text{Log}(P_s - P_w^*)$.

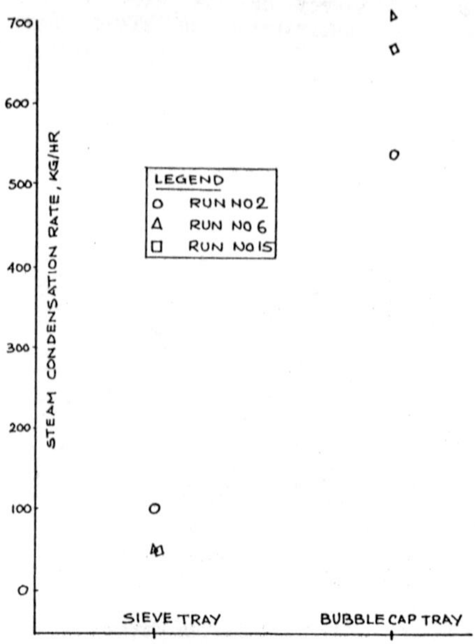

FUGURE 7 : COMPARISON OF MAXM. CONDENSATION ON SIEVE TRAY WITH CONDENSION ON BUBBLE CAP TRAY FOR TRANSPORT LIMITED RUNS

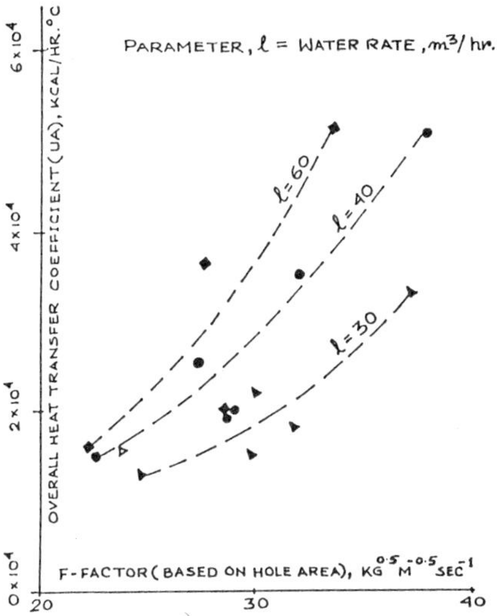

FIGURE 8 : OVERALL HEAT TRANSFER COEFFICIENT VS F-FACTOR FOR SIEVE TRAY

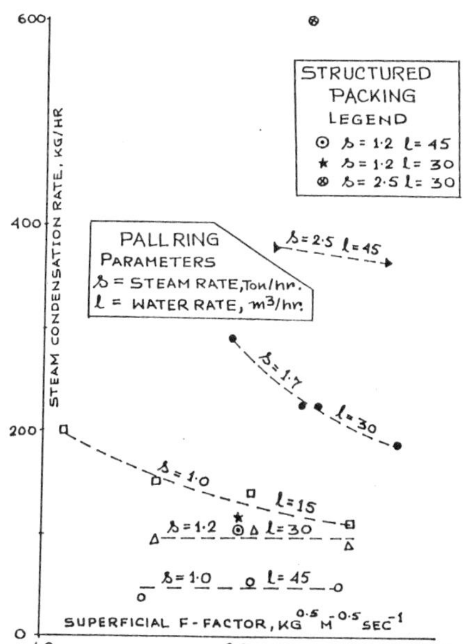

FIGURE 9 : STEAM CONDENSATION RATE VS F-FACTOR FOR PACKING

INFLUENCE OF WIND LOAD ON THE EFFICIENCY OF DISTILLATION COLUMNS

C.F. Hong
4th Design Dept, Shanghai Phamaceutical Industry Design Institute, 1856 Nanjing Road (West Part), Shanghai 200040, P.R. China

This paper laid stress on analyzing the effects of erection technique and wind load on packed columns and plate columns. An ethylbenzene/styrene distillation tray column was taken for example. The author had recorded the wind scale for two different months. And the relationship between the styrene loss in column top stream and the wind scale was given in the form of diagrams. To control the effect of wind load some measures on erection technique by strengthening column rigidity were taken. Therefore the tray efficiency increased by 13%. Thus the recovery of products raised and energy saving was also obvious.

INTRODUCTION

Distillation process is a very important means for separating liquid mixture. It is widely used in chemical, refining, petrochemical and pharmaceutical industry. In order to achieve good results the cruxes of the matter are not only the precise computation of technology, reasonable design of columns, skillful operation on them, but also the strict demands on erection of columns. It is most important especially for the high columns installed in the open air. Besides the consideration of the horizontal deviation of trays caused by the deviation of erection that makes column body inclined, the inclination of trays induced by external loads, eg wind load exerted on column body, must be carefully under consideration either. Under the action of wind force the column deflection occurs and the trays inside the column will be inclined. This will cause unstability of column operation. In such situation the tray efficiency discussed here decreases by about 13%. In this paper some problems dealing with column efficiency were investigated.

THE INFLUENCE ON COLUMN EFFICIENCY BY WIND LOAD

For columns with high ratio of height to diameter and installed in the open air, besides the deflection in the direction of wind load, at both sides of column back area to the wind the vortexes are alternately formed, which will force column body to vibrate in the direction of perpendicular to wind load[1]. High column fixed merely on base tends to be affected by wind load. The stronger the wind load, the greater the deflection of column body and the inclination of trays. This affects the operation of column unfavourably, and the column efficiency will be reduced.

PACKED COLUMNS

In case of packed cloumn, normally the inclination or deflection of column induced either by wind load or by installing on a non-stationary base like off-shore barge may form serious maldistribution of reflux flow in packed bed, even some parts of packing are partly wetted or totally dry and a great deal of wall flow is formed. Local resistance for rising vapour are unequal in whole packed bed. There appears maldistribution of vapour phase as well. This results in insufficient or no contact at all between liquid and vapour phases. The ratio of liquid to vapour differs from point to point in the same cross section, so the packing efficiency falls off rapidly.

H.V. Svensson[2] introduced a set of barge-mounted methanol plants. When process plants were operated off-shore, the methanol column was inevitably affected by the wave-induced motions of the barge. The author indicated that except for the most severe operating conditions, the operating efficiency of distillation columns on floating plants is about 10% below that of an on-shore based column. The author had also tested a column on a motion simulator which had 3 degrees of freedom: roll, pitch and heave motion. The diameter of test column was 1000 mm, height of packed bed 2210 mm, No. 50 IMTP, loading factor Cs = 0.03–0.08 m/s. The results were shown in Figure 1 and Figure 2.

R.C. Plüss and P. Bomio[3] indicated that when column was subjected to external loads, it is better to choose packed column before plate column for mass transfer. Since the liquid flows as a very thin film on the surface of the packing, which adheres well to the packing such, that it is but slightly influenced by external loads.

PLATE COLUMNS

Routine design for plate column normally takes the height of clear liquid on tray h_L = 50–100 mm. If the design of tray parameter is reasonable, the requirement for tray levelness may not be so strict. For some distillation columns operated in vacuum such as ethylbenzene/styrene distillation column, which needs very low pressure drop, the design height of clear liquid is only about 25 mm. Therefore the tray levelness is strict. The corresponding deflection because of the inclination of column body caused by erection and wind load should be carefully controlled.

The inclination of column body results inevitably in the local heights of liquid on tray unequal, and then the local resistance of rising vapour unequal neither. The thicker the liquid on tray, the less the vapour passing through it. The leakage of liquid tends to occur at this part of tray. The thinner the liquid on tray, the higher the vapour velocity and the more the entrainment. It is even more serious for split-flow trays. In the case of inclination of column body, maybe one half of tray is operated under unstable conditions, while the other half of tray has excessive liquid load and leads to flood in downcomer. The inclination of column body will also reduce the operating turndown ratio of a column. The above problems can be sometimes solved by means of increasing weir height, enlarging downcomer area, installing baffles along the liquid flow path and enhancing tray spacing. But the best way is to take measures to strengthen the column rigidity.

There was an ethylbenzene/styrene distillation column with Linde sieve trays built in a factory in China. The column diameter was 2600 mm for rectifying section, 2200 mm for stripping section and the total height of this column including the skirt base was 47,000 mm. It was installed in a five-store frame and fixed with the frame at the height of 15 meters from ground.

Figure 3 and Figure 4 show the amount of styrene loss from column top stream versus local daily wind scale in August and December, respectively, in a same year. It is obvious from the figures that the loss of styrene increases with the increasing of wind scale. The maximum calculated deflection of this column y = 217 mm caused by wind load, the maximum inclined gradient of tray Δt = 12.3 mm, which causes maldistribution of liquid flow on tray and not sufficient contact between vapour and liquid. The local residence time of liquid in tray are also unequal. Liquid back mixing on tray will reduce the driving force of mass transfer and thus tray efficiency decreases. Styrene loss from top stream raised from normal value 3–5% up to 8% or so.

Afterwards when a new set of styrene project was designed, the above problems were considered. Some measures were taken to control the column amplitude caused by wind load. According to the scale of the styrene plant, the calculated diameters of ethylbenzene/styrene distillation column were 1200 mm and 1000 mm for upper and lower section respectively, the total height was 46 000 mm. Since the ratio of height to diameter was too high to be stable for column itself, in consideration of rigidity of column and reduction of deflection caused by wind load the design diameter was taken as 1600 mm for both sections, the height of column was kept the same as calculated, ie 46 000 mm. The hydraulic calculation of tray was still based on design scale. If the column had been installed on the base alone in the open air, the maximum deflection y would have been 960 mm and the maximum inclined gradient $\Delta t = 28.0$ mm, which was greater than weir height (25 mm), when the wind pressure was 4.0 MPa.

To reduce the inclined gradient of trays induced by wind load, the united triangle platforms were adopted, which combined this column with other two columns together in order to strengthen the rigidity of column against wind load. In case of the same wind pressure (4.0 MPa), the maximum calculated deflection $y = 120$ mm, the maximum inclined gradient $\Delta t = 4.2$ mm.

After several years' operation it was proved that the design of this column was successful. Both the quality and quantity of styrene product were prior to that produced from other similar plants in China. By preliminary estimation the yearly increased revenue was about one third of the investment of the column. The economic benefits were very satisfied. The comparison is shown in Table 1.

The first plan in Table 1 was not built and the second one was taken in design.

CONCLUSION

By controlling the deflection of column body caused by wind load and reducing the inclined gradient of trays, the contact between liquid and vapour can be improved. The tray efficiency discussed in this paper raised by about 13%.

In condition of the same purities of both ends of column respectively, the increased theoretical plates may reduce reflux ratio from 7.0–7.5 down to 6.5–7.0.

The reduced reflux ratio results in not only energy saving but also the reduction of column investment.

Reduction of styrene loss from top stream means the enhancement of recovery of styrene product.

From above mentioned items the yearly increased revenue was equivalent to about ⅓ of the column investment. Its economic benefits were obvious.

REFERENCES

1. Sheng, Z.W. and Sun, Y.L., 1991, *Petrochemical Equipment (China)*, 20 (2): 35–39.
2. Svensson, H.V., 1982, *CEP*, 78 (11): 43–46.
3. Plüss, R.C. and Bomio, P., 1987, *IChemE Symposium Series No. 104*, A259–A271.

TABLE 1

Item	A similar column in China	Column mentioned in this paper	
		first plan	second plan
Column diameter (rectifying section/stripping section) (mm)	ϕ2600 / ϕ2200	ϕ1600	ϕ1600
Column height (mm)	47,000	46,000	46,000
Way of fixation	standing on base and fixed with the frame at the height of 15 meters from ground	standing on base itself	combined with other two columns by united triangle platforms
Wind pressure(MPa)	4.0	4.0	4.0
Maximum deflection y (mm)	217	960	120
The maximum inclined gradient Δt of trays (mm)	12.3	28.0	4.2
Reflux ratio	7-7.5	7-7.5	6.5-7
Styrene loss from column top stream (wt%)	3.0-5.0	3.0-5.0	≤1.0
Ethylbenzene fraction in bottom product(wt%)	0.2	0.2	<0.14
Calculated number of theoretical plates	50	50	60
Real number of plates	75	75	75
Total tray efficiency(%)	67.0	67.0	80.0

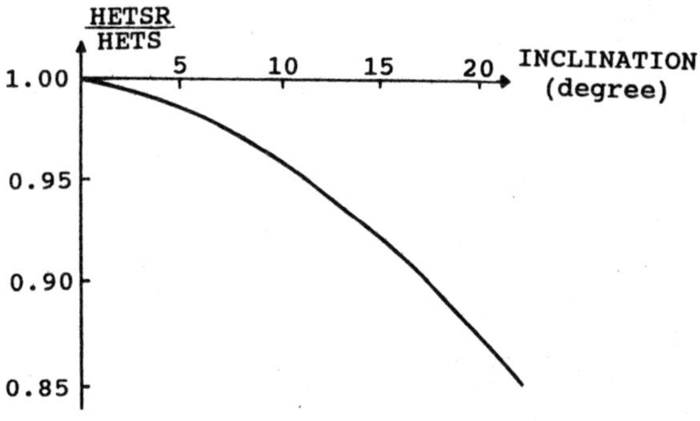

Figure 1. Efficiency vs. static column inclination

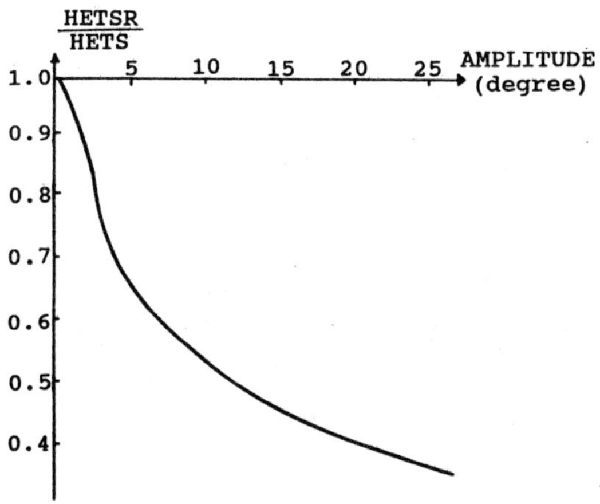

Figure 2. Efficiency vs. motion amplitude

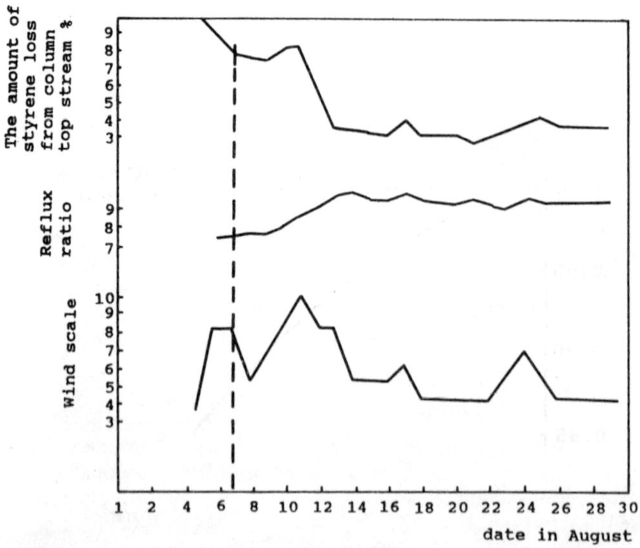

Figure 3. Influence of wind scale on the amount of styrene loss from column top stream and reflux ratio

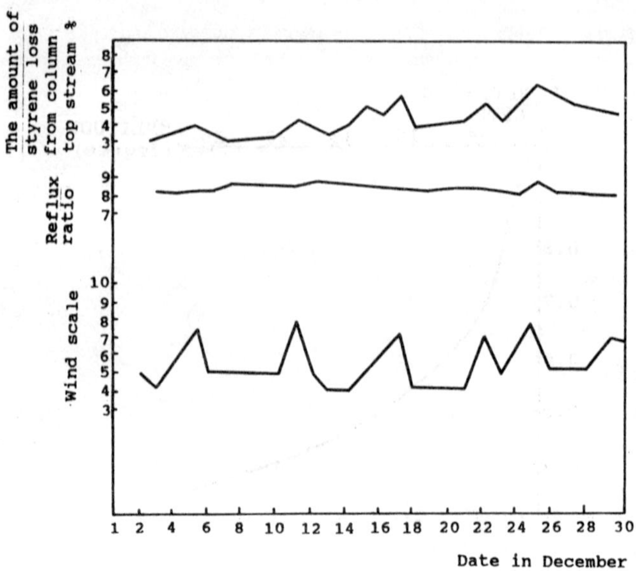

Figure 4. Influence of wind scale on the amount of styrene loss from column top stream and reflux ratio

MODELLING OF SHORT-PATH HIGH VACUUM DISTILLATION

Z. Kawala
Technical University of Wroclaw, Institute of Chemical Engineering, 50–373 Wroclaw ul. Norwida 4/6, Poland

Heat and mass transfer resistances in the liquid phase of a liquid film evaporating into vacuum were studied. Results of the research into the influence of the resistances on the hydrodynamics, evaporation rate and separation efficiency of the film are presented. The appropriate equations of heat and mass transfer for a thin-layer falling film and centrifugal molecular still were solved numerically, so that the radial and axial distributions of temperature and concentration in a liquid film were obtained.

INTRODUCTION

During molecular distillation in falling and centrifugal stills, a high rate of liquid film evaporation from the surface into the high vacuum combined with a low rate of heat transfer from the film's interior to the interface causes a self-cooling effect on the interface[1,2]. This accounts for the axial and radial temperature gradients in the liquid layer that flows in a vertical falling film apparatus or horizontal centrifugal still.

In the case of distillation of a multi-component mixture, a high rate of surface evaporation and low rate of diffusion in the liquid layer additionally influence the radial and axial concentration distributions[3,4].

The gradients produce changes in the surface evaporation rate and separation efficiency along the flow path. A study of this problem is presented here for pure liquid and binary mixtures.

1. HYDRODYNAMICS OF EVAPORATING FILM

1.1 FALLING FILM APPARATUS

The apparatus diagram and velocity distribution in a falling film are shown in Figure 1. The main part of the installation consists of a cylindrical evaporator surrounded by a condenser jacket. A substance to be distilled passes from a storage tank F on to the surface of the evaporator where it flows down in the form of a thin film, evaporating partly into vacuum. The velocity distribution for a laminar film is described by the well-known equation[5]:

$$w_r = \frac{g\rho}{\eta} S^2 \left[\frac{r-R}{S} - \frac{1}{2}\left(\frac{r-R}{S}\right)^2 \right] \tag{1}$$

with $R \le r \le R + S$

To calculate the film thickness at a given point in the apparatus, the amount of liquid evaporated so far should be taken into account. The formula for the thickness at a point z is given by the following equation:

$$S_z = \left[3\eta \left(\frac{m_o}{2\pi R g \rho^2} - \frac{1}{g\rho^2} \int_{z_o}^{z} G \mathrm{d}z \right) \right]^{1/3} \tag{2}$$

1.2 CENTRIFUGAL APPARATUS

In a centrifugal apparatus, the main part is a rotating evaporator disc fed in the centre with a stream perpendicular to its surface, as shown in Figure 2.

Due to centrifugal forces the liquid changes its flow direction from axial to radial. For the radial direction, the velocity distribution in a liquid film on the disc can be calculated from the equation[6]:

$$w_r = \frac{\omega^2 r \rho S}{\eta} z - \frac{1}{2} \frac{\omega^2 r \rho}{\eta} z^2 \tag{3}$$

and for the axial direction from

$$w_z = \frac{\omega^2 \rho z^2}{\eta} \left[\frac{z}{3} - \left(S + \frac{r}{2} \frac{\mathrm{d}S}{\mathrm{d}r} \right) \right] \tag{4}$$

The evaporating film thickness at the distance r from the middle of the disc may be calculated from the following formula:

$$S_r = \left[\frac{3\eta \left(m_o - \int_{r_o}^{r} 2\pi r G \, \mathrm{d}r \right)}{2\pi \omega^2 r^2 \rho^2} \right]^{1/3} \tag{5}$$

2. RATE OF EVAPORATION IN VACUUM

Considering liquid evaporation into high vacuum, it is necessary to take into account processes that occur in both the vapour phase and the liquid layer. Highly rarefied vapour cannot be treated as a continuous medium and its description should apply the kinetic gas model that allows for effects of mutual interactions between individual particles.

The effective evaporation rate G in the case of evaporation into high vacuum is given by the formula[7]:

$$G = P^o \left(\frac{M}{2\pi R_u T} \right)^{1/2} \left[1 - (1 - F)\left(1 - e^{-Kn_e}\right)^n \right] \tag{6}$$

where

$$Kn_e = \frac{h}{k\beta}$$

and

$$F = \frac{A_k}{A_k + A_p}$$

The correction coefficient k can be obtained from the empirical equation[7]:

$$\log k = 0{,}2F + 1{,}38(F + 0{,}103)^4 \tag{7}$$

3. HEAT AND MASS TRANSFER EQUATIONS FOR A LIQUID FILM

During liquid flow down the vertical surface or over the rotating disc of an evaporator, continuous heat transfer from deeper layers of the film towards the surface occurs. This is a result of conduction and convection. The temperature variation in an evaporating film may be described by the Fourier-Kirchhoff equation. By an analogy between heat and mass transfer, change in the concentration of each component of the film liquid may be given by the equation of steady-state diffusion with superimposed convection.

3.1 FALLING FILM APPARATUS

On the assumption that heat transfer is steady; liquid flows down vertically only; the temperature distribution is axisymmetrical; and axial heat conductivity is negligible, the Fourier-Kirchhoff equation has the following form in cylindrical coordinates:

$$w_z \frac{\partial T}{\partial Z} = a\left[\frac{1}{r}\frac{\partial T}{\partial r} + \frac{\partial^2 T}{\partial r^2}\right] \tag{8}$$

The initial and boundary conditions for this equation are as follows:

1. $T = T_o$ for $z = 0$ and $R \leq r \leq R + S$
2. $T = T_w$ for $r = R$ and $0 \leq z \leq L$
3. $\lambda \frac{\partial T}{\partial r} = r_p G$ for $r = R + S$ and $r_o \leq z \leq L$

This third boundary condition describes the heat flux used to evaporate the liquid from the surface at rate G.

The equation for the distribution of more volatile component concentration in a distilled binary mixture is:

$$w_z \frac{\partial C_A}{\partial z} = D_{AB}\left(\frac{1}{r}\frac{\partial C_A}{\partial r} + \frac{\partial^2 C_A}{\partial r^2}\right) \tag{9}$$

The initial and boundary conditions are:

1. $C_A = C_{A0}$ for $z = 0$ and $R \leq r \leq R + S$
2. $\frac{\partial C_A}{\partial r} = 0$ for $r = R$ and $0 \leq z \leq L$
3. $D_{AB} \frac{\partial C_A}{\partial r} = \frac{G_A}{M_A}$ for $r = R + S$ and $0 \leq z \leq L$

3.2 CENTRIFUGAL APPARATUS

For this case the simplified Fourier-Kirchhoff equation is of the following form:

$$w_r \frac{\partial T}{\partial r} + w_z \frac{\partial T}{\partial z} = a \frac{\partial^2 T}{\partial z^2} \tag{10}$$

where initial and boundary conditions are as follows:

1. $T = T_o$ for $r = r_o$ and $0 \le z \le S$
2. $T = T_w$ for $z = 0$ and $r_o \le r \le R_t$
3. $\lambda \dfrac{\partial T}{\partial r} = r_p\, G$ for $z = S$ and $r_o \le r \le R_t$

and the diffusion equation has the following form

$$w_r \frac{\partial C_A}{\partial r} + w_z \frac{\partial C_A}{\partial z} + D_{AB} \frac{\partial^2 C_A}{\partial z^2} \tag{11}$$

where the initial and boundary conditions are:

1. $C_A = C_{A0}$ for $r = r_o$ and $0 \le z \le S$
2. $\dfrac{\partial C_A}{\partial z} = 0$ for $z = 0$ and $r_o \le r \le R_t$
3. $D_{AB} \dfrac{\partial C_A}{\partial z} = \dfrac{G_A}{M_A}$ for $z = S$ and $r_o \le r \le R_t$

Analytical solution of any of the above equations of heat and mass transfer is impossible, as both temperature and concentration on the film surface are not constant, but vary along the flow path in all the cases. Therefore, differential equations were solved numerically. Substituting finite-difference ratios with respect to central difference[8] for partial derivatives in equations (8) and (9), the following forms of these equations result for a falling-film apparatus:

$$T_{r,\,z+\Delta z} = T_{r,\,z}\left[1 - \frac{2r}{K}\right] + T_{r+\Delta r,\,z}\left[\frac{\Delta r}{2K} + \frac{r}{K}\right] + T_{r-\Delta r,\,z}\left[\frac{r}{K} - \frac{\Delta r}{2K}\right] \tag{12}$$

where

$$K = \frac{r\,\Delta r^2\, W_{z,\,r}}{a\,\Delta z}$$

and

$$C_{A_{r,\,z+\Delta z}} = \left[1 - \frac{2r}{N}\right] C_{A_{r,\,z}} + \left[\frac{\Delta r}{2N} + \frac{r}{N}\right] C_{A_{r+\Delta r,\,z}} + \left[\frac{r}{N} - \frac{\Delta r}{2N}\right] C_{A_{r-\Delta r,\,z}} \tag{13}$$

where

$$N = \frac{r\,\Delta r^2\, w_{z,r}}{D_{AB}\,\Delta z}$$

For a centrifugal apparatus a transformation of equations (10) and (11) yields the following finite difference equations:

$$T_{z,r+\Delta r} = (A - B)\,T_{z+\Delta z,r} + (1 + B - 2A)\,T_{z,r} + A\,T_{z-\Delta z,r} \tag{14}$$

where

$$A = \frac{a\,\Delta r}{w_r\,\Delta z^2}; \qquad B = \frac{w_z\,\Delta r}{w_r\,\Delta z}$$

and

$$C_{A_{z,r+\Delta r}} = (E - B)C_{A_{z+\Delta z,r}} + (1 + B - 2E)C_{A_{z,r}} + E\,C_{A_{z-\Delta z,r}} \tag{15}$$

where

$$E = \frac{D_{AB}\,\Delta r}{w_r\,\Delta z^2}$$

4. NUMERICAL CALCULATION RESULTS

Numerical calculations were carried out for a molecular falling film apparatus with the evaporator radius of $R = 0.1$ m and distance between the evaporator and condenser surfaces of $h = 0.02$ m. The distance h had the same value for a centrifugal apparatus. It was assumed that the substances distilled were di-n-butyl phthalate and its mixture with di-n-butyl sebacate. The calculations were done for the devices with an isothermal evaporator wall. The variation and profiles of temperature, concentration, film thickness, the velocity distribution, and evaporation rate were determined, as well as the composition of the distillates obtained in these installations. Only the most representative results are given in the paper.

The calculations for the distributions of particular parameters were carried out up to the film dry-out point.

Respective points are marked in the figures with vertical broken line.

Figure 3 shows the influence of the initial feed rate on the surface temperature distribution when evaporating di-n-butyl phthalate in a falling film apparatus. After small cooling of the surface at the initial stage, a rapid growth of T_s follows. Then the temperature difference $T_w - T_s$ decreases gradually along the flow path, so that T_s approaches T_w at the end stage.

The surface temperature distribution influences directly the evaporation rate distribution in this apparatus, which is shown in Figure 4. In Figure 5, the di-n-butyl phthalate concentration profiles are shown for various values of dimensionless film length L during binary mixture distillation. The X-axis represents dimensionless film thickness, while the Y-axis shows the dimensionless local-initial concentration ratio.

Figure 6 shows the rate of mixture evaporation in a centrifugal apparatus. A decrease in the evaporation rate for higher values of the disc radius is brought about by a decrease in the concentration of a more volatile component in the surface layer. The profile of changes in the composition of the distillate obtained in this kind of apparatus as a function of the initial composition of feed is shown in Figure 7. Relatively high stability of the distillate composition for small disc radii in this case results from the low initial temperature of the substance. This thickness distribution of an evaporating liquid film, up to the point of its dry-out, is given in Figure 8.

CONCLUSION

High-vacuum distillation analysis as presented here accounts for phenomena in both liquid film and vapour phase, thus providing for a better design of distillation installations. With the assistance of computer programs developed for numerical calculations, it is possible to estimate the evaporator height needed in a falling film apparatus or its diameter in a centrifugal one. It is also possible to monitor changes in stream, distillate composition, temperature, liquid film thickness, and its dry-out zone.

LIST OF SYMBOLS

A_K	condensation surface area, m^2
A_p	evaporation surface area, m_2
$a = \lambda/\rho c_p$	temperature diffusivity, m^2/s
C	concentration, $kmol/m^3$
D_{AB}	diffusion coefficient of component A through component B, m^2/s
F	surface ratio
G	rate of surface evaporation, kg/m^2s, kg/m^2h
g	gravitational acceleration, m^2/s
h	distance between evaporation and condensation surface, m
k	anisotropy grade of vapour
M	molar mass, kg/kmol
m	mass flow, kg/s, kg/h
n	number of intermolecular collisions after which the vapour reaches isotropic state
P^o	vapour pressure, Pa
R_u	universal gas constant, J/kmol/K
r	coordinate in radial direction, m
S	film thickness, m
T	temperature, K
w	velocity in film, m/s
z	coordinate in axial direction, m

GREEK SYMBOLS

β	mean path of vapour molecule, m
η	dynamic viscosity, Pa.s
$\Delta = z/L$	dimensionless film length
λ	thermal conductivity, W/mK
ρ	density, kg/m^3

SUBSCRIPTS

A	more volatile component
B	less volatile component
r	in relation to radius
z	distance from inlet

REFERENCES

1. Greenberg, D.B., 1972, *AIChE J*, 18: 269–279
2. Kaplon, J., Kawala, Z. and Skoczylas, A., 1986, *Chem Eng Sci*, 41: 519–522.
3. Bhandarkar, M. and Ferron, J.R., 1988, *Ind Eng Chem Res*, 27: 1016–1020.
4. Kawala, Z. and Stephan K., 1989, *Chem Engn Technol*, 12: 406–414.
5. Stephan, K., 1988, *Warmeubertragung beim Kondensieren und beim Sieden* (Springer-Verlag, Berlin).
6. Emslie, A.G., Bonner, F. and Peck, L., 1958, *J Appl Phys*, 5: 858–861.
7. Kawala, Z., 1983, *Kinetics of Surface Evaporation under Molecular Distillation Conditions* (Wydawnictwo Politechniki, Wroclawskiej, Wroclaw).
8. Rosenberg, D.U. 1975, *Methods for the Numerical Solution of Partial Differential Equations* (Elsevier Publ Comp Inc, New York).

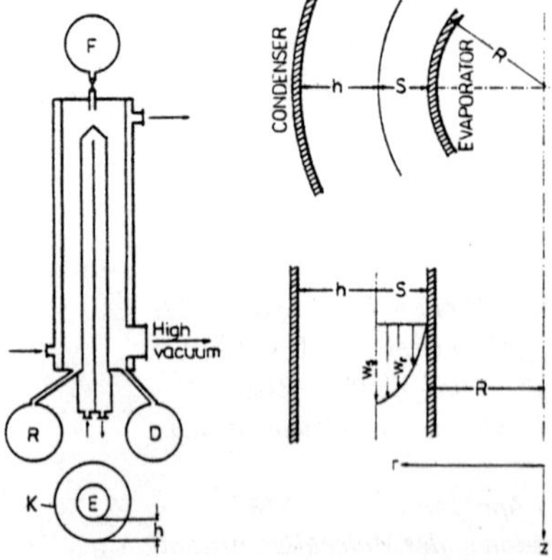

Fig.1. A scheme of falling film apparatus and velocity distribution in the film
D-distillate, E-evaporator, F-feed, K-condenser, R-residue

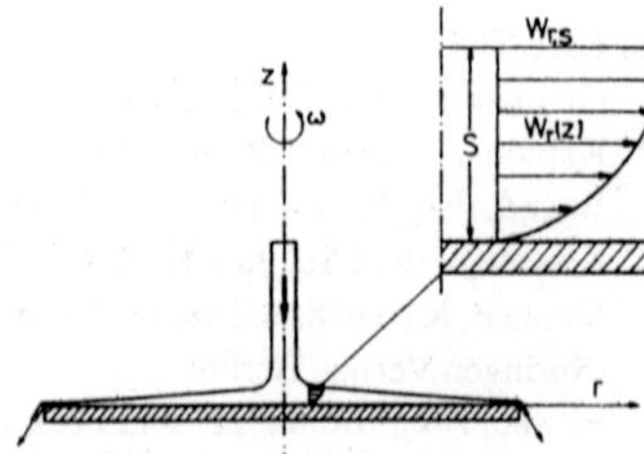

Fig.2. Scheme of liquid flow on rotating disc

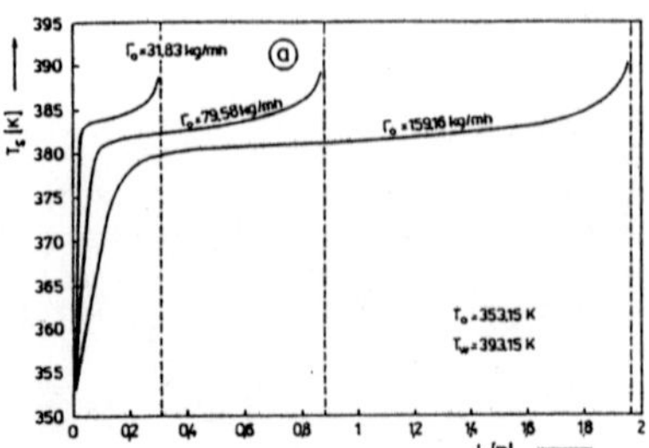

Fig.3. Effects of initial feed rate on distribution of surface temperature and dry-out position of liquid

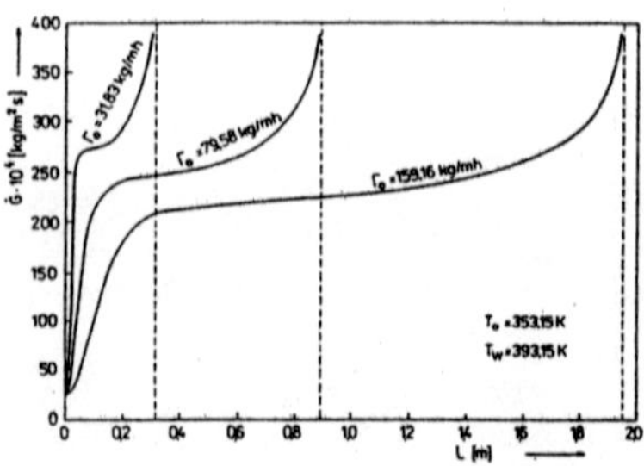

Fig.4. Effects of initial feed rate on evaporator rate

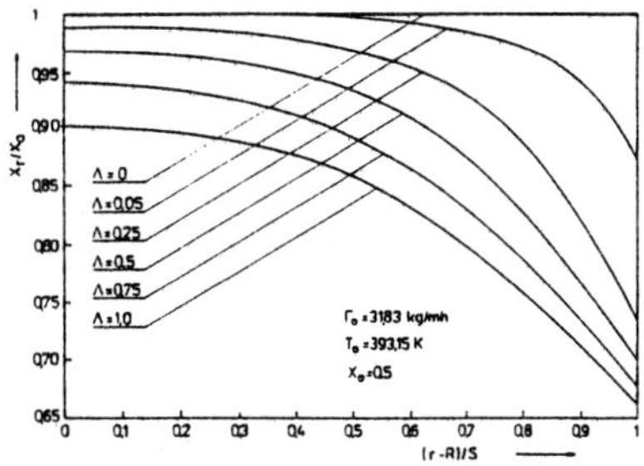

Fig.5. Concentration profiles in the film for different feed rates

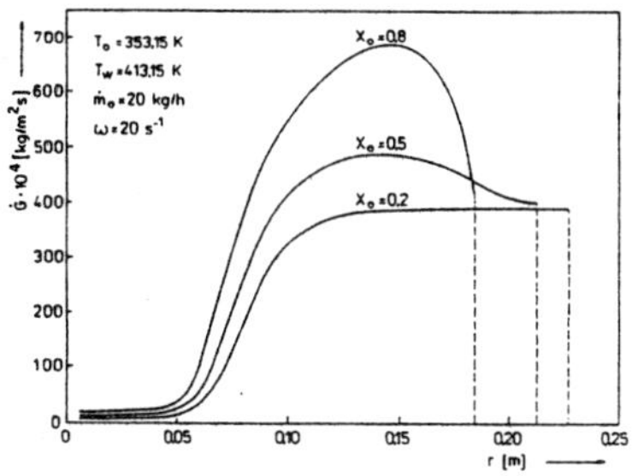

Fig.6. Distribution of evaporation rate as a function of initial concentration of the feed

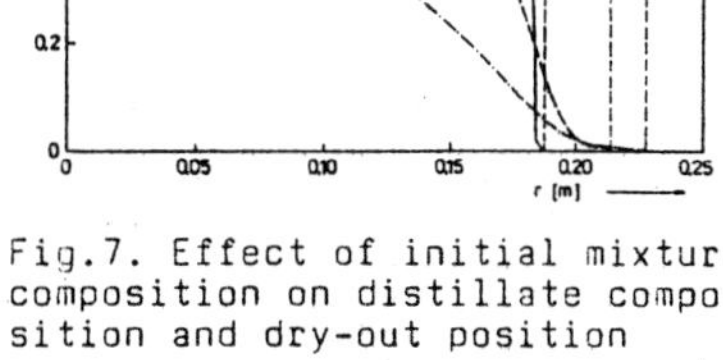

Fig.7. Effect of initial mixture composition on distillate composition and dry-out position

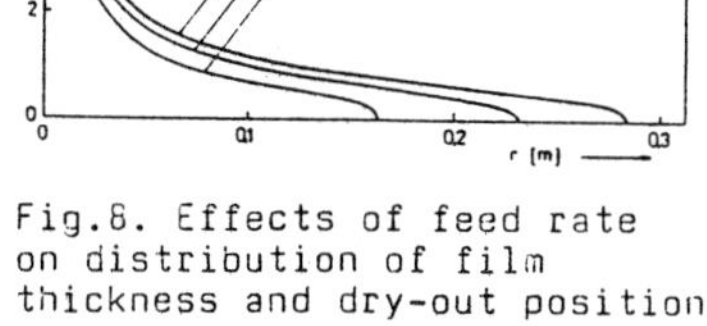

Fig.8. Effects of feed rate on distribution of film thickness and dry-out position

SIMULATION OF THREE-PHASE DISTILLATION

S. Wasylkiewicz
Institute of Chemical Engineering and Heating Equipment, Technical University of Wroclaw, Wroclaw, Poland

The model of multicomponent, multistage, three-phase distillation column is presented. In the proposed calculation method, the VLL flash algorithm is applied to calculate parameters of streams leaving the stage on the ground of known properties of all streams enternig the stage. The method is extremely stable and usually converges to the final solution. The versatility of the simulation algorithm is demonstrated with a practical distillation problem.

Keywords: Three-phase distillation, simulation.

INTRODUCTION

A lot of publications have been devoted to modelling and simulation of multistage separation processes in which only two phases (vapour and liquid or two liquids) exist. Far less attention has been paid to three-phase distillation systems (vapour-liquid-liquid) because they rarely appear in chemical industry and make the simulation procedure far more difficult. It is important to point out that this systems always admits the trivial solution (compositions of both liquid phases are identical) which sometimes is unreal. Therefore no wonder that most computations methods, which are robust in vapour-liquid distillation, fail for three-phase systems.

In some cases three-phase distillation can be very economically attractive because in this instance there usually appears an additional effect of a substance separation between two liquid phases. For example, using properly a side stream phase separator and removing only one of two liquid phases it is possible to separate three-component mixture in one distillation column (Block, 1976). Similar conception applied to a separation flowsheet of fermentation butyl beer (Pucci, 1986) has made it possible to reduce considerably the number of distillation columns (two columns in comparison to the standard four-column flowsheet).

In the vicinity of extraction critical point, a small change in feed composition can cause disappearence of the second liquid phase and it is more and more likely to get the false trivial solution. As a rule liquid-liquid extraction is carried out far from the critical point. On the other hand in the case of three-phase distillation compositions of liquids on some plates can be very close to the composition at critical point. Some of the plates can work as three-phase stages, some as two-phase ones. The type of a plate can not be properly determined before calculations. These facts cause additional difficulties and considerably decrease stability of simultaneous correction methods (Ferraris, 1981).

Block and Hegner (1976) have applied Newton-Raphson method where mean mole fractions in liquid phase were used as iterative variables. Ross and Seider (1980) and Walraven, Van Rompay and Luyten (1988) have increased convergence of the method by means of inside-out technics. In an inner loop vapour and average liquid composition have been calculated. In the outer loop liquid-liquid equilibrium compositions have been computed by means of phase-splitting-algorithm (Walraven, 1988). Schuil and Bool (1985) have implemented a mixed K-value model into existing computer programs for the simulation of vapour-liquid equilibrium stages. Pucci, Mikitenko and Asselineau (1986) have used a plate-to-plate calculation for simulation of three-phase distillation.

The proposed method of simulation of technological systems makes it possible to include three-phase distillation plates in computations of distillation columns. Theoretical stages of this type are calculated by means of FlasVLL procedure which realizes one of six basic flashes in vapour-liquid-liquid systems (Wasylkiewicz, 1992). In FlshVLL procedure it is checked up whether there are two or three phases in equilibrium on the plate. So at the begining of simulation there is no need to known which plates are two-phase stages and which are three-phase ones. If FlshVLL procedure is applied to computations of vapour-liquid distillation columns, the results will be the same as in the instance of FlshVL procedure (six basic flashes in vapour-liquid systems (Wasylkiewicz, 1992)). But the computation time can be far greater because of continuous checking exit liquid phase stability by means of phase-splitting-algorithm (Wasylkiewicz and Jamka, 1992).

MULTIFLASH METHOD

A schematic diagram of a single-stage multicomponent flash in a vapour-liquid-liquid system is presented in Figure 1. Parameters of a feed supplying the stage: a molar flow rate F, overall composition $Z = [z_i]$ (i = 1,...,NC) and molar enthalpy H_f are given. In the most general case the feed can be divided in the equilibrium stage into three streams: a vapour stream (composition — $Y = [y_i]$ (i = 1,...,NC), molar enthalpy — H_v) which is a fraction e_v of the feed, a liquid stream (composition — $X1 = [x_i^I]$ (i = 1,...,NC), molar enthalpy — H_l^I) which is a fraction $e_l\ (1-e_v)$ of the feed, and a second liquid stream (composition — $X2 = [x_i^{II}]$ (i = 1,...,NC), molar enthalpy — H^{III} which is a fraction $(1-e_l)\ (1-e_v)$ of the feed. All of the streams are in thermodynamic equilibrium at temperature T and pressure P. A heat flux Q is supplied into the stage. If $Q < 0$ the heat is withdrawn from the stage.

A set of equations describing the equilibrium distillation in the vapour-liquid-liquid system consists of NC mass balances (for each component i = 1,...,NC), NC equations of vapour-liquid equilibrium, NC equations of liquid-liquid equilibrium, a heat balance and three summation equations for each of the phases. Relationships $\gamma_i(T,X)$ and $P_i^s(T)$ for each component, $H_v(T,Y)$ and $H_l(T,X)$ for each of the phases are also needed. They are accessible in the course of computations by means of suitable functions or procedures from physicochemical property data bank (Wasylkiewicz, 1990). If one assumes that quantities P, Z and H_f are given, another value of one state variable (eg, Q, e_v or T) should be assigned additionally in order that the set of equations describing the equilibrium distillation stage be solved univocally. It is important to point out that if concentrations z_i ($\Sigma z_i = 1$) are given, the set of mass balances for each component i = 1,...,NC and three summation equations are linearly dependent. If one additionally assumes a value of the heat flux Q, a problem of general flash will be obtained (adiabatic flash when $Q = 0$) (NFl = 34), if a value of e_v — a partial vaporization problem (NFl = 35), if a value of stage temperature T — an isobaric-isothermal flash (NFl = 36). Detailed algorithms of the three aforementioned flash problems have been elaborated by Wasylkiewicz (1992). In computations, a typical adiabatic distillation theoretical stage can be represented by the adiabatic flash problem (NFl = 34, $Q = 0$), an evaporator and a condenser — by the partial vaporization problem (NFl = 35).

EXAMPLE PROBLEM

An industrial distillation operation (Block, 1976) has been selected to illustrate the simulation program SimFS. A three-component feed of butyl alcohol (B), water (W) and propyl alcohol (P) is to be separated into a propyl alcohol rich top product and a butyl alcohol rich bottom product. The feed composition is $x_B = 0.13$, $x_W = 0.65$ and $x_P = 0.22$ and its flow rate is $F = 50.0$ mol/s. The feed is introduced at its boiling temperature $T_f = 363.25$ K at stage 8 of a twelve-stage distillation column (Figure 2).

The first stage represents an equilibrium evaporator (NFl = 35, $e_v > 0$) and the twelfth stage is a total condenser (NFl = 35, $e_v = 0$). There is an atmospheric pressure in the column ($P = 101.325$ kPa). Distillate flow rate is $D = L_s^{12} = 29.0$ mol/s.

Nonidealities in the liquid phases were described by means of NRTL model (Renon, 1968). Parameters of binary interactions have been taken from two sources (Table 1).

TABLE 1

Binary interaction parameters of NRTL and UNIQUAC models for the butanol — water — propanol system.

NRTL model, Gmehling (1981)				
789.995	-491.499	0.317	propanol	butanol
18.196	1345.412	0.316	butanol	water
690.991	38.313	0.295	water	propanol
NRTL model, Block (1976)				
-220.53	257.90	0.30	propanol	butanol
324.17	1264.71	0.48	butanol	water
987.30	-25.74	0.30	water	propanol
UNIQUAC model, Gmehling (1977)				
543.486	-305.960		propanol	butanol
55.865	253.683		butanol	water
-82.262	349.242		water	propanol

Calculated liquid-liquid equilibrium diagrams for the butanol (B) — water (W) — propanol (P) system at a temperature of 298.15 K are shown in Figures 3 and 4. They have been calculated using NRTL parameters from Gmehling (1981) and Block and Hegner (1976) publications, respectively. It is evident that the description of liquid-liquid equilibrium is not correct for the second set of NRTL parameters, eg mutual solubility of propanol and water is less than in the butanol — water system (compare the experimental data (Gmehling, 1981)).

Using Gmehling's (1981) NRTL binary interaction parameters, stationary profiles of concentrations, temperatures and molar flow rates of vapour and two liquid phases (Figure 5) in the rectification column have been calculated by means of a flowsheet simulation program SimFS (Wasylkiewicz, 1986, 1992). For comparison analogous calculations have been worked out for NRTL parameters from Block and Hegner (1976) article. A printout of the column simulation results for these parameters is shown in Table 2. The first two numbers denote a number of theoretical stages (NSt = 12) and a type of a column (RKol = 3). In this instance it is the three-phase distillation column (VLL). Afterwards, for successive stages (from the twelfth stage to the first

TABLE 2
Printout of results of rectification column simulation with two- and three-phase distillation plates. NRTL parameters (Block, 1976)

12 3 NSt, RKol: VLL

I	NFl	Iv	Il	Il2	T	P	1	2	3	L/L2/V	;
12	35	11	0	0	361.44	101.325	0.030559	0.618995	0.350446	8.80736E+01	Ls: 2.90000E+01
							0	0	0	0	
							0.014053	0.599489	0.386458	0.00000E+00	Q:-4.89527E+06
11	34	10	12	12	362.03	101.325	0.058552	0.649314	0.292134	8.79920E+01	
							0	0	0	0	
							0.030559	0.618995	0.350446	1.17074E+02	
10	34	9	11	11	362.71	101.325	0.086271	0.677083	0.236646	8.79392E+01	
							0	0	0	0	
							0.051456	0.641694	0.306849	1.16989E+02	
9	34	8	10	10	363.32	101.325	0.108964	0.696046	0.194990	8.78720E+01	
							0	0	0	0	
							0.072059	0.662401	0.265540	1.16934E+02	
8	34	7	9	9	363.77	101.325	0.131821	0.693760	0.174418	1.37880E+02	
							0	0	0	0	
							0.088898	0.676425	0.234678	1.16897E+02	F1
7	34	6	8	8	364.21	101.325	0.147268	0.707688	0.145044	1.37874E+02	
							0	0	0	0	
							0.106848	0.693244	0.199909	1.16895E+02	
6	34	5	7	7	364.63	101.325	0.161662	0.720050	0.118288	1.37870E+02	
							0	0	0	0	
							0.125101	0.709521	0.165378	1.16885E+02	
5	34	4	6	6	364.99	101.325	0.174491	0.730206	0.095303	1.37867E+02	
							0	0	0	0	
							0.142235	0.724051	0.133714	1.16880E+02	
4	34	3	5	5	365.27	101.325	0.188773	0.734095	0.077132	1.37791E+02	
							0	0	0	0	
							0.157527	0.736012	0.106461	1.16866E+02	
3	34	2	4	4	365.52	101.325	0.244821	0.684412	0.070767	1.09806E+02	
							0.024923	0.960082	0.014995	2.79565E+01	
							0.174517	0.740620	0.084863	1.16791E+02	
2	34	1	3	3	365.73	101.325	0.270275	0.674171	0.055554	1.04002E+02	
							0.024084	0.965210	0.010707	3.37199E+01	
							0.188191	0.748069	0.063740	1.16755E+02	
1	35	0	2	2	365.90	101.325	0.292963	0.665365	0.041672	1.88184E+01	
							0.023459	0.969112	0.007429	2.18158E+00	
							0.199945	0.754155	0.045900	1.16726E+02	Q: 4.89610E+06

one), there are given: a stage number (*I*); type of the stage (NFl — a number determining the type of flash calculations); numbers of stages which feed the stage *I* with the vapour phase (*Iv*), the first liquid phase (*Il*) and the second liquid phase (*Il2*); temperature (*T*); pressure (*P*); compositions of both liquid phases and the vapour phase for components 1, 2 and 3 (butanol, propanol and water); molar flow rates of the first liquid phase (*L*), the second liquid phase (*L2*) and the vapour phase (*V*). *Q* is a heat flux supplying a stage, *Ls* — a side stream of the first liquid phase, and *F1* — the feed entering the eighth stage of the column.

CONCLUSIONS

In the proposed method of simulation of technological systems, a modular approach has been adopted, which however differs fundamentally from the classical approach. Whole apparatuses or individual stages of mass exchange has been taken as modules. Calculations of the system of modules has been organised also quite differently. It would be most profitable to apply simultaneous computation of all modules on a multiprocessor computer.

The vapour-liquid-liquid flash problems has been solved quite otherwise than in the classical approach (Prausnitz, 1980). It makes it possible to elaborate a reliable procedure for separation stage calculations and then to include it in the proposed method of simulation of technological systems. The simulation program SimFS is also very robust and makes it possible to get convergence of calculations even for initial approximations far from the final solution.

Calculated compositions of the products (see Table 2) are almost the same as the results obtained by Block and Hegner (1976). A small difference most likely results from application of different methods for calculation of physicochemical properties of pure substances and their mixtures (Wasylkiewicz, 1990, 1992).

REFERENCES

Block, U. and Hegner, B., 1976, *AIChE J*, 22: 582.

Ferraris, G.B. and Morbidelli, M., 1981, *AIChE J*, 27: 881.

Gmehling, J. and Onken, U., 1977, *Vapor-Liquid Equilibrium Data Collection, Aqueous-Organic Systems, Chemistry Data Series, Vol. I, Part 1* (DECHEMA, Frankfurt).

Gmehling, J., Onken, U. and Arlt, W., 1981, *Vapor-Liquid Equilibrium Data Collection, Aqueous-Organic Systems, Supplement 1, Chemistry Data Series, Vol. I, Part 1a* (DECHEMA, Frankfurt).

Prausnitz, J.M., Anderson, T.F., Grens, T.A., Eckert, C.A., Hsieh, R. and O'Connell, J.P., 1980, *Computer Calculations for Multicomponent Vapor-Liquid and Liquid-Liquid Equilibria* (Prentice-Hall, New York).

Pucci, A., Mikitenko, P. and Asselineau, L., 1986, *Chem Eng Sci*, 41: 485.

Renon, H. and Prausnitz, J.M., 1968, *AIChE J*, 14: 135.

Ross, B.A. and Seider, W.D., 1980, *Comp Chem Eng*, 5: 7.

Schuil, J.A. and Bool, K.K., 1985, *Comp Chem Eng*, 9: 295.

Walraven, F.F.Y. and Van Rompay, P.V., 1988, *Comp Chem Eng*, 12: 777.

Walraven, F.F.Y., Van Rompay, P.V. and Luyten, R., 1988, *Chem Eng Process*, 24: 133.

Wasylkiewicz S., 1986, Computation of multistage separation processes by transient-state method, *World Congress III of Chemical Engineering, Tokyo, 1986*, 4: 777.

Wasylkiewicz, S. and Królikowski, L., 1990, Physico-chemical property data bank for chemical process design, *10th International Congress of Chemical Engineering, Chemical Equipment Design and Automation, CHISA'90, Praha 1990*, paper no 1034.

Wasylkiewicz, S., 1992, Modelling and simulation of multistage separation processes, *Scientific Papers of the Institute of Chemical Engineering and Heating Equipment of Wroclaw Technical University, No 65, Monographs No 40*, Wroclaw 1992.

Wasylkiewicz, S., 1992, Vapour-liquid-liquid flash distillation, *XIV Polish Conference of Chemical and Processing Engineering, Muszyna (Poland) 1992*, in press.

Wasylkiewicz, S. and Jamka, D., 1992, Phase splitting algorithm, *XIV Polish Conference of Chemical and Processing Engineering, Muszyna (Poland) 1992*, in press.

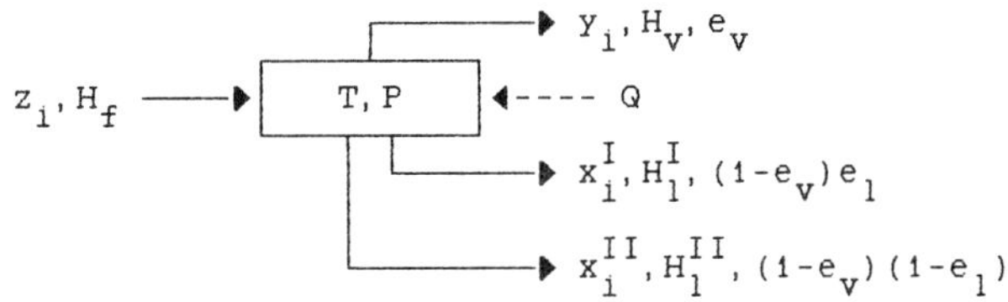

Fig. 1. Schematic diagram of a single-stage VLLE flash

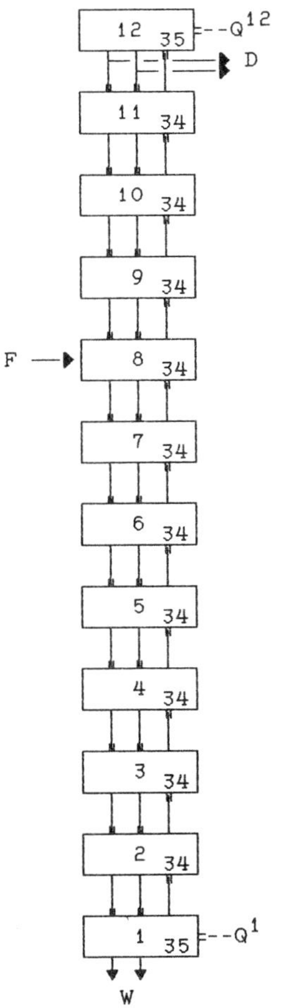

Fig. 2. Schematic diagram of three-phase distillation

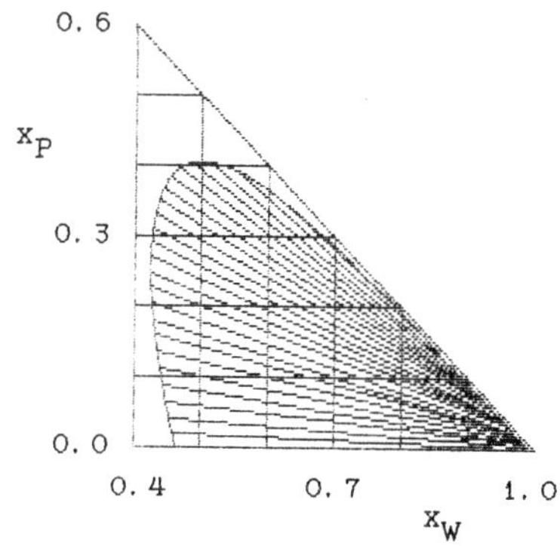

Fig. 3. Liquid - liquid equilibrium diagram for the butanol (B) - water (W) - propanol (P) system at 298 K. NRTL parameters (Gmehling, 1981)

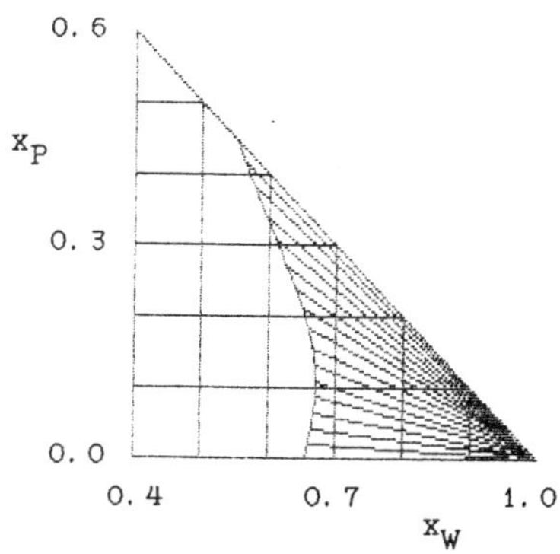

Fig. 4. Liquid - liquid equilibrium diagram for the butanol (B) - water (W) - propanol (P) system at 298 K. NRTL parameters (Block, 1976)

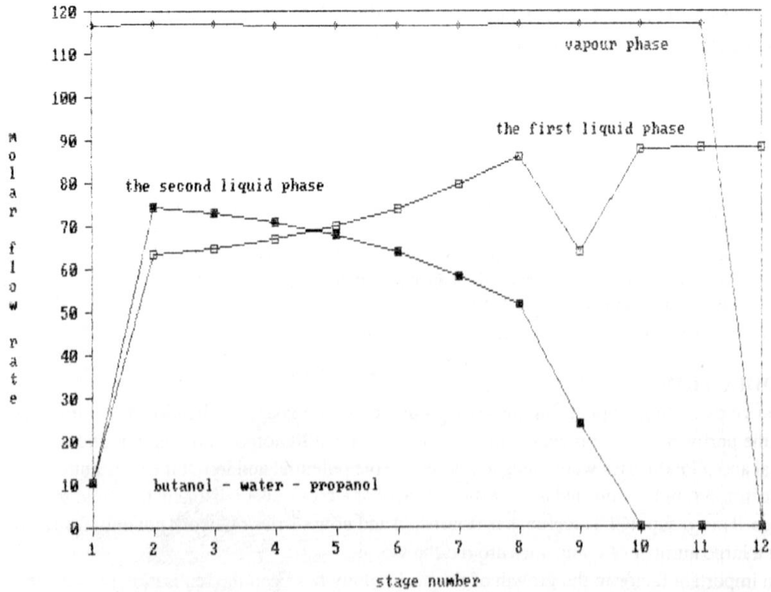

Fig. 5. Stationary profiles of molar flow rates of vapour and two liquid phases in the distillation column. NRTL parameters (Gmehling, 1981)

INVESTIGATION OF PROBLEMS IN DISTILLATION COLUMN OPERATION USING RADIOISOTOPE TECHNOLOGY

Dr T.L. Jones
ICI Tracerco

Radioisotope technology has the unique ability, conferred by the properties of ionising radiations, to investigate process problems while the process is on line and without disruption of the process in any way. This paper reviews the current techniques available for the investigation of distillation operations and provides examples of this technology by means of case studies.

INTRODUCTION

In recent years, radioisotope techniques have found ever-increasing application to the investigation of the performance of process plant in general and distillation operations in particular. The chemical and oil industries were quick to appreciate the potential and in ICI the substantial scope for radioisotope applications led to the establishment of a group specialising in this field some 30 years ago. This group, ICI Tracerco, has flourished and now carries out work not only for ICI but also for a large number of companies around the world.

An important factor in the growth of this technology has been the realisation on the part of production management of the enormous savings which can result from the successful application of radioisotope technology. Most measurements can be made while the process is on-line without disrupting the operating conditions thus saving down-time.In addition, the rapidity and convenience with which the measurements can be made (utilising as they do equipment external to the process) leads to a direct reduction in service costs.

Radioisotope techniques can be broadly divided into two categories, depending upon the way in which the radioactive material is employed:

- sealed source applications and
- tracer studies.

Throughout the application of a sealed-source technique, the radioactive material remains encapsulated within a sealed container; penetrating radiations from the source are directed at the vessel of interest. Variations in the transmitted or scattered radiation are detected and are related to the contents of the vessel. In many ways, these methods are analogous to the use of X-rays for medical diagnosis, though there are significant differences in the technologies that are involved.

Radiotracer techniques are fundamentally different. In this case radioactive material in a suitable form is injected into the process stream, thereby rendering a portion of it radioactive. The subsequent movement of this 'labelled' process material (as it flows through pipes or up or down distillation columns) can be monitored by using radiation detectors mounted at appropriate positions on the plant.

Because of the general construction of processing plant, the radioisotopes that are of greatest use are those whose radiations are capable of penetrating the (often substantial) walls of vessels and piping. Apart from a few specialised applications, this means that this section of the technology is concerned with gamma-ray and neutron emitters.

SEALED SOURCE APPLICATIONS

Industrial X-Ray or gamma radiography is carried out by directing a beam of radiation through a pipe or vessel and measuring the transmitted radiation by means of photographic film. Apart from the practical difficulties of doing this on a 5 metre diameter, 50 metre high distillation column, photographic film is not particularly sensitive. A better measuring technique is to use radiation detectors incorporating scintillation crystals and photomultiplier tubes which are approximately 1000 times more sensitive than film and to reduce the radioactive source size by a corresponding factor.

The basic technique involves positioning a gamma ray source and a radiation detector on opposite sides of the vessel under inspection and relating the level of transmitted radiation to the mass of material in the radiation beam. For narrow beam attenuation the intensity of the radiation beam after traversing the vessel is given by:

$$I = I_{o*e}^{-\mu o x}$$

where I is the transmitted radiation, I_o is the incident radiation, μ is the mass absorption coefficient (constant for a particular gamma energy and a particular absorber), x is the thickness of the absorber and o is the density. Thus if the separation of the source and detector is kept constant the intensity of the transmitted radiation is a function of the density of the medium.

The principle of using this technique to examine a distillation column is to select a scan line across the column and to make a large number of measurements in a vertical plane thus building up a density profile of a vertical slice through the column. In practice the source and detector are positioned diametrically opposite to each other on the column. They are then moved vertically downward and the intensity of the transmitted radiation is recorded relative to the elevation on the column.

As the source and detector scan through a 'normal' vapour space in the column a transmission peak is observed and when the scan line intersects a tray with liquor on it an absorption trough is observed. In practice it is found that the changes experienced in moving from regions of low intensity to high intensity are not perfectly sharp because wide beam instead of narrow beam radiation is used. This is necessary because the shielding required to produce narrow beam conditions would make the source container too heavy and unwieldy to use. Consequently there is a contribution to the radiation field at any point due to radiation scattered indirectly from the internal structure of the column to the detector.

Comprehensive interpretation of the data usually requires a great deal of experience but more obvious faults in operation such as flooding or collapsed trays are relatively easy to detect. Some of the range of operational faults are shown in the somewhat idealised representation in Figure 1. Most of the data can be acquired by just one scan but to identify more subtle faults on columns with complex internals or to quantify more accurately the density changes in various parts of the column, a reference scan obtained when the column is operating satisfactorily or a blank scan on an empty column is desirable. Also a flooded column will mask any further problems within the flooded section and additional scans at lower throughputs may be necessary.

Because all of the equipment is external to the vessel, the technique is applicable to any process material and is not impaired by conditions of high or low temperature or high pressure. The technique has been used on trayed and packed columns of diameters varying from a few tens of millimetres to 9 metres or more with wall thickness up to 100 millimetres of steel. It has also been used to monitor the operation of cryostatic columns within the cold box.

NEUTRON SOURCE APPLICATIONS

Several types of compact, portable, fast neutron sources are commercially available. The properties of neutrons which are useful in process investigation measurements are:

a. Fast neutrons are capable of penetrating substantial thicknesses of metal.
b. Fast neutrons are moderated principally by collision with hydrogen nuclei.
c. The resulting thermal neutrons are capable of being detected by any of a range of commercially available detectors.

One of the simplest yet most useful applications of neutron sources is the detection of the level of hydrogenous material inside a process vessel — see Figure 2.

A portable measuring head consisting of a fast neutron source and a slow neutron detector is positioned against the wall of the vessel and moved up or down over the surface. Fast neutrons penetrate the wall of the vessel and become thermalised by collisions with hydrogenous material in the adjacent medium. Slow neutrons diffusing back out of the vessel are detected. The response of the detector is thus a function of the hydrogen concentration of the material adjacent to the measuring head and the position of the level may be inferred from the large signal change which is observed as the measuring head moves through the position of the liquid/vapour interface.

On distillation columns this technique is particularly useful for measuring base levels and levels in downcomers as well as operating levels in stock tanks, receivers and reboilers.

RADIOACTIVE TRACER TECHNIQUES

Unsealed sources as solid, liquid or gaseous materials labelled with an appropriate radioisotope form the basis of a range of techniques which are important in studying the dynamics of process streams both within and around distillation columns.

The radioactive material is injected into the process stream thereby rendering a portion of it radioactive.The subsequent movement of the 'labelled' material as it flows through the distillation process is monitored using radiation detectors mounted at appropriate positions around the process.

The general features of radiotracer techniques which make them important as investigational and measurement tools are:

a. Great sensitivity (following from the sensitivity with which small amounts of radioactivity can be measured with modern detection equipment).
b. Unequivocal information — the radiotracer is usually the only radioisotopic species present.
c. Relative insensitivity to 'difficult' conditions, eg high temperature or pressure, corrosive materials, size of process. Tracers commonly used in process investigation are predominantly gamma ray emitters of short half-life. Gamma radiation offers the possibility of detection external to the medium under investigation and hence leads to experimental simplification in that sampling and complex analysis can often be avoided.

The range of radioactive tracer techniques in current use has been well documented[1]. Some of the techniques which have specific application in distillation operations are:

a. Flowrate measurement for material balance eg feed, products, reflux, steam to boilers and ejectors and cooling water to condensers.

b. Distribution studies to measure carry over, carry up, residence times and channelling in packed beds.

c. Investigative tests to measure leaks of steam/water/process material in reboilers, condensers, ejectors and internal leakage from draw off trays.

Some examples of the application of these techniques are illustrated in the following case studies.

DENSITY PROFILE OF AN ABSORBER COLUMN TO INVESTIGATE THE EFFECTIVENESS OF ANTIFOAM ADDITION

An oil absorber column on a refinery was experiencing severe foaming problems resulting in a high level of liquid entrainment in the vapour overhead line. In an attempt to provide sufficient space at the top of the column for liquid to disengage, the top two trays had been removed, but the problem still persisted. It was decided to incorporate an antifoaming additive to the feed to the column. In order to ascertain the effectiveness of this treatment; gamma-ray absorption scans were carried out before and after the addition of the antifoam.

The results of the two sets of measurements are shown in Figure 3. Without antifoam, very heavy foaming — approaching flooding — was observed at the top of the column. Further down the column the foaming problem became progressively smaller though all trays, with the exception of Tray 1, were heavily laden. In contrast, the scan carried out during the antifoam addition indicated that the column was operating normally with well defined liquor levels and vapour spaces. These data provided positive evidence that the antifoam addition was successful and subsequent scans, carried out with different concentrations of antifoam, enabled production management to assess the minimum level of additive required to prevent foaming.

INVESTIGATION OF TRAY FOULING IN A DISTILLATION COLUMN

A fractionation column on a phenol plant was producing material of poor specification. The column was examined using the gamma-ray absorption technique to investigate the source of the problem. The results indicated that in the upper part of the column there was good phase separation. However in the lower part of the column progressively high vapour densities and tray loadings were observed. From Figure 4 it can be seen that Trays 38 and 39 in particular were carrying large amounts of liquid and were approaching the 'flooded' condition.

On the basis of these results, the column was shut down, opened and the suspect region examined. Heavy deposits of an insoluble salt were found on the the trays in this part of the column. Clearly, these deposits were impeding the transfer of vapour and liquor up and down the column, leading to inefficient fractionation. After washing the trays with appropriate solvent and bringing the column back on line, a second scan was carried out which confirmed that the liquor hold-up in the lower part of the column had cleared.

After several weeks operation, however, the product quality progressively deteriorated and a further column scan indicated that liquor was being held-up in the same region as before. It was clear that the problem could only be resolved by a change in the operating conditions or a change in the column design. The problem was ultimately solved by changing the design of the trays but in the meantime a regular schedule of scanning was introduced to provide precise information on the rate of performance deterioration.

There are many examples of this technique being used as part of a debottlenecking process. On one fractionator on a refinery in Europe the internals had been modified to increase throughput by 20%. When this increase was not achieved because of instability, the column was scanned and the area in the vicinity of the top pump around draw-off tray was shown to be flooded although the process conditions did not indicate a problem. At the next shutdown modifications were made to the orientation of the flow baffles associated with the top pump around tray. Further scans after start-up confirmed that the column was operating satisfactorily at an increased feed rate close to the expected design conditions.

DETECTING TRAY DAMAGE

Figure 5 is a gamma-ray absorption scan of part of a fractionating column. Examination of the data clearly indicates that;Trays 49 and 50 are missing; Tray 48, though present, is damaged and holding little liquid; Tray 31 has fallen and the debris is lying on top of Tray 32.

Scans such as this are frequently carried out prior to a planned shutdown and, on the strength of the results, replacement trays ordered in advance thus preventing a possible extension of the shutdown.

This, specifically, was the case in relation to an atmospheric tower on a refinery in North America. Stability and operational problems encountered in the vicinity of the gas oil product draw-off tray led the operator to suspect tray damage. The questions effectively were — how many trays were damaged and to what extent, was the bottom tray safe for a man to stand on or was additional scaffolding required?

A gamma scan provided data to enable a detailed tray damage report to be prepared. Some of the trays were only damaged on one side of the centre downcomer, others were tilted or buckled or out of position resulting in unbalanced liquor levels. The bottom tray was mechanically sound and operating normally so additional scaffolding was not necessary.

In spite of the irregular density profile, the interpretation of the gamma scan represented the actual internal tray conditions in over 90% of all cases.

LIQUID RETENTION IN A PACKED COLUMN

The gamma ray density profiling technique is also applicable to packed columns. The procedure for collecting data is essentially the same as for trayed columns. In order to define an overall density profile, however, it will be necessary to select a number of different scan lines. To differentiate between increased density due to fouling or to liquor holdup or both, for example, it is necessary to carry out scans over a range of throughputs. In one particular case the specified throughput could not be achieved and the differential pressure was higher than expected.

The density profiles indicated fouling of the column packing but the deposits did not affect the liquid retention in the packing. At a subsequent shut down the fouling, caused by polymerisation, was confirmed. The operator was interested in monitoring liquid retention at various feed rates and also during the life of the column. By comparison of a series of scans at different times and feed rates with a blank scan of the dry packing a great deal of information on the hydrodynamic behaviour of the column was obtained.

VAPOUR DISTRIBUTION STUDY IN A PACKED BED

Tracer techniques were used on a column in a fractionation train to investigate the effectiveness of a new deflector plate at the inlet of the vapour return. The purpose of the plate was to distribute the vapour evenly throughout the column. When the anticipated improvement in product quality was not achieved, the distribution of the vapour phase was monitored using a gaseous radioisotope tracer.

The test was performed by injecting the radioisotope Krypton 85 into the vapour return line. This is a chemically inert gas so the tracer did not affect product quality and its low gamma ray energy did not present any radiological hazard to personnel. Sensitive radiation detectors were placed on the column around the circumference and just below each packed section as shown in Figure 6. The output from each of the 13 detectors was collected and analysed.

The test results showed that there was a significant maldistribution of the vapour flow throughout the column. The new deflector plate was forcing most of the flow up the side of the column opposite to the vapour inlet. The data also showed that the maldistribution existed in the first and second packed sections and the vapour only became well distributed in the third section.

This information provided the engineers with proof that it was necessary to redesign the vapour inlet. The deflector plate was replaced by a vapour sparger and the column performance improved to meet the product specification.

DETECTION OF WATER LEAKAGE INTO PRODUCT

In the distillation process there are several possible sources of water ingress including condensers, coolers and steam heated reboilers. In one particular example of water contamination it was necessary to investigate all three possible sources. The system is shown diagrammatically in Figure 7.

Three separate injections were made around the batch distillation unit. Bromine 82 as potassium bromide was injected into the cooling water to the product cooler and samples of the product assayed for tracer content. Bromine 82 was injected into the condenser cooling water feed and samples of the product exit pump A were removed and assayed for activity

Tritiated water was injected into the steam feed to the reboiler and samples from both the column bottoms and overheads were taken and assayed for tracer content.

The minimum detectable limit was evaluated for each of these samples, positive results were obtained from the bottoms samples and, when the reflux was operating, the overhead samples. From this data it was clear that any leakage in the cooling or condensing systems was very small and the only significant leak — 0.05% of the overheads rate — was from the reboiler.

ACKNOWLEDGEMENT

All investigations described above have been carried out by personnel from ICI Tracerco whose cooperation in preparing the paper is hereby acknowledged.

REFERENCE

1. Charlton, J.S., 1986, *Radioisotope Techniques for Problem Solving in Industrial Process Plants*, Leonard Hill, London.

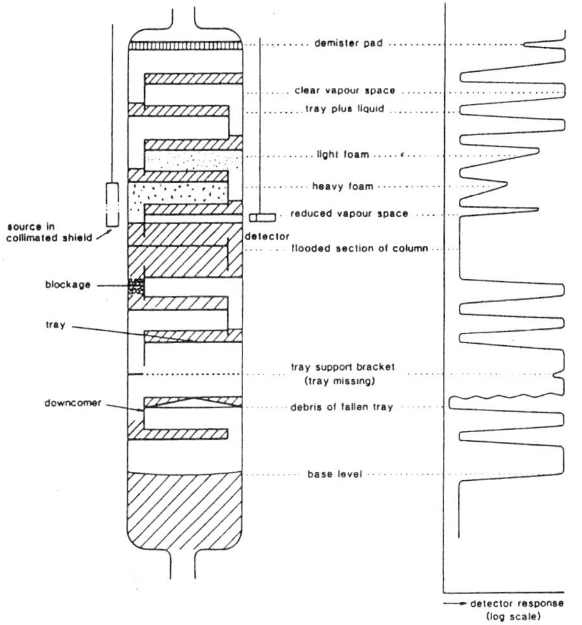

Figure 1. Gamma-ray absorption scan of a distillation column

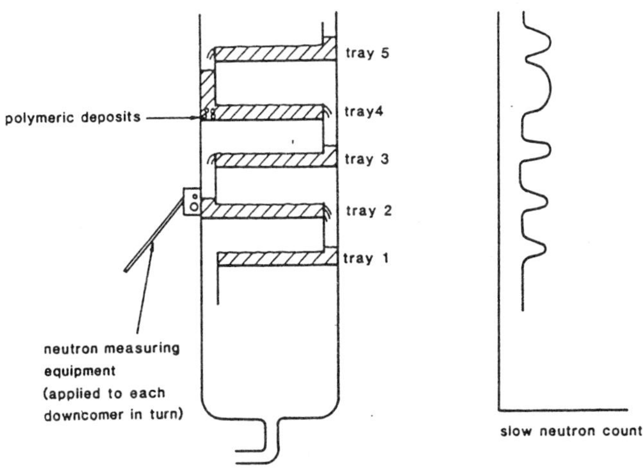

Figure 2. Inspection of distillation column downcomers using a neutron moderation technique

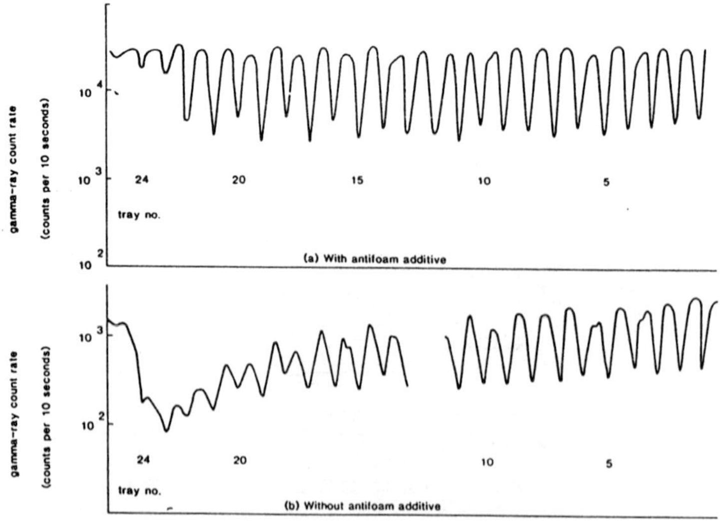

Figure 3. Gamma-ray transmission characteristics of an oil absorber column: (a) with antifoam additive; (b) without antifoam additive

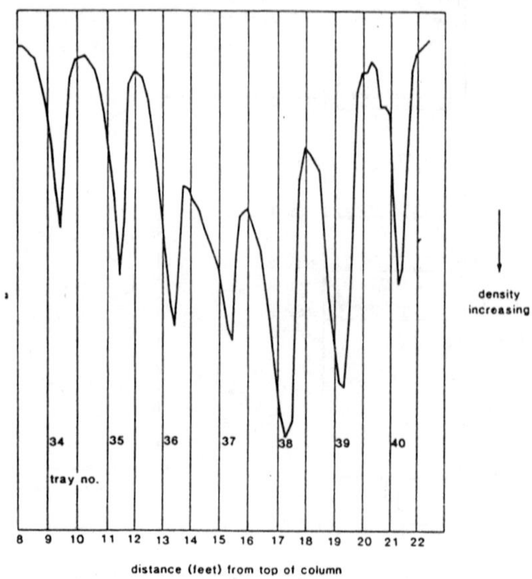

Figure 4. Gamma ray absorption inspection of a fractionation column (note heavy vapour density and high liquid loading of trays 38 and 39)

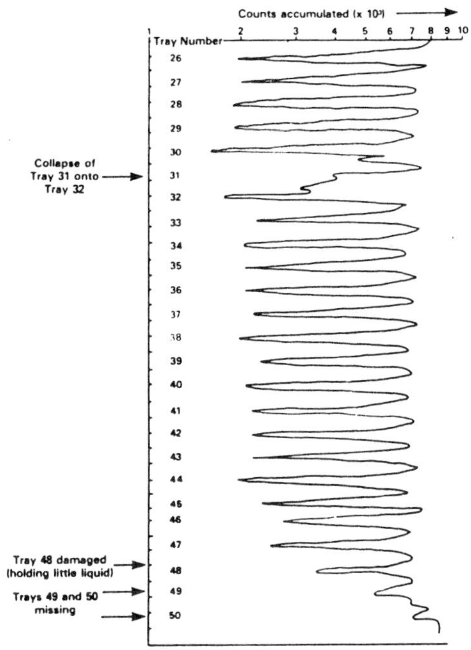

Figure 5. Gamma-ray absorption scan of a fractionating column

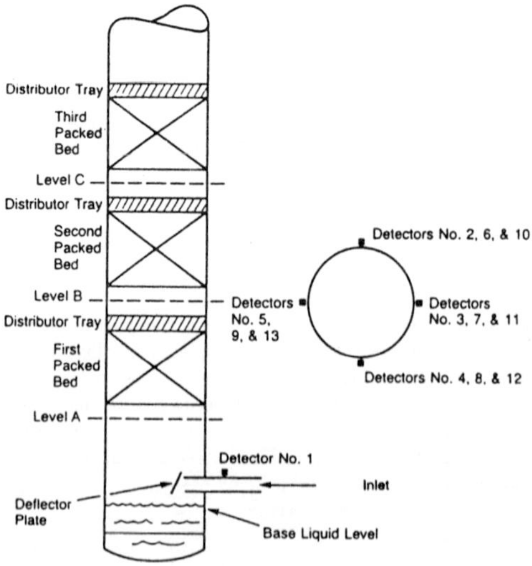

Figure 6. Vapour distribution study in a packed bed

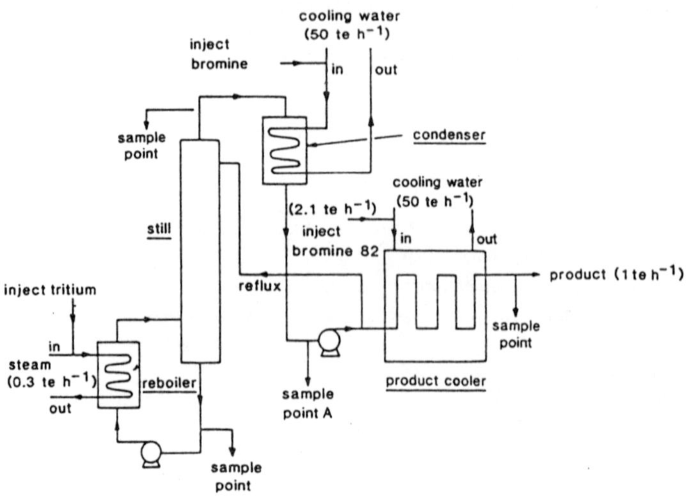

Figure 7. Detection of water into product leak using 'direct' tracer sampling technique

NEW ENHANCED CAPACITY MD TRAY DEBOTTLENECKS DEETHANIZER

M.R. Resetarits, R.J. Miller*, J.L. Navarre**, M. Linskeseder and P. Reich-Rohrwig***
* UOP, Tonawanda, New York, USA
** UOP NV Antwerp, Belgium
***OeMV Aktiengesellschaft, Schwechat, Austria

Since their inception in 1964, Multiple Downcomer (MD™) Trays have been employed hundreds of times in grass-roots and revamped columns wherein high liquid-to-vapor ratios were encountered. MD Trays have higher capacities than crossflow trays, eg, 4-pass trays. After years of laboratory work, a new tray, having an even higher capacity than an MD Tray, has been developed. Enhanced Capacity MD Trays were recently employed by OeMV engineers in a front-end deethanizer to debottleneck their ethylene plant in Schwechat, Austria. That deethanizer was initially equipped with 1975-vintage MD Trays. Operating data collected after the 1990 ECMD Tray revamp showed that a capacity increase of at least 15% was effected.

INTRODUCTION

A new enhanced capacity Multiple Downcomer (ECMD) Tray has been successfully commercialized by UOP at the OeMV Ethylene Plant in Schwechat, Austria. This new tray was used to revamp a front-end deethanizer to effect a capacity increase of at least 15%.

Prior to the revamp, the subject deethanizer was the primary bottleneck in the ethylene plant. This column originally contained 1975-vintage MD Trays. As the throughput demands to the column increased, the maximum capacity (ie flood point) of the existing MD Trays was inevitably reached. This situation prompted OeMV to contact UOP regarding the possible debottlenecking of the subject column. After much discussion, OeMV agreed to be the first company in the world to use ECMD Trays in a commercial column. As a result, the existing trays were replaced and a 15% capacity increase was effected.

MD TRAYS

The MD Tray has been used in nearly 400 columns throughout the world[1,2]. Figure 1 is a typical MD Tray with multiple downcomers, long outlet weirs and no receiving pans. These trays have often been used to revamp conventionally-trayed (2- and 4-pass) columns to effect capacity increases[3,4]. In addition, they have been used in grass roots applications when either the column size (height and diameter) can be reduced appreciably or the number of columns required for a given separation can be reduced[5].

In the past, UOP has encountered difficulties when revamping these high-capacity columns to increase capacities. However, with the development of ECMD Trays, capacity increases in MD-Trayed columns are now possible. The OeMV deethanizer was the first column to demonstrate this revamp scenario. Companies which currently operate MD-Trayed columns could consider ECMD Trays for future capacity enhancements.

OEMV REVAMP

The OeMV deethanizer is shown in Figures 2 and 3. The overall column configuration remained the same both before and after the revamp. The number of trays remained unchanged with 56 MD Trays being replaced by 56 ECMD Trays. Also, the two feed points remained unchanged with the vapor feed entering at Tray 23 and the liquid feed entering at Tray 19.

Table 1 contains operating data showing the process conditions before and after the revamp. Prior to the revamp, the deethanizer was operating in the incipient stages of flooding at a total feed rate of 108.9 Metric Tons Per Hour (MTPH).

OeMV sought to increase the capacity of the deethanizer to a minimum total feed of 120 MTPH — a 10% capacity increase. In addition, a feed rate of 132 MTPH was sought for future feed rate increases. As a result, the ECMD Trays were designed at a feed rate of 120 MTPH with the capacity to operate at a feed rate of 132 MTPH. During start-up, a feed rate of 125.3 MTPH was attained in the revamped deethanizer. At this feed rate (15% higher than the maximum MD Tray feed rate), no hydraulic instabilities were observed.

The maximum capacity of the ECMD Trays has not been reached. Since start-up, sufficient feed to the deethanizer has not been available to determine the ultimate capacity of the deethanizer. Thus, the primary conclusion from OeMV is that ECMD Trays were able to provide a column capacity improvement of at least 15% over MD Trays.

ECMD TRAY PERFORMANCE

In addition to the observed capacity increase, the ECMD Trays were found to operate at the same efficiency as the MD Trays. As shown in Table 1, column performance after the revamp revealed a better propene purity in the Make and a lower ethane purity in the Tails. The result was a good overall efficiency.

Also, after the revamp, a decrease in the average tray pressure drop was observed. Specifically, at 15% higher loadings, the ECMD Trays were observed to have a lower pressure drop than the MD Trays. The lower observed pressure drop was expected but the magnitude of the decrease was somewhat unexpected. In laboratory tests at the same loadings, ECMD Trays had been observed to operate at pressure drops up to 10% less than MD Trays. This low tray pressure drop is even more significant when it is considered that an MD Tray operates at roughly a 40% lower pressure drop than a typical conventional tray.

It was also observed that the revamped deethanizer was meeting product purity specifications within two hours after start-up. The ability to start-up quickly could prove to be an additional benefit of the ECMD Tray.

TRAY DESIGN

Precise details regarding the augmentations that are employed to increase the capacity of MD Trays to ECMD levels can not be divulged at this time. Laboratory work and field testing are still underway in support of a patent application.

LABORATORY STUDIES

The successful operation of ECMD Trays in the OeMV deethanizer was the culmination of a distillation tray capacity enhancement development program at UOP. The primary objective of this program was to develop a distillation tray with a significantly higher capacity than an MD Tray.

The ECMD Tray was a direct result of the aforementioned program. This tray was developed in the UOP distillation laboratory located in Tonawanda, New York. The first positive results were obtained in a 0.6 m × 0.6 m air-water test simulator. This work marked the inception of the ECMD development program. Several variations were tested in this small column. The capacity results were compared to prior MD Tray data and a significant capacity increase was observed.

A 'best' version of the ECMD Tray was then placed in a 2.4 m diameter air-water simulator for additional testing. The capacity tests in this column were done using three trays at 305 mm tray spacings. A matrix of data at various liquid and vapor loadings was collected and the ultimate capacity of the trays was determined.

After these capacity tests, the trays in the large test column were replaced by conventional MD Trays with the same active area and downcomer area. The capacity tests revealed that, again, ECMD Trays have a significantly higher tray capacity than MD Trays. Specifically, they appear to have roughly a 20% higher tray capacity than MD Trays. Flood data from these capacity studies are shown in Figure 3. The results corroborated the earlier flood data obtained in the 0.6 m × 0.6 m test column.

On a side note, ECMD Trays were found to have a significanlty higher tray capacity than MD Trays in the laboratory. However, this does note necessarily correspond to an equally higher column capacity. Many factors ultimately determine the capacity of a given column. Tray capacity is just one of those factors.

CONCLUSIONS

A new enhanced capacity Multiple Downcomer Tray has been developed by UOP and, after extensive testing within a laboratory environment, successfully tested at OeMV. At OeMV, this tray proved to have a higher capacity than, and an efficiency equal to, the MD Tray. It should prove useful in many future revamps to attain capacity increases and in many new towers to attain investment decreases.

ACKNOWLEDGEMENTS

The authors would like to acknowledge Michael J. Lockett for his contributions to the ECMD Development Program, Joseph Agnello for his field support during the column revamp, John R. Ulmer for the mechanical design of the trays, and the business and technical managements of OeMV and UOP for their support and innovativeness.

REFERENCES

1. Delnicki, W.V. and Wagner, J.L., 1970, *Chem Eng Progress*, 66 (3): 50.
2. O'Neill, P.S., Wisz, M.W., Ragi, E.G., Page E.H. and Antonelli, R., 1985, Paper presented at AIChE Spring National Meeting, Houston, Texas, March 1985.
3. Lockett, M.J. and Wisz, M.W., 1983, paper presented at European Federation of Chemical Engineering, Working Party Meeting on Distillation, Absorption and Extraction, Winterhur, Switzerland, June 1983.
4. Resetarits, M.R., Agnello, J., Lockett, M.J. and Kirkpatrick, H.L., 1988, paper presented at the 1988 AIChE National Meeting, New Orleans, Louisiana, March 1988.
5. Kirkpatrick, R.D., 1978, *Oil and Gas Journal*: 72

TABLE 1: PROCESS CONDITIONS

	Before Revamp (MD)	After Revamp (ECMD)
Expected Feed, mtph	110.0	132.0
Design Feed, mtph	100.0	120.0
Actual Feed, mtph	108.9	125.3
Feed 1, mtph	98.4	109.3
Temperature, deg C	-15.0	-14.4
Quality	1.0	1.0
Feed 2, mtph	10.5	16.0
Temperature, deg C	17.6	15.0
Quality	0.0	0.0
Make Rate, mtph	72.0	81.9
Propene, ppm V	1242	526
Tails Rate, mtph	36.9	43.4
Ethane, vol%	0.63	1.26
Column Pressure Drop, mBar	240	230
Average Delta P per Tray, mBar	4.29	4.11

TABLE 2: MD TRAY DESIGN COMPARISON

	Before Revamp (MD)	After Revamp (ECMD)
Diameter, mm	2736	2736
No. Trays	56	56
Tray Spacing, mm	450	450
Average Active Area	78.9	77.2
Receiving Pan Area, m^2	0.0	0.0
Average Weir Height, mm	90.0	88.6

Photographs of Typical MD Trays.

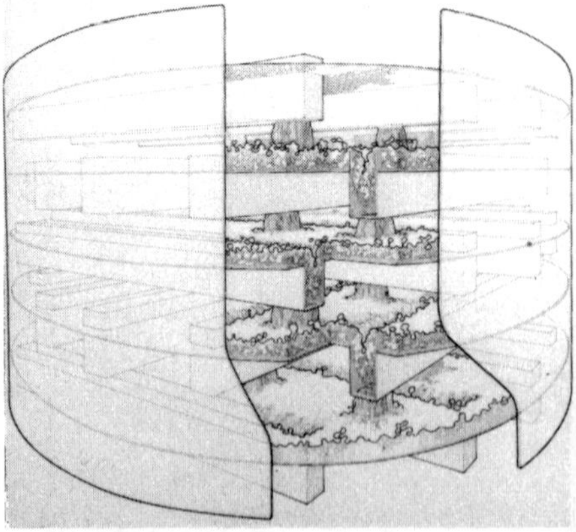

Figure 1: MD Tray

FIGURE 2: OeMV Deethanizer

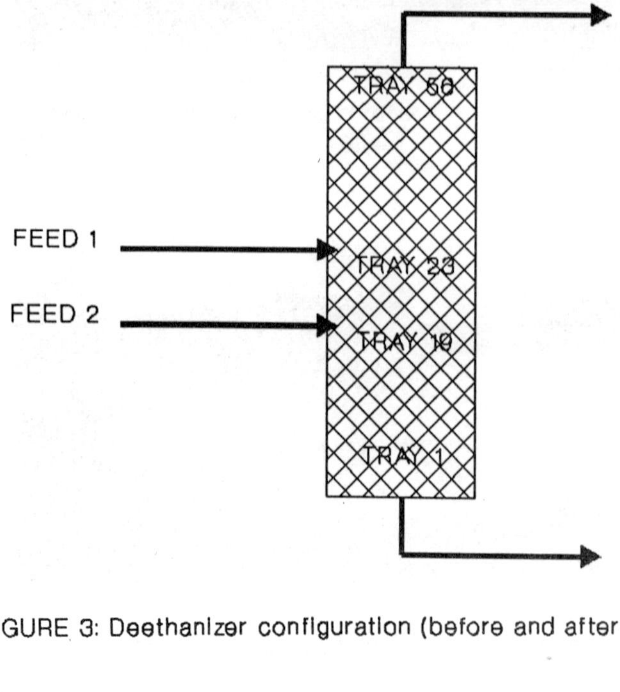

FIGURE 3: Deethanizer configuration (before and after revamp)

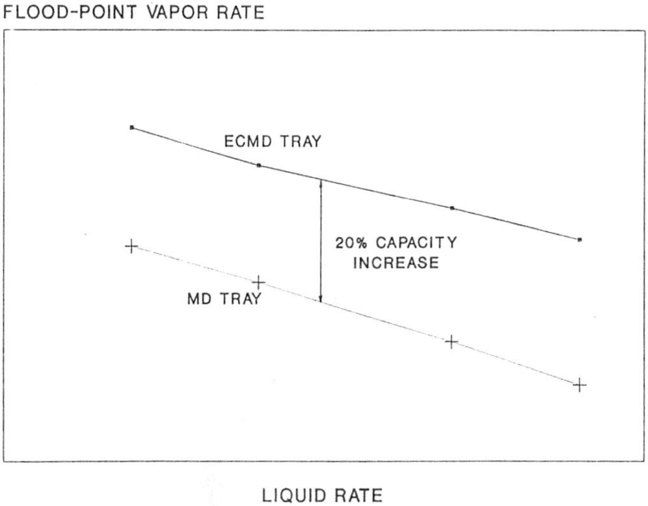

FIGURE 4: 2.4 m Test Column Results

INTRODUCTION

INFERENTIAL CONTROL STUDIES ON AN INDUSTRIAL COLUMN

D. Alkaya, C. Üzgen, T. Gürkan
Chemical Engineering Department, Middle East Technical University, 06531, Ankara, Turkey

Brosilow's inferential control scheme is utilized for an industrial, multicomponent, high-purity distillation column. A dynamic model called DAL is developed which includes all the nonidealites in VLE and enthalpy calculations. The steady-state simulation of the column is done with SIMSCI PROCESS package and the output obtained is used for the initial rating of DAL. The plant data is considered to correct dynamic simulation results for any model/plant mismatch. The selection of sensor location necessary for inferential control is done by using SVD (Singular Value Decomposition) analysis and TVA (Temperature Variation Analysis). The SVD analysis also revealed that the column can be controlled with single-end.

The transfer function relationships between inputs and outputs are found using DAL program by step testing. These are further used in the design of the estimator and IMC controller. The results are checked by the simulation studies on the column working with the control configuration based on 'eigenstructure' concept.

Keywords: Inferential control, unsteady-state modelling, distillation column control

INTRODUCTION

Inferential Control (Brosilow and Joseph, 1978) is used when there are problems with on-line measurement of controlled variables. Its implementation depends on the nature of the process and it requires a mathematical unsteady-state model of the process. In a distillation column composition profiles are closely related to the temperature profile. Thus the most important variables are the tray temperatures which provide an estimate of the product compositions. Accurate models are needed to select the trays at which the temperature measurement will be taken.

In recent years, a generalized model for the dynamic simulation of distillation columns is presented by Gani, Ruiz and Cameron (1986). The model is capable of solving problems related to operability of single and multiple columns and to column instability for different types of operations. Lagar, Paloschi and Romagnoli (1987) using the eigenvalues of the Jacobian corresponding to the equations describing the dynamic behaviour of a multicomponent distillation column implemented a policy to switch between nonstiff and stiff integration methods. Carling and Wood (1988) studied the dynamic behaviour of depropanizer by dynamic simulation program developed by Wong (1985) and modified by Carling (1986). Gani and Cameron (1989) coupled the steady-state simulation problem to an existing dynamic model for distillation columns. In the dynamic model of Luyben (1989) mass, and energy balances were solved for multicomponent distillation columns while the nonlinearities and the nonidealities on the VLE and enthalpy calculations are not considered.

In this study, the program of Luyben is used by including nonlinearities and nonidealities in VLE and enthalpy correlations considering the appropriate correlations given in the PROCESS package. The inferential control of the column is studied both by steady and unsteady state approaches.

EXPERIMENTAL STUDIES

The distillation column (Figure 1) studied is an industrial one existing in the Alkylation section of a Monomer Plant. It is aimed to separate EB (EthylBenzene) from MEB (MethylEthylBenzene) and DEB (DiEthylBenzene) with desired product compositions of 0.9980 and 0.01 with respect to EB at the top and at the bottom of the column respectively. The main specifications of the column are given in Table 1.

Steady-state and dynamic plant data were obtained from the industrial column by experimentation. Steady-state plant data were obtained under condition of typical operation. The dynamic plant data were obtained by giving positive or negative step and pulse changes to the feed flow rate, reflux and boilup rates. The experimental responses obtained are shown in Figure 2.

SIMULATION STUDIES

STEADY-STATE SIMULATION

It is done by using PROCESS Package. Chao-Seader correlation is chosen for phase equilibrium calculations due to its accuracy for light hydrocarbon mixtures. Curl-Pitzer method is used to predict both liquid and vapour enthalpies and the enthalpy deviations from ideality. The output from the simulation package consists of the temperature, pressure, compositions, vapour and liquid rate profiles along the column, and the heat duties.

The efficiency of the column is determined by trying to obtain the same compositions at the top and bottom of the column with those of the design case while changing the number of trays and feed tray location. The efficiency of the column is found to be 100 % by using the design data as inputs (Alkaya, 1990).

The steady-state simulation of the column is done by using column design data and column operating data. The inputs for simulation are selected to be feed temperature, its flow rate and compositions, reflux rate, number of trays and heat duties. The outputs are the flow rates and compositions of top and bottom products and all variable profiles throughout the column. The design and operating data and the outputs obtained by steady-state rating of DAL using some column operating and design data as inputs are given in Table 4. All the results of simulation are in good agreement with both design and operating data.

UNSTEADY-STATE SIMULATION

The unsteady-state computer program, DAL, which solves general equations of unsteady-state material, energy and momentum balances utilizes data file obtained from PROCESS package for the column as; feed flow rate, temperature, pressure, compositions; T, x, L profiles along the column, volumetric hold-ups on the column base and reflux drum, efficiency, pressure at the top and bottom of the column, reflux flow rate and heat input to the reboiler. Data for chemical components includes coefficients for the correlation used in the estimation of vapour pressure, liquid enthalpy, vapour enthalpy and density obtained from the property estimation section of the PROCESS package (Alkaya, 1990). The enthalpy calculations used in DAL are given below.

ENTHALPY CALCULATION

The enthalpy of a liquid mixture can be evaluated by using the equation:

$$h^{L} = h_{I}^{L} + \sum_{i=1}^{m} x_i \, \overline{\Delta h_i} \tag{1}$$

Here $\overline{\Delta h_i}$ is the partial molar enthalpy of component *i*, at temperature *T*, pressure *P*, and composition *x*, relative to the molar enthalpy of pure, ideal vapour *i* at the same temperature. The partial molar enthalpy for every component *i* is found from an appropriate form of the Gibbs-Helmholtz equation (Prausnitz *et al*, 1980). Thus, equation (1) can be written as

$$h^{L} = h_{I}^{L} - \sum_{i=1}^{m} x_i \, RT^2 \left(\frac{\mathrm{d} \ln f_i^{po}}{\mathrm{d}T} \right)_{pure} \tag{2}$$

where f_i^{po} is fugacity of pure liquid at zero pressure, at temperature T, ϕ_i^{s} is the fugacity coefficient at saturation temperature, T_s respectively.

The enthalpy of the real vapour is given by

$$h^{V} = h_{I}^{V} + \sum_{i=1}^{m} y_i \, \Delta h_{I}^{V} \tag{3}$$

where Δh_I^{V} is the molar enthalpy correction for a vapour at temperature T and pressure P relative to the ideal vapour at the same T and composition. This correction is given exactly by

$$\Delta h_{I}^{V} = \int_0^P \left(V - T \frac{\partial V}{\partial T}_{P,y} \right) \mathrm{d}P \tag{4}$$

where V is the vapour-phase molar volume. To evaluate the integral in equation 4, for non-associating mixtures, the virial equation is used truncated after the second coefficient.

The ideal enthalpies of the liquid and vapour phases are determined by the equation fitted by the package as

$$h_I = c_1 + c_2 T + c_3 T^2 + c_4 T^3 + c_5 T^4 \tag{5}$$

by using the appropriate constants for each phase given by the package.

The nonidealities in VLE calculations are handled using Chao-Seader correlation and for density and vapour pressure evaluation correlation similar to equation (5) are used.

In the program, first, pressure profile of the column is determined and then using the appropriate Subroutines, corrected temperatures and vapour compositions are calculated. Energy balance and enthalpy Subroutines are used to calculate vapour rates. Initial holdups are determined by a Subroutine which makes use of the Francis Weir Formula.

The differential hold-up values indicates whether there is a deviation between the calculated vapour rates and those obtained from the PROCESS package or not. When there are positive or negative deviations in the hold-ups, then the steady-state rating is initiated which comprises the first section of the code. In the steady-state rating, liquid rates are calculated by using total mass and enthalpy balances.

Although all the coefficients for the property estimation of components and initial steady-state rating values are obtained from PROCESS package when they are used in DAL a small difference in initial temperature values can be seen. This causes some deviations in the results. Therefore, the initial steady-state rating is further processed in DAL to generate its own steady-state data. The final steady-state attained is checked by giving a disturbance to Q (–20 % as given in experiment) in both DAL and in PROCESS Package. The final steady-state values obtained are in good aggreement as shown in Figure 3. This proves that dynamic simulation works as well as steady-state PROCESS simulation considering initial and final steady-state values.

Liquid compositions are calculated using component balances and calculated values of liquid and vapour rates, temperature and vapour compositions. If the error between composition values calculated by PROCESS package and DAL program are less than 0.0001 the program goes to the unsteady-state section of the program. Otherwise, it returns to the initial estimation of the temperature and vapour compositions.

In the dynamic analysis, changes to selected inputs should be given before the start of the unsteady-state simulation. Then, for each input, vapour and liquid rates, hold-ups on each tray, vapour and liquid compositions and temperature profiles are obtained with respect to time until they reach their final steady-state values.

MODEL-PLANT MISMATCH

The last comparison is made between the response obtained by DAL and the plant data. This is the most important check usually missing in most of the previous studies (Finco and Luyben, 1989). The response data obtained by using DAL program for bottom temperature is drawn with respect to time on Figure 4 for a 20 % decrease in Q together with the response obtained from the industrial column The response of the bottom of the column is considered for comparisons because of difficulty of experimental detection of changes in the upper part of the high purity column. It is observed that the responses obtained by using DAL are faster than the response of the experiments. Thus, the hydraulic time constant has to be adjusted to match the plant dynamic data as it was done by Finco and Luyben (1989). It was observed that the time constant of the bottom hold-up drum has more affect on the system response. Different values of the time constants for the bottom hold-up drum were used to match the curves as shown in Figure 4. The best match is obtained with the bottom hold-up drum time constant of 3 hour.

Using SVD analysis, 2nd and 8th trays are selected for sensor location. This selection is based on the lowest CNs obtained for different pairs. The CN values for different pairs are shown in Table 3. In the table, some pairs such as 2-7, 2-6, 3-7 also have low CN values. However, TVA analysis which necessitates the information about the tray temperature on the face of disturbances indicate that trays 6 and 7 have inverse response behaviour as shown in Figure 5. (The disturbance is given to the EB concentration as –3 %.) Thus, the final selection is based on both SVD and on TVA (Alkaya, 1991)and trays 2 and 8 are selected.

The application of the eigenstructure concept of Luyben (1988) to the column showed that the single end control of the column is possible by changing reflux with feed flow rate and controlling the column by manipulating the heat input on the face of the feed concentration changes. It is found that the top section temperatures of the column are not sensitive to the composition changes introduced to the column as shown in Figure 5. So, in the design of the estimator and controller based on the IMC concept, EB concentration change at the bottom is considered and control is achieved accordingly. Since the transfer functions representing the column are necessary for the design of both estimator and the controller, they are obtained by step testing and are shown in Table 4.

The estimator is designed to infer the effect of unmeasured disturbances on the primary output from frequent measurements of the secondary outputs. The equations needed for the estimator design are given below (Desphande, 1985):

$$x_B(s) = b^T(s)z_F(s) + c(s)Q(s) \tag{7}$$

$$T(s) = A^T(s)z_F(s) + P(s)Q(s) \tag{8}$$

and using equations (7) and (8) the following equation can be written

$$x_B(s) = \alpha_i^T(s)\,(T(s) - P(s)Q(s)) + c(s)m(s) \tag{9}$$

where $\alpha_i(s) = \alpha_i(0)\,\dfrac{\tau_{1g}(s) + 1}{\tau_{g1}(s) + 1}$ and $\alpha_i(0) = (A^TA)^{-1}A^Tb$.

Here $\alpha_i(0)$ is the static estimator, τ_{1g} the average of all the time constants in the matrix A corresponding to the measurement i, τ_{g1} the average of all time constants in matrix b that corresponds to primary process output x_B. By using the corresponding matrices from Table 4 the estimator design takes the following form

$$\alpha(s) = \begin{vmatrix} \dfrac{-0.0302(3.62\,s+1)}{(2.984\,s+1)} & \dfrac{0.0199(3.642\,s+1)}{(3.4835\,s+1)} \\ \dfrac{-0.0426(1.178\,s+1)}{(2.984\,s+1)} & \dfrac{0.0285(1.178\,s+1)}{(3.4835\,s+1)} \end{vmatrix} \tag{10}$$

The controller that is used in the inferential scheme, is designed using IMC rules as suggested by Desphande (1985) and is given below

$$G_c(s) = -c(s)^{-1} = (1.897)^{-1}\,\frac{1.156\,s+1}{0.1156\,s+1} \tag{11}$$

The response of the single-end controlled column to feed concentration change of 3 % is given in Figure 6 in terms of bottom EB concentration using estimator and controllers given in equations (10) and (11).

CONCLUSIONS

A good representative model for the process is obtained using the dynamic model DAL by considering nonlinearities and nonidealities in VLE and enthalpy calculations. It is found that the column control can be achieved by using the 2nd and 8th tray temperatures, which are selected by the utilization of SVD and TVA analysis together. The estimator design is done with the help of transfer functions of the column and the IMC controller worked well for the single-end controlled column suggested by the 'eigenstructure' concept.

REFERENCES

1. Alkaya, D., 1990, MS Thesis, Middle East Technical University.
2. Alkaya, D., Üzgen, C. and G ürkan, T., 1991, *AIChE Annual Meeting, Los Angeles.*
3. Brosilow, C. and Joseph, B., 1978, *AIChE J*, 24: 485–508.
4. Carling, C., 1986, *Proc IFAC Symp on Dynamics and Control of Reactors and Distillation Columns*: 167–173.
5. Desphande, B.P., 1985, *Distillation Dynamics and Control*, Instrument Society of America: 443–459.
6. Gani, R., Ruiz, C.A. and Cameron, I.T., 1986, *Computers and Chem Engng.*
7. Gani, R. and Cameron, I.T., 1989, *Computers and Chem Engng.*
8. Lagar, G., Paloschi, J. and Ramagnoli, J.A., 1987, *Computers and Chem Engng*, 4: 383–394.
9. Luyben, W.L. and Finco, M.V., 1989, *Ind Eng Chem Res*, 28: 75–83.
10. Luyben, W.L., 1988, *Ind Eng Chem Res*, 27: 208–210.
11. Luyben, W.L., 1989, *Process Modelling Simulation and Control for Chemical Engineers*, McGraw-Hill, New York.
12. Prausnitz *et al*, 1980, *Computer Calculations for Multicomponent Vapour-Liguid and Liquid-Liquid Equilibria*, Prentice Hall.

Table 1. Column Specifications and Operating Conditions

flow rates, kgmol/sec x 10^3		compositions , mol fraction of EB	
feed	7.40	distillate	0.9980
distillate	7.05	bottoms	0.2625
bottoms	0.35	feed	0.9513
reflux	6.75		
feed temperature, K	439.1	number of trays	50
feed pressure, kPa	206.8	feed tray location	21
reboiler duty, W x 10^5	4.556	diameter,m	1.066
condenser duty, W x 10^5	-4.816	weir height,m	0.076

Table 2 Simulation Results with Design and Operating Data

	DESIGN		OPERATING	
	Design	SIMULATOR	Experiment	SIMULATOR
$F^{\#}$(kgmol/sec 10^3):	10.166		7.408	
B (kgmol/sec 10^3):	0.526	0.530	0.347	
$D^{\#}$(kgmol/sec 10^3):	9.639		7.061	
$L^{\#}$(kgmol/sec 10^3):	7.042		6.750	
$y_{D,EB}$:	0.9985	0.9983	0.9980	0.9980
$y_{D,MEB}$:	0.0013	0.00017	0.002	$2.372\ 10^{-3}$
$y_{D,DEB}$:	0.0001	4.515 1	—	$6.826\ 10^{-7}$
$x_{B,EB}$:	0.0024	0.0007	$1\ 10^{-4}$	$2.417\ 10^{-3}$
$x_{B,MEB}$:	0.1391	0.1324	0.2084	0.2009
$x_{B,DEB}$:	0.8585	0.8608	0.7915	0.7965
T_B (K) :	485.8	474.4	475.3	473.9
Q_C (10^5 W) :	-8.140	-8.240	—	-4.816
Q_R (10^5 W) :	7.380	6.210	—	4.556

inputs of the program

Table 3 Condition Numbers Between Corresponding Trays

Tray	1	2	3	4	7	8
1				40.1477		
2						
3						
4		21.1730	48.6160			
5	28.3902	21.1730	25.1126	49.9796		
6	22.1893	16.4406	18.3786	29.0079		
7	18.4204	13.8738	15.7138	23.5378		
8	15.0754	12.4146	14.9465	22.6355		
9	12.8280	12.0292	15.9855	25.2853		
10	11.8084	13.3715	15.9724	33.6443	1361.0	442.056
11	12.9795	18.0976	30.1785	55.5672		

Table 4 Transfer Function Relations

	T2	T8	$x_{B,EB}$	$x_{B,DEB}$
Z_F	$\frac{410.657\ e^{-1.53s}}{3.753\ s + 1}$	$\frac{800.0\ e^{-0.126s}}{1.21\ s + 1}$	$\frac{-0.396\ e^{-1.449s}}{2.34\ s + 1}$	$\frac{2.154\ e^{-0.542s}}{4.544\ s + 1}$
Q_R	$\frac{833.33\ e^{-0.365s}}{0.663\ s + 1}$	$\frac{928.3}{0.359\ s + 1}$	$\frac{-1.897\ e^{-0.794s}}{1.156\ s + 1}$	$\frac{2.462\ e^{-0.313s}}{0.015\ s + 1}$

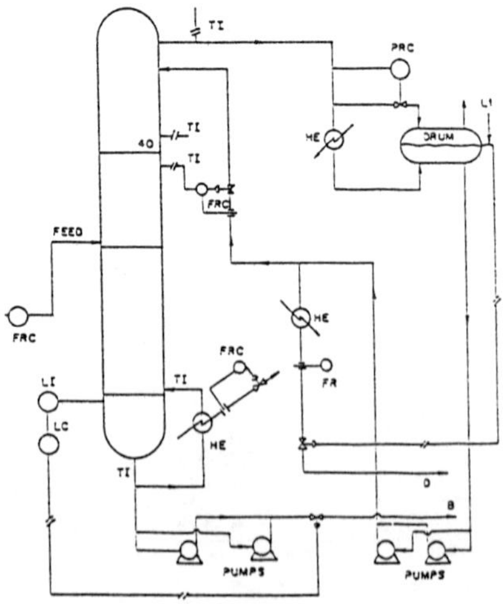

Figure 1. Experimental Ethylbenzene (EB) column

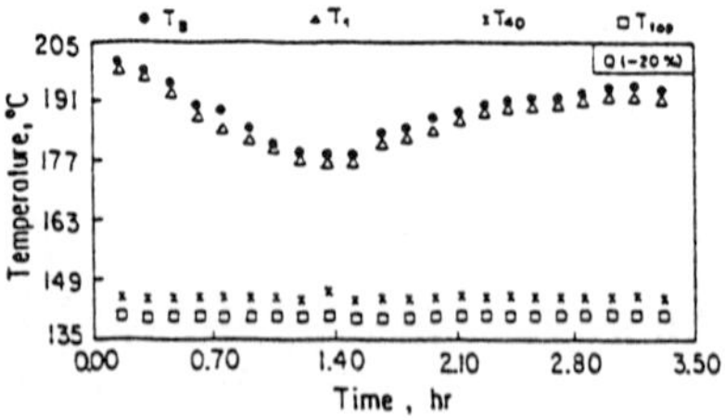

Figure 2. Column experimental response to pulse change in Q

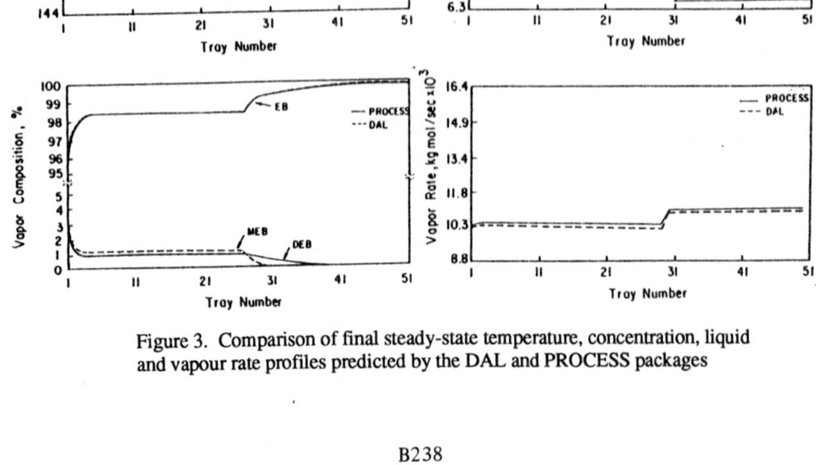

Figure 3. Comparison of final steady-state temperature, concentration, liquid and vapour rate profiles predicted by the DAL and PROCESS packages

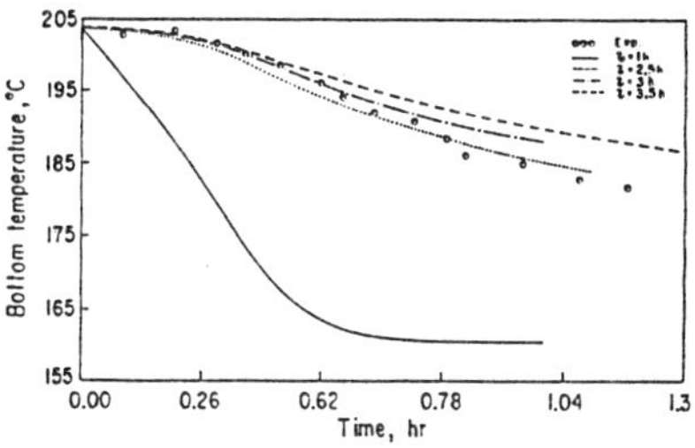

Figure 4. Model/plant mismatch with bottom temperature

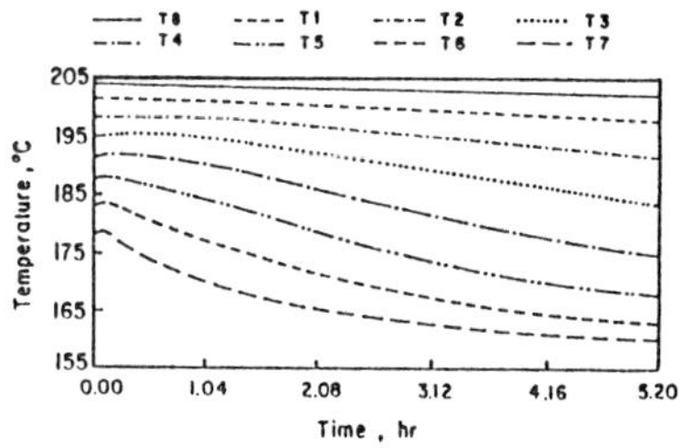

(a)

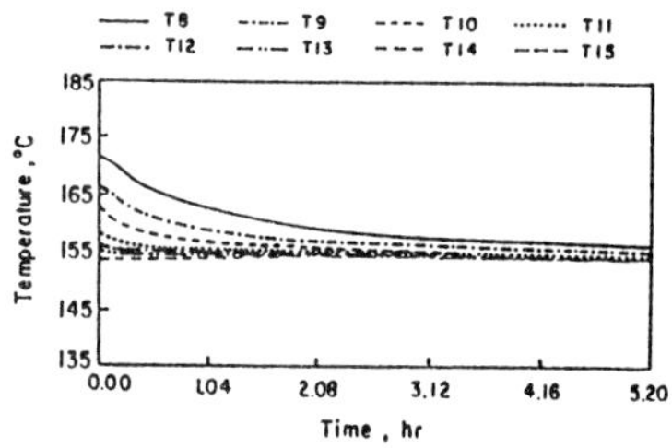

(b)

Figure 5. Temperature profiles for feed disturbance

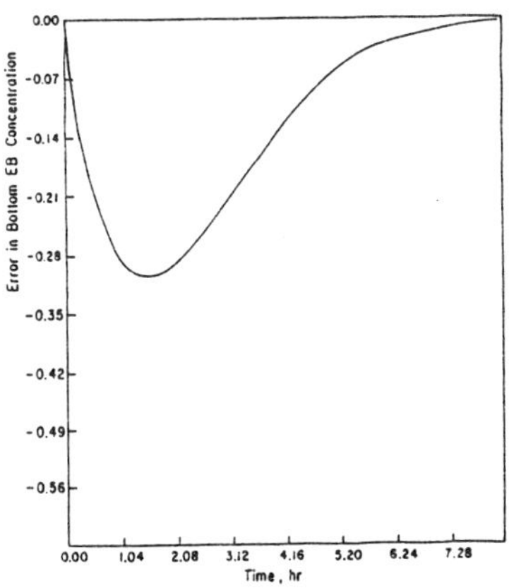

Figure 6. Responses of the single-end controlled column

BENEFITS FROM THE RIGOROUS MASS TRANSFER RATE APPROACH TO MODELLING ACID GAS ABSORPTION USING MIXED AMINES

D.J. Vickery*, J.T. Adams* and R.D. Wright**
* ChemShare Corporation, Houston, Texas, USA
** Amoco Production Company, Houston, Texas, USA

The solution of absorption and distillation problems has traditionally used the equilibrium stage concept, in which the bulk vapour and liquid phases are assumed to be in equilibrium. Tray efficiency is the empirical parameter used to relate this model to the actual equipment. This paper describes the technical issues involved in using the more fundamental and rigorous Mass Transfer Rate method for distillation and absorption design and modelling and compares performance predictions using this method with actual gas sweetening plant operating data.

INTRODUCTION

In recent years, interest has grown in a more fundamental and rigorous method — the 'Mass Transfer Rate' approach in which mechanical, chemical and thermodynamic equilibrium are assumed to exist only at the fluid interface. Rates of heat transfer and component mass transfer are determined taking into account the mass and thermal diffusion resistances as well as any chemical reactions taking place in the liquid and vapour phases. Clearly, the increased rigour of the Mass Transfer Rate method is only achieved at the expense of providing or predicting additional physical and transport properties as well as heat and mass transfer coefficients. However, there has been hitherto relatively little published on the practical benefits of applying this more fundamental approach to separation modelling.

Design and specification of absorption and distillation processes almost always involves application of commercial or proprietary models of separation process equipment. Using today's technology, that separation model is generally based upon the equilibrium stage model. In the equilibrium stage model, vapour and liquid are assumed to enter a tray or cross section of packing in a column, exchange matter and energy and leave in equilibrium with each other. However, columns rarely, if ever, operate at equilibrium. In fact, distillation and absorption are, by their nature, non-equilibrium processes: mass and energy are transferred across the fluid interface at rates which depend upon the extent to which the phases are **NOT** in equilibrium with each other. Historically, this departure from equilibrium conditions has led to the use of efficiencies. Efficiencies are meant to compensate for the deviation of actual column operation from the equilibrium assumed in the column model.

It is possible to look at the processes occurring in trayed and packed columns with a higher degree of rigour. Consider a stage model where vapour and liquid enter a tray and material and energy transfer between the two fluids through the interface between those fluids. Each phase provides resistance to that transfer which can be modelled using a Fickian type rate expression. In this 'Mass Transfer Rate' model, it will be assumed that equilibrium exists only at the fluid interface. This interface is a singular surface which provides no resistance to mass or heat transfer and does not accumulate energy or matter. The resistance to mass transfer is limited to a thin film on each side of the fluid interface. Mass and heat transfer through this region can be modelled using the film theory of heat and mass transfer. Chemical reaction may occur in either or both of the two fluid phases.

The equations describing this model of separation processes may be termed the **MERQ** equations. That is:

- **M**aterial balance equations
- **E**nergy balance equations
- **R**ate equations
- e**Q**uilibrium relations

The model uses detailed information about the column construction and characteristics along with fluid composition, flow and physical property information determined during the simulation of the separation process to calculate mass transfer coefficients for each component in each fluid phase, heat transfer coefficients, interfacial area available for material and energy transfer, phase equilibrium ratios at the interface and column hydrodynamic parameters such as degree of flooding and pressure drop. When chemical reactions are occurring, the model is capable of describing the effect the reactions have on mass and heat transfer rates. Complete descriptions of the 'Mass Transfer Rate' or 'Non-equilibrium Stage' Model can be found in Krishnamurthy *et al* (1985) and Sivasubramanian (1985).

AMINE BASED GAS TREATING

Amine treating of natural gas to remove acid species CO_2 and H_2S is an important industrial application of absorption technology. Unfortunately, significant differences in both the equilibrium and the kinetic effects for H_2S and CO_2 can make accurate design and modelling by traditional methods difficult. Furthermore, the magnitude of these effects varies substantially from primary to secondary to tertiary amines. One positive consequence of these effects is the potential to use a blend of amines (usually two), which is tailored to obtain the degree of separation while maintaining minimum regeneration energy needs. Such an optimization of both the amine blend and the detailed equipment configuration can only be realistically performed with the help of a rigorous Mass Transfer Rate approach to the calculations.

Removal of H_2S and CO_2 from natural gas streams is an important application of chemically reactive solvents in absorption practice. There are often limits placed upon the levels of H_2S and CO_2 which may be in gas streams. These restrictions may be due to economic considerations such as reducing the level of corrosion of gas pipelines, to environmental regulations such as limiting the amount of sulphur released during combustion of the gas or to consumer requirements such as minimum gas heating values. Unfortunately, both H_2S and CO_2 tend to have low physical solubilities in most acceptable (physical) solvents. To compensate for this, two general approaches are available. The first approach is to operate under conditions where physical solubility is enhanced — eg high pressure and low temperature, the second is to use a solvent which enhances the acid gas solubility by reacting with it — providing a sink for the acid gases.

Among the chemically reactive solvents is the commercially important family of alkanolamines. These solvents can be thought of as substituted ammonia molecules. The type of alkanolamine is determined by the number of substitutions on the nitrogen atom. For example, if one hydrogen atom on the nitrogen is replaced with a functional group, the molecule is termed a primary amine. If two hydrogens are replaced the molecule is a secondary amine, and if all three hydrogens are replaced, it is a tertiary amine. Examples of primary amines are monoethanolamine (MEA) and di-glycolamine (DGA); diethanolamine (DEA) and di-iospropanolamine (DIPA) are examples of secondary amines, while methyl-diethanolamine (MDEA) and triethanolamine (TEA) are examples of tertiary amines.

The most important single difference between primary or secondary amines and tertiary amines is that tertiary amines do not directly react with CO_2 to form carbamates. At most, they catalyze the hydrolysis of CO_2 to bicarbonate. This fact leads to varied applications of the classes of amines. When deep removal of CO_2 is required, primary or secondary amines such as MEA and DEA are typically applied. On the other hand, when CO_2 concentrations are already within or near tolerance, tertiary amines can selectively remove H_2S while picking up a minimum of CO_2.

This selectivity arises from two sources. First, MDEA is more **thermodynamically selective** for H_2S removal than primary and secondary amines because MDEA does not react directly with CO_2. That is if we allow a mixture of equal moles of H_2S and CO_2 to sit over an aqueous solution of MDEA for a very long period of time, more of the H_2S will be absorbed than CO_2. This can be seen by examining the data of Jou *et al* (1982). Furthermore, MDEA is **kinetically selective** for the removal of H_2S. That is if we plot the amount of H_2S and CO_2 in the liquid phases of the system described above, we will see that initially far more H_2S will be absorbed than CO_2. In fact, the amount of H_2S absorbed at a given intermediate time may be greater than the final equilibrium concentration it achieves. As the CO_2 is absorbed, it may replace some of the H_2S and cause its desorption, though less CO_2 will be absorbed than with primary or secondary amines. Thus an equilibrium competition between kinetic and thermodynamic effects arises.

To a degree, the competition between chemical kinetics and thermodynamics allows one to exercise some control over the selectivity of MDEA towards hydrogen sulphide over carbon dioxide by adjusting the gas-liquid contact time. This can be done through appropriate choice of tower hydraulics including liquid depths on trays, solvent circulation rates, packing versus trays and so on. In fact, the very short liquid residence times in packed towers recommends the use of packed equipment for increased selectivity when these packed columns are designed using reliable methods. Higee promises to give degrees of selectivity well beyond what can be reached using conventional technology (Bucklin *et al*, 1987). However, in many applications one finds that MDEA is too selective, ie, it allows too much carbon dioxide to slip through the absorber, while diethanolamine (DEA) and monoethanolamine (MEA) are not selective enough. Lack of sufficient selectivity causes too much carbon dioxide to be removed and pushes up solvent regeneration costs. The best solvent is one with exactly the right selectivity. This can be produced by blending a primary or secondary amine with a tertiary amine such as MDEA in just the right proportions.

Another advantage of tertiary amines is the relatively weak bonding that occurs with CO_2. This means that regeneration to remove CO_2 from MDEA is more economical than with commercially used primary and secondary alkanolamines.

THE RATE APPROACH TO COLUMN MODELLING

Traditionally, all tower calculations produce either the number of equilibrium stages needed to perform a specified separation or the separation achieved by a specified number of ideal trays. The translation from real trays to equilibrium stages and back again depends on tray efficiency information.

MATERIAL BALANCE

Equilibrium stage methods apply component and energy balance calculations around each complete ideal stage in a tower. In contrast, Mass Transfer Rate based modules such as GasPlant and GasPlant-Plus consider component and energy balances around each phase present on every real stage of the column (or, in the case of packed towers, in each horizontal slice of the tower). The individual phase balances are linked by the requirements that (i) the mass transfer rate of each chemical component and energy transfer rate from one phase must equal the respective transfer rate into the other and (ii) the phases are in mechanical, thermal and chemical equilibrium at the interface between them. There is no requirement that the bulk phases themselves be in equilibrium; in fact, departures from equilibrium are required for mass transfer (ie, for a separation) to take place.

MASS AND ENERGY TRANSFER RATES

Mass and energy are transferred from one phase to another at rates that depend on individual film coefficients for mass and heat transfer and on concentration and temperature driving forces for the transfer. Both the gas and liquid phases are important in determining the overall transfer rates because both phases offer resistance to transfer. This approach is completely analogous to the one used in determining heat transfer rates in heat exchangers. There, one uses correlations for shell-side and tube-side heat transfer coefficients and logarithmic mean temperature differences to calculate heat transfer.

Mass-transfer-coefficient correlations are available for various types of trays; for dumped packing of various sizes, materials and styles; and for certain structured packings. These coefficients depend on physical and transport properties, flow rates, temperature, pressure and type of equipment, including such details as the packing type, size, material and number of redistributors and the tray type and its construction details. These coefficients are then used with the differences in chemical potential in each film to determine the rates of mass transfer in the separation equipment.

Additionally, energy is transferred between the contacting phases. Two phenomena contribute to the energy transfer; conductive transfer due to temperature gradients and convective transfer of energy within the transferring molecules. As was the case with the phase equilibrium, thermal equilibrium is assumed to exist only at the fluid interface. The bulk vapour and liquid may differ significantly in temperature. This is commonly observed at the bottom of absorption towers and on the feed tray of the regenerator. The difference in temperature of the vapour and liquid exiting the bottom tray of an MEA or DGA contactor can be 80°F or more.

PHASE EQUILIBRIUM

Phase equilibrium plays an important role in rate models just as it does in equilibrium stage calculations. In a rate model, phase equilibrium is assumed at the fluid-fluid interface, not between the bulk phases leaving a tray or packed section. Thus, phase equilibrium is one of the key factors in setting driving forces for mass transfer and, therefore, in determining the separation that is actually achieved on a real tray or within an actual cross section of packing. The liquid-phase activity coefficient model used within GasPlant and GasPlant-Plus is based on the extended Debye-Huckel model of Edwards, Newman and Prausnitz (1975) as described by Deshmukh and

Mather (1981). Because the model is thermodynamically sound, it has been possible to extend it to mixed amine systems. These extensions have been described in some detail by Chakravarty *et al* (1985) and Vickery and Weiland (1986, 1987). Predictions of the thermodynamic model have been checked against limited experimental data.

REACTION EFFECTS AND ENHANCEMENT FACTORS

Liquid-phase chemical reactions always increase mass transfer rates. Reaction effects are usually expressed through enhancement factors. Unlike tray efficiencies and HETPs, enhancement factors are not empirical constants to correct for the effects of chemical reaction. They are very well defined parameters that are calculated from the knowledge of the chemical kinetics, the reaction stoichiometry, liquid composition and a few physical and transport properties of the components in the liquid. Enhancement factors depend, therefore, on such parameters as temperature, amine concentrations, acid gas loadings, kinetic constants and so on. Reaction models that correctly account for the way that two amines interact with each other during the process of absorbing the acid gases are used for blended amines. These models are not empirical; they are based on knowledge of the actual kinetics involved.

The major reactions which occur within the H_2S - CO_2 - alkanolamine system are:

1)	Dissociation of water;	H2O	$\rightleftarrows H^+ + OH^-$
2)	Protonation of amine;	Am + H2O	$\rightleftarrows AmH^+ + OH^-$
3)	Formation of bisulfide;	H2S	$\rightleftarrows H^+ + HS^-$
4)	Dissociation of bisulfide;	HS^-	$\rightleftarrows H^+ + S^=$
5)	Formation of bicarbonate;	H2O + CO2	$\rightleftarrows H^+ + HCO3^-$
6)	Dissociation of bicarbonate;	$HCO3^-$	$\rightleftarrows H^+ + CO3^=$
7)	Formation of carbamate;	2Am + CO2	$\rightleftarrows AmH^+ + Carb^-$

It is reaction (7), the formation of carbamate, which occurs only with primary and secondary amines.

Reactions (1), (2), (3), (4) and (6) involve only the transfer of a proton and are much faster than reactions (5) and (7). In general, these reactions can be assumed to have achieved equilibrium in the bulk liquid phase. The formation of the carbamate is a fast reaction and also tends to achieve equilibrium in the bulk liquid. Thus in each of these cases, the resistance to mass transfer lies primarily in the vapour film. Absorption under these conditions is said to be vapour phase controlled.

However, reaction (5) is a slow reaction. For tertiary amine systems, it is generally this reaction which determines the transfer rate of the CO_2. For this reason, CO_2 absorption in tertiary amines is generally liquid phase controlled.

EXAMPLE 1 — AMOCO'S WASSON CO_2 RECOVERY PLANT

Amoco's Wasson CO_2 removal plant is a part of their gas treating facilities in Wasson, Texas. CO_2 is used for tertiary recovery in fields of West Texas. The gas from the fields is brought to the plant, compressed and H_2S is removed using an iron chelate process. The remaining gas is dried by contacting it with a TEG solution. The dry gas then proceeds to a Ryan-Holmes column. The bottoms stream forms an NGL product and the resulting gas which is upwards of 50% CO_2 is sent

	Sour Gas	Lean Solvent
Temperature	73° F	133° F
Pressure	814 psia	820 psia
Composition		
Nitrogen	2.28 %	
Carbon Dioxide	51.80 %	0.08 loading
Methane	36.96 %	
Ethane	8.96 %	
MDEA		32 wtpct

Table 1. - Typical Feed Conditions for Wasson MDEA Absorber

to an MDEA loop for bulk CO_2 removal. The recovered CO_2 is used for reinjection into the fields.

The bulk CO_2 removal system employs MDEA to remove the large amounts of CO_2 from the gas stream. The lower regeneration requirements of MDEA as opposed to MEA or DEA means that the typical amine stripper can be replaced with a two stage flash process. The pressure drop within the flashes is sufficient to regenerate the amine solution.

The absorption column is a 60 foot packed tower, 9 feet in diameter. Two redistributors are located in the column, one at 20 feet and one at 40 feet. The packing is a 1.5 inch plastic ring. Inlet pressure of the CO_2 rich gas is approximately 800 psi. Typical feed stream conditions to the absorber are given in Table 1.

The lean solvent is already fairly rich in CO_2. Thus the exiting gas is not 'sweet' in the traditional sense. Typical sweet gas CO_2 content is 5% CO_2. The column operates so that it is equilibrium limited at the bottom of the column. The idea is to bring the rich CO_2 loading up until it is in equilibrium with the entering sour gas. In order to achieve this, the column must have many equivalents of a theoretical plate. Also, the higher the column operating pressure the greater the CO_2 pickup. Thus low pressure drop and a large total liquid holdup are desired. These considerations lend themselves well to packed column operation.

Since the column is equilibrium limited, the major benefits of the rate model for this application are to determine the minimum amount of packing required to achieve equilibrium, the ability to look at column profiles and being able to model a packed column rather than using HETP and/or efficiencies. An equilibrium stage model will do almost as well predicting rich solvent and sweet gas compositions. However, due to the assumption of equal vapour and liquid temperatures in a cross sectional area of packing, the equilibrium model may miscalculate the exiting temperatures.

If one were to try to optimize this process in some way, there are three variables which strongly influence the CO_2 pickup in the absorber. The first is the exit pressure of the second stage flash; the second variable is the temperature of the lean solvent stream and the last is the solvent capacity.

The two flashes act as the 'regenerator' for the process. The first flash acts to remove a significant amount of the CO_2 but it is the second flash that finally determines the amount of CO_2

released. Both flashes yield a flash gas of approximately 93% CO_2 so the strongest influence of the solvent loading is the flash pressure. The lower the pressure, the lower the back pressure on CO_2 over the solution and, hence, the lower the CO_2 load in the solvent. Doubling the flash pressure can significantly increase the solvent loading.

The second factor, the lean solvent temperature, affects the results by shifting the bicarbonate and carbonate equilibrium points and Henry's constant. For example, the following data for the Henry's constant of CO_2 in pure water and the reaction dissociation constants for the bicarbonate and carbonate formation reactions are taken from Edwards *et al* (1978).

$$\ln(H_{CO2}) = -6789.04/T - 11.4519*\ln(T) - 0.010454*T + 94.491$$

$$\ln(K_5) = -12092.1/T - 36.7816*\ln(T) + 235.482$$

$$\ln(K_6) = -12431.7/T - 35.4819*\ln(T) + 220.067$$

where T is Kelvin.

This gives $\ln(K_5)$ a maximum at 132°F, $\ln(K_6)$ a maximum at 171°F and $\ln(H_{CO2})$ a maximum at 308°F. These factors interact and it is not immediately obvious how inlet lean solvent temperature will affect equilibrium.

The final factor, the solvent capacity is determined by the circulation rate and the solvent strength. Increasing either will give a greater solvent capacity and allow increased CO_2 pickup. Increased circulation rate will also affect the energy balance by increasing the amount capacity of the solvent to accept heat.

The effect of changing each of these parameters on sweet gas CO_2 content for a given circulation rate and sour gas feed is shown in Table 2. Note that contrary to what may have been expected, the CO_2 pickup increases as the lean solvent temperature increases. Since the absorber

Solvent Strength (wtpct)	Lean Solvent Temperature (°F)	2nd Stage Flash Pressure (psia)	Sweet Gas CO2 (Mole pct.)
30	133	33	22.8
30	133	20	7.8
31	133	33	21.8
31	133	20	5.9
32	133	33	20.9
32	133	20	3.5
32	128	33	22.5
32	128	20	7.1
32	138	33	18.9
32	138	20	3.3

Table 2. - Effect of Process Operating Parameters on Sweet Gas CO2 Content

achieves equilibrium, the increase is not due to faster reaction kinetics but rather to the shift in equilibrium conditions for reactions (5) and (6). This shift is enough to compensate for the reduced solubility of the gases in the solvent. For other operating conditions, this may not be the case.

EXAMPLE 2 — MARATHON'S CODY WYOMING PLANT

During the 1987 Laurance Reid Gas conditioning conference, Joseph Harbison of Marathon Oil Company and Glenn Handwerk presented operating data on a gas plant Amoco was operating in Cody, Wyoming. They had a selective gas treating system which normally ran with 25–30% MDEA. Somehow, the system had become contaminated with DEA. While this had an adverse effect on the selectivity of the process, it provided an excellent source of operating data for the mixed amine system.

The detailed contactor balance from June 21, 1986 in Table 3 shows the following sour and sweet gas component mole fractions for the sour and sweet gas streams on a dry basis.

While the corresponding rich and lean amine compositions do not quite close a material balance, they are good enough to show the affect of the column operating parameters on process conditions. Among the parameters which will be examined are the tray active area, weir height, number of functioning trays and solvent composition.

Figure 1 shows the affect of relative DEA and MDEA solvent strength on the sweet gas composition. The simulations were run using a total solvent strength of 26 wtpct. The most obvious feature of the figure is that as DEA strength increases, CO_2 content of the sweet gas decreases but H_2S content increases. This is due to competition between reaction kinetics and equilibrium as described earlier. The lean solvent entering the column encounters a gas which is fairly low in H_2S content (1000 ppm) but rich in CO_2 (14%). The fast kinetics of the H_2S reactions with both DEA and MDEA allow the H_2S to be absorbed just as fast as it can diffuse through the gas side

Component	Sour Gas	Sweet Gas
N2	1.145	1.303
CO2	19.202	14.383
H2S	2.078	0.022
C1	58.891	64.489
C2	8.551	9.472
C3	4.938	5.259
i-C4	1.163	1.324
n-C4	2.098	2.408
i-C5	0.765	0.905
n-C5	0.634	0.771
C6+	0.535	0.664

Table 3. - Cody Wyoming Contactor Gas Stream Compositions - 6/21/86

film. On the other hand, the rate at which the CO_2 can be absorbed is controlled not only by the gas phase diffusion, but by a combination of the kinetics of the slower bicarbonate formation reaction and the amount of free DEA available to react with CO_2 in the faster carbamate formation reaction. As the DEA strength increases, the capacity for carbamate formation increases and it is possible to remove CO_2 at a faster rate.

This increased rate of CO_2 removal has a negative impact on the H_2S pickup. Since the solvent was able to pick up more of H_2S initially, it achieved a relatively high loading of H_2S quickly. Then as more CO_2 is picked up, there is greater competition between the CO_2 and H_2S to remain in solution. Eventually, some of the H_2S actually desorbs and returns to the vapour phase. This shows up as a maximum in the H_2S composition in both the vapour and liquid. In essence, an internal recycle of H_2S develops. In the case of the Cody plant, entering sour gas H_2S content is 2.078% on a dry basis or 2.066% on a wet basis. The internal maximum H_2S content is 2.18% two trays above the bottom tray. In the same way, the maximum H_2S loading in the solvent occurs 3 trays above the bottom tray and is a value of 0.167 mole H_2S / mole amine compared to a rich loading of 0.156 mol/mol. This internal recirculation reduces the number of trays available for H_2S removal. With no DEA present, the CO_2 pickup drops and the internal H_2S recirculation is not observed.

Similar effects can be seen if straight MDEA is run through the same column configuration but with a varying number of trays. Figure 2 shows how the number of trays affects the H_2S and CO_2 content in the sweet gas. With 8 trays, no internal H_2S recirculation is noted. However, with 18 trays, the H_2S content of the gas reaches a maximum of 3.32 mole percent — a 60 percent increase over the inlet composition. Note that to reach the same H_2S content as the 10 tray mixed amine absorber requires 16+ trays. The same level of CO_2 removal as seen in the mixed amine example can be achieved with approximately 12 trays.

The amount of CO_2 pickup can also be controlled by adjusting the active area of the tray or the weir height. Reducing the active tray area or lowering the weir height has the effect of reducing the 'volume' of the tray and, in turn, the residence time of the liquid on the tray. This decreases the degree to which the bicarbonate formation reaction goes to completion and the overall CO_2 pickup. Conversely, increasing active area and/or weir height increases residence time and CO_2 pickup and reduces H_2S pickup. The effect of the weir height and active area on the dry sweet gas content for the mixed amine system are shown in Tables 4 and 5.

Active Area (ft^2)	H2S (ppm)	CO2 (mole %)
4.88	466	14.03
5.135	496	14.00
5.39	535	13.97

Table 4. - Affect of Active Tray Area on Sweet Gas Composition

Weir Height (Inches)	H2S (ppm)	CO2 (mole %)
1.75	364	14.13
2.00	496	14.00
2.25	666	13.89

Table 5. - Affect of Weir Height on Sweet Gas Composition

CONCLUSIONS

The use of rate-based technology in the simulation of separation processes such as absorption and distillation has proven itself to be technically superior to the traditional equilibrium stage model. This rate-based technology is employed in the GasPlant and GasPlant-Plus programs for the simulation of alkanolamine based sour gas treating. The programs are useful not only in the design stage of plant operation where they can be used to determine the ability of a given column to meet operating requirements, to evaluate alternate types of equipment for separation and to analyze the effect of process and equipment parameters on plant operating conditions; but also in the maintenance of those plants. Specifications allow the determination of tray diameters based upon desired flooding as well as direct specification of diameter. This sizing ability removes the necessity of iterating between a simulation tool and tower sizing programs.

Plant profiles can be compared to model profiles to locate discrepancies in operation. For example if a tray has lost active area due to a physical obstruction or if a solvent stream has been contaminated by another alkanolamine. The model can also be used to evaluate possible alternatives for correcting the problem.

The ability to evaluate process conditions before actually building the plant appears to be especially important in cases where selective treatment is required. Overdesign is not a viable option in these cases. We have shown that for selective gas treating too many trays, too large a tray or too high a weir can all increase CO_2 pickup at the cost of lost H_2S capacity. The results can be off spec gas products or unacceptably dilute sulphur plant feed.

REFERENCES

Bucklin, R.W., Buckingham, P.A. and Smelser, S.C., 1988, The Higee demonstration test of selective H_2S removal with MDEA, Addendum to *Proceedings of the 41st Laurance Reid Gas Conditioning Conference, Norman, Oklahoma, March 7–9, 1988.*

Chakravarty, T., Phukan, U.K. and Weiland, R.H., April 1985, Reaction of acid gases with mixtures of amines, *Chemical Engineering Progress.*

Deshmukh, R.D. and Mather, A.E., 1981, A mathematical model for equilibrium solubility of hydrogen sulfide and carbon dioxide in aqueous alkanolamine solutions, *Chem Eng Sci*, 36: 355.

Edwards, T.J., Mauer, G., Newman, J. and Prausnitz, J.M., 1978, Vapor-liquid equilibria in multicomponent aqueous solutions of volatile weak electrolytes, *AIChE Journal*, 24: 966.

Harbison, J.L. and Handwerk, G.E., 1987, Selective removal of H_2S utilizing generic MDEA, *Proceedings of the 40th Laurance Reid Gas Conditioning Conference, Norman, Oklahoma, March 2–4, 1987.*

Jou, F., Mather, A.E. and Otto, F.D., 1982, Solubility of H_2S and CO_2 in aqueous methyldiethanolamine solutions, *Ind Eng Chem Process Des Dev*, 21: 539.

Krishnamurthy, R. and Taylor, R., 1985, A nonequilibrium stage model of multicomponent separation processes, *AIChE Journal*, 31: 449–455. Ibid., 1985, 456–464.

Sivasubramanian, M.S., 1985, The heat and mass transfer rate approach for the simulation and design of acid gas treating units, *MS Thesis* (Clarkson University, Potsdam, New York).

Vickery, D.J. and Weiland, R.H., 1986, Solubility of acid gases in blends of amines, paper 44c, *American Institute of Chemical Engineers Annual Meeting, Miami Beach, Florida, November 2–7, 1986.*

Vickery, D.J. and Weiland, R.H., 1987, Simulating gas treating with blended amines, paper 44f, *American Institute of Chemical Engineers Spring National Meeting, Houston, Texas, March 29–April 2, 1987.*

Vickery, D.J., Campbell, S.W. and Weiland, R.H., 1988, Gas treating with promoted amines, Presented at the *Laurance Reid Gas Conditioning Conference, Norman, OK, March 7–9, 1988.*

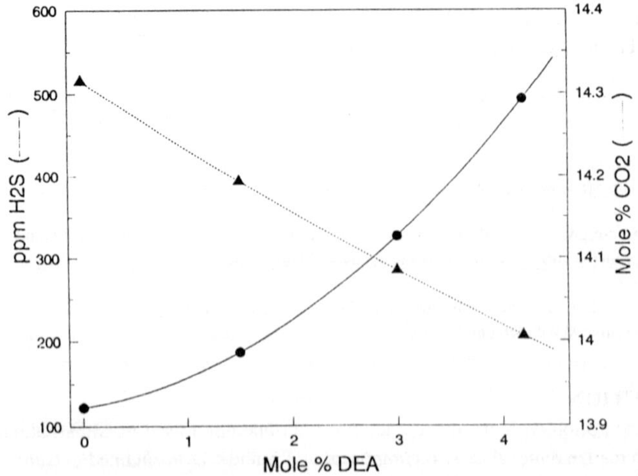

Figure 1. Effect of Solvent Strength on Sweet Gas Composition

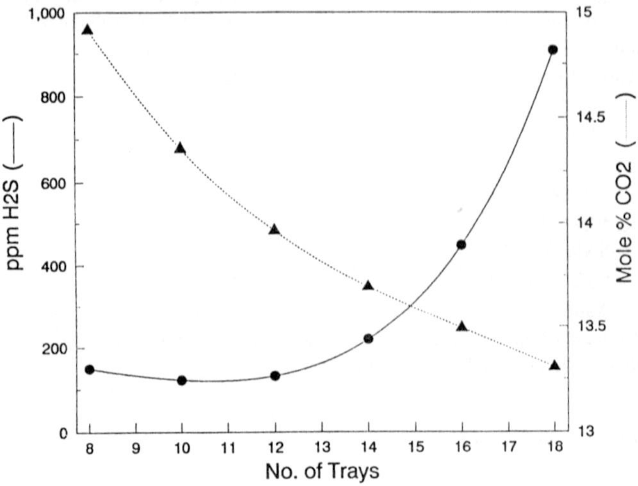

Figure 2. Effect of Trays on H2S and CO2 Sweet Gas Content

HYDRODYNAMICS OF A GAS-LIQUID COLUMN EQUIPPED WITH A STRUCTURED PACKING

C. Pondebat*, J.L. Peytavy* , M.H. Huor**, M. Prevost**, A. Laurent***
*ELF-AQUITAINE, BP 34, 64170 Lacq, France
**LEAP-ENSIGC, Chemin de la Loge, 31078 Toulouse, France
***LSGC-CNRS-ENSIC, BP 451, 1 Rue Grandville, 54001 Nancy Cedex, France

This study was carried out in a pilot column of 0.3 m inner diameter, working in counter-current operation and equipped with structured packing. It allowed:
- to establish the influence of respective gas and liquid flow rates and of the liquid properties upon pressure drop and dynamic liquid holdup,
- to obtain the hydrodynamic diagram.

Experimental results and established correlations have been compared to some existing in the literature.

INTRODUCTION

Gas-liquid absorption with chemical reactions is an operation used in ELF-AQUITAINE processes which realize removal of acid components (H_2S and CO_2) contained in natural gases with aqueous solutions of DEA (diethanolamine), MDEA (methyldiethanolamine) or Activated-MDEA. The mastership of these processes is especially based upon a good knowledge of industrial contactors; for this purpose, hydrodynamic behaviour and efficiency, in terms of mass transfer coefficients, of various industrial trays have been particularly studied. Some studies in progress aim at extending this knowledge to contactors more recently appeared: structured packings.

The results presented in this paper register in this context; they refer to the hydrodynamic of a pilot column of 0.3 m inner diameter, packed with Mellapak 250 HC and working in counter-current operation of gas and liquid. The main point of experimental work consisted in measuring pressure drop and dynamic liquid holdup of this packing versus gas and liquid flow rates and in determining the loading and flooding curves. The originality of this study lies on the extend of the scale of liquid flow rate and on the choice of the following 5 gas-liquid systems:

- Nitrogen — Aqueous solution of MDEA (3.78 N),
- Nitrogen — Aqueous solution of DEA (3.56 N),
- Nitrogen — Soda (5.10 N),
- Nitrogen — Soda (2.84 N),
- Nitrogen — Aqueous solution of Potassium Carbonate (0.297 N) and Bicarbonate (0.884 N).

The empirical correlations which are proposed to account for experimental measurements are compared to some existing in the literature.

EXPERIMENTAL APPARATUS

APPARATUS

The main part of experimental device is a pilot column of 0.3 m inner diameter packed with 16 elements of Mellapak 250 HC on 3.6 m total height. Each element is composed of a set of stainless steel corrugated sheets which are perforated and jointed by two flanges of steel gauze.

The characteristics of the packing are as follows:

- geometric area : 250 m^2m^{-3},
- void fraction : 97.5 %,
- angle of flow channel based on horizontal: 60°.

The packed part of the column is interrupted by pressure and temperature gauges. The whole packing is supported by a grid which has a void fraction higher than that of the packing.

The column operates at room temperature, under 1.05×10^5 Pa; the explored ranges have been:

- for gas: F=0 to 4 $\sqrt{Pa}$,
- for liquid: u_L=0 to 0.045 ms^{-1},
  μL=1.34×10-3 to 9.10×10^{-3} Pas,
  ρL=1042 to 1188 kgm^{-3},
  σL=42.9×10^{-3} to 76.7×10^{-3} Nm^{-1}.

EXPERIMENTAL MEASUREMENTS OF PRESSURE DROP

For each liquid flow rate, pressure drop has been measured, increasing gas flow rate, with two manometers; one amplifier for pressure drop up to 400 Pa, another one for greater values.

Scavenging with nitrogen (between two measurements) avoids any problem of plugging of the pressure gauges with liquid, particularly near flooding. Measurement accuracy is about 10 Pa, with good reproducibility.

EXPERIMENTAL MEASUREMENTS OF DYNAMIC LIQUID HOLDUP

For these measurements, we have used the method of liquid collection at the bottom of the column: after starting and detecting the liquid level at the bottom, the liquid inlet and outlet valves are closed and the gas flow rate is decreased, the surpressor shutdown or the flow rate decrease being ineffective on the reproducibility of measurements; then the liquid drains off (10 minutes) and the amount recovered is weighed.

To calculate the dynamic liquid holdup, the amount of liquid held in the distributor is deduced from the amount recovered; the packed volume is considered equal to the volume taken up by the 16 elements of Mellapak. The accuracy of the measurement is about 10 %.

EVIDENCE OF LIQUID-SIDE HYDRODYNAMIC REGIMES

Whatever the liquid used, a kind of bubbling appeared for values of $u_{L,Eff}$ higher than 0.3 ms^{-1}: ripples on the liquid surface become very large, are uplifted and so the gas flows pocket-shaped.

INFLUENCE OF REGIMES ON PRESSURE DROP (BELOW THE LOADING POINT)

If, for each liquid and each liquid flow rate, the experimental results are correlated according to the model $\Delta P/H=\Delta P_0/H+K_0FA_0$, and if the variation of the constants K_0 (Figure 1) and A_0 (Figure 2) is plotted versus $u_{L,Eff}$, three working zones on liquid side are evidenced.

For A_0, the following points are observed:

- for $u_{L,Eff} \leq 0.255$ ms^{-1} (zone 1): A_0 remains constant and near 2
- for $0.255 \leq u_{L,Eff} \leq 0.36$ ms^{-1} (zone 2): A_0 decreases quickly with $u_{L,Eff}$
- for $u_{L,Eff} \geq 0.36$ ms^{-1} (zone 3): A_0 becomes constant again and almost null.

For K_0, no discontinuity in the values is noted.

INFLUENCE OF REGIMES ON DYNAMIC LIQUID HOLDUP (BELOW THE LOADING POINT)

Below the loading point, the dynamic liquid holdup is almost independent on gas flow rate.

If for a given gas flow rate (F=0 for example) the values of liquid holdup are plotted versus $u_{L,Eff}$ (Figure 3), the three previous zones are evidenced: the slope of the variation of β versus $u_{L,Eff}$ increases from one zone to the other.

RESULTS

PRESSURE DROP IN DRY PACKING

The experimental measurements have been smoothed by the following empirical correlation with ±1 % accuracy:

$$\frac{\Delta P_0}{H} = 13.908.F^{1.700} \qquad (1)$$

PRESSURE DROP IN IRRIGATED PACKING

The influence of operating parameters upon pressure drop is given by the following correlations with ±15 % accuracy:

* below the loading point:

$$\frac{\Delta P}{H} = \frac{\Delta P_0}{H} + K_1.\left[\frac{\mu_L}{\mu_{Water}}\right]^{A_1}.\exp(K_2.u_L).F^{A_2 + A_3.\left[\frac{\mu_L}{\mu_{Water}}\right] + A_4.u_L} \qquad (2)$$

The values of the constants K_1, A_1, K_2, A_2, A_3, A_4 are summed up in Table 1 for the different liquid-side hydrodynamic regimes:

TABLE 1
Values of constants K_1, A_1, K_2, A_2, A_3, A_4

	$u_{L,Eff} \leq 0.255$ (r=0.979)	$0.255 \leq u_{L,Eff} \leq 0,36$ (r=0.833)	$u_{L,Eff} \geq 0.36$ (r=0.968)
K_1	1.5386	39.177	39.984
A_1	0.018673	-1.1022	-0.12183
K_2	140.10	96.012	76.320
A_2	1.5614	0.25132	-0.45405
A_3	0.13303	0.39616	0.043704
A_4	-28.096	-36.316	18.974

* above the loading point and whatever the value of $u_{L,Eff}$ (r=0.905):

$$\frac{\Delta P}{H} = \left[\frac{\Delta P}{H}\right]_{\text{below loading point}}.\exp\left[2.3823.(F-F_1)\right] \qquad (3)$$

where F_1 may be calculated from correlation (6).

Figure 4 shows the influence of the kinetic factor upon pressure drop for one value of liquid flow rate: both experimental and calculated values are given.

DYNAMIC LIQUID HOLDUP

The influence of operating parameters upon dynamic liquid holdup is given by the following correlations with ±10 % accuracy:

* below the loading point:

$$\beta = K_3 \cdot \left[\frac{\nu_L}{\nu_{Water}}\right]^{A_5} . u_L^{A_6} . \exp(K_4 . F) \tag{4}$$

The values of the constants K_3, A_5, A_6, K_4 are summed up in Table 2 for the different liquid-side hydrodynamic regimes:

TABLE 2
Values of constants K_3, A_5, A_6, K_4

	$u_{L,Eff} \leq 0.255$ (r=0.990)	$0.255 \leq u_{L,Eff} \leq 0.36$ (r=0.986)	$u_{L,Eff} \geq 0.36$ (r=0.926)
K_3	0.38548	1.0467	1.8229
A_5	0.17954	0.056825	-0.04219
A_6	0.36739	0.5867	0.7191
K_4	0.02048	0.03956	0.04258

* above the loading point and whatever the value of $u_{L,Eff}$ (r=0.931):

$$\beta = [\beta]_{\text{below loading point}} \exp[0.40401.(F-F_l)] \tag{5}$$

HYDRODYNAMIC DIAGRAM

For a given liquid flow rate, the loading and flooding points have been determined graphically from the curves representing pressure drop or dynamic liquid holdup versus gas kinetic factor. A comparison between these two methods shows a very good agreement.

The loading and flooding curves obtained may be plotted in the Souders' coordinates defined hereafter:

$$X = \frac{L}{G} \cdot \sqrt{\frac{\rho_G}{\rho_L}} \quad \text{and} \quad Y = \frac{F}{\sqrt{\rho_L - \rho_G}}$$

The experimental points have been smoothed by the following empirical correlations with ±10 % accuracy:

* loading curve (r=0.970):

$$Y_l = \exp(-3.0287.X^{0.1070}) \tag{6}$$

* flooding curve (r=0.969):

$$Y_f = \exp(-2.9777.X^{0.1336}) \tag{7}$$

DISCUSSION — COMPARISON WITH LITERATURE

Before going further, two remarks have to be made:

- the large majority of the results in the literature are only available below loading, so we have limited the comparison to this zone;
- this comparison is also limited to explicit models cited in the literature.

This comparison has shown clearly the difficulty of extrapolating:

- to high liquid flow rates; in our study, the range of variation of u_L was from 0.003 to 0.045 ms^{-1}, whereas in the literature the common maximum value is about 0.02, which corresponds approximately in our study to the zone 1 ($u_{L,Eff}$0.255 ms^{-1});
- to different liquids, especially for the high liquid flow rates: it was difficult to gather our experimental results, very different as regards liquid nature, in correlations which did not take into account the physico-chemical properties of liquid.

PRESSURE DROP

For pressure drop in dry packing, an excellent agreement between our results and those of the literature is noted: for example, Billet *et al*[1] point out a variation of $\Delta P_0/H$ with $F^{1.814}$.

For pressure drop in irrigated packing, agreement of our results with the literature remains satisfactory if the comparison is limited to zone 1; on the other hand, if these models are applied to our complete range of liquid flow rate, results become worse and worse when liquid viscosity decreases: this is illustrated by the two first references of Table 3, considering our two extremes: MDEA ($\mu L=9.10\times10^{-3}$ Pas) and Carbonates ($\mu_L=1.34\times10^{-3}$ Pas).

Only the model developed by Mackowiak[5] allowed us to regroup all the experiments, although it considers two liquid-side hydrodynamic regimes, large variations between experimental and calculated values are observed (more than 50 %).

TABLE 3
Comparison of our results with the literature (pressure drop)

Results of literature	Application on our experiments
Billet and al. [2] $\frac{\Delta P}{H} \sim F^{A_7}.\left[1-K_5.u_L^{2/3}\right]^{-5}$ $A_7=1.735$ - $K_5=2.38$	- for MDEA (r=0.986) : $A_7=1.79$ - $K_5=4.26$ - for Carbonates (r=0.983) : $A_7=0.27$ - $K_5=4.33$
Bravo and al. [3-4] $\frac{\Delta P}{H} \sim \left[K_6+\frac{K_7}{Re_G}\right].F^2.\left[1-K_8.Fr_L^{1/2}\right]^{-5}$ Re_G and Fr_L are based on S dimension $K_6=0.171$ - $K_7=92.7$ - K_8 depends on packing	- for MDEA (r=0.989) : $K_6=0.115$ - $K_7=76.2$ - $K_8=4.73$ - for Carbonates (r=0.951) : $K_6=-0.191$ - $K_7=910.2$ - $K_8=4.67$
Mackowiak [5] $\frac{\Delta P}{H} \sim 5.4.K_9.Re_L^{A_8}.function(F,\beta)$ if $0.3\leq Re_L\leq 12.3$ $\frac{\Delta P}{H} \sim 3.8.K_9.function(F,\beta)$ if $Re_L>12.3$ $K_9=0.091$ - $A_8=-0.14$	- for all the liquids (r=0.262) : $K_9=0.037$ - $A_8=-0.66$

DYNAMIC LIQUID HOLDUP

For dynamic liquid holdup, the comparison between our results and those cited in the literature (Table 4) shows a rather good agreement; nevertheless a weakest influence of the liquid flow rate is worth noting.

TABLE 4

Comparison of our results with the literature (dynamic liquid holdup)

Results of literature	Application on our experiments
Billet and al. [1] $\beta = K_{10} \cdot u_L^{A_9}$ $K_{10}=1.55$ - $A_9=2/3$	- for MDEA : * A_9 unadjusted (r=0.883) : $K_{10}=1.80$ - $A_9=2/3$ * A_9 adjusted (r=0.989) : $K_{10}=0.74$ - $A_9=0.42$ - for Carbonates : * A_9 unadjusted (r=0.983) : $K_{10}=1.55$ - $A_9=2/3$ * A_9 adjusted (r=0.991) : $K_{10}=1.1$ - $A_9=0.58$
Mackowiak [5] $\beta = K_{11} \cdot B_L^{A_{10}}$ $K_{11}=2.2$ - $A_{10}=1/2$ where $B_L = \left[\frac{v_L}{g^2}\right]^{1/3} \cdot \left[\frac{u_L}{\varepsilon_0}\right] \cdot \left[\frac{a_0}{6.\varepsilon_0}\right]$	- for all the liquids : * A_{10} unadjusted (r=0.965) : $K_{11}=2.30$ - $A_{10}=1/2$ * A_{10} adjusted (r=0.978) : $K_{11}=1.45$ - $A_{10}=0.42$

CONCLUSION

Measurements of pressure drop and dynamic liquid holdup have been carried out with five different gas-liquid systems in a column packed with Mellapak 250 HC. Some correlations giving the influence of operating parameters have been developed and compared to results of literature. Liquid-side hydrodynamic regimes have been evidenced and hydrodynamic diagram has been established.

This study has led to an improvement of knowledge of hydrodynamics of structured packings at high liquid flow rates.

REFERENCES

1. Billet, R. and Mackowiak, J., 1988, *Chem Eng Technol*, 11: 213–227.
2. Billet, R. and Mackowiak, J., 1985, *Chem Ing Tech*, 57 (11): 976–978.
3. Bravo, J.L., Rocha, J.A. and Fair, J.R., 1986, *Hydroc Proc*, 65 (3): 45–49.
4. Fair, J.R. and Bravo, J.L., 1987, *IChemE Symposium Series No 104*: A183–A201.
5. Mackowiak, J., 1991, *Chem Eng Process*, 29 (2): 93–105.

NOMENCLATURE

a_0	geometric area of packing	m^2m^{-3}
A_i	constant used in a model (i=0 to 10)	
B	base dimension of flow channel	m
B_L	non dimensionnal liquid load	
F	gas kinetic factor ($F=u_G\sqrt{\rho_G}$)	$\sqrt{Pa}$
F_l	gas kinetic factor at loading	$\sqrt{Pa}$
Fr_L	Froude number	
g	acceleration due to gravity	ms^{-2}

G	mass velocity of gas	$kgm^{-2}s^{-1}$
h	height dimension of flow channel	m
H	height of packing	m
K_i	constant used in a model (i=0 to 11)	
L	mass velocity of liquid	$kgm^{-2}s^{-1}$
Per	available perimeter based on tower cross section	mm^{-2}

$$Per = \frac{4\ S + B}{B\ h}$$

r	regression coefficient	
Re_G	gas phase Reynolds number	
Re_L	liquid phase Reynolds number	
S	side dimension of flow channel	m
u_G	superficial gas velocity	ms^{-1}
u_L	superficial liquid velocity	ms^{-1}
$u_{L,Eff}$	effective liquid velocity	ms^{-1}

$$u_{L,Eff} = \left[\frac{3\ u_L}{2\ Per}\right]\left[\frac{g\ Per}{3\ \nu_L\ u_L}\right]^{1/3}$$

X	Souders' abscissa	
Y	Souders' ordinate	ms^{-1}
Y_f	Souders' ordinate at flooding	ms^{-1}
Y_l	Souders' ordinate at loading	ms^{-1}

Greek letters

β	dynamic liquid holdup	m^3m^{-3}
ΔP	pressure drop in irrigated packing	Pa
ΔP_0	pressure drop in dry packing	Pa
ε_0	void fraction of packing	m^3m^{-3}
μ_L	liquid dynamic viscosity (subscript Water for water)	Pas
ν_L	liquid kinematic viscosity (subscript Water for water)	m^2s^{-1}
ρ_G	gas density	kgm^{-3}
ρ_L	liquid density	kgm^{-3}
σ_L	liquid surface tension	Nm^{-1}

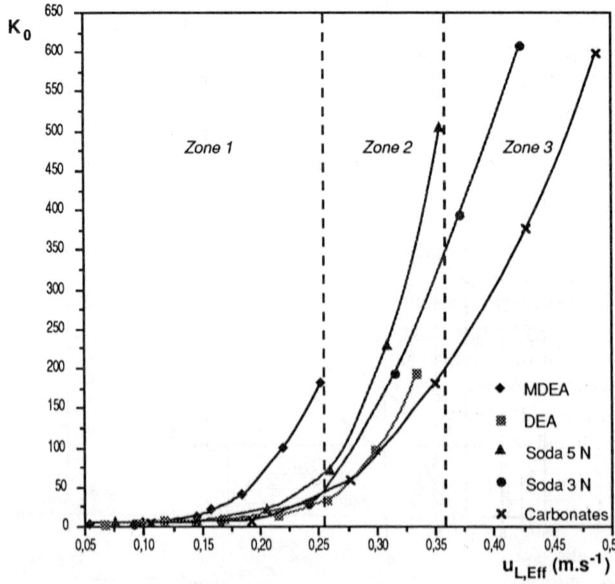

Figure 1 : Influence of hydrodynamic regimes upon K_0

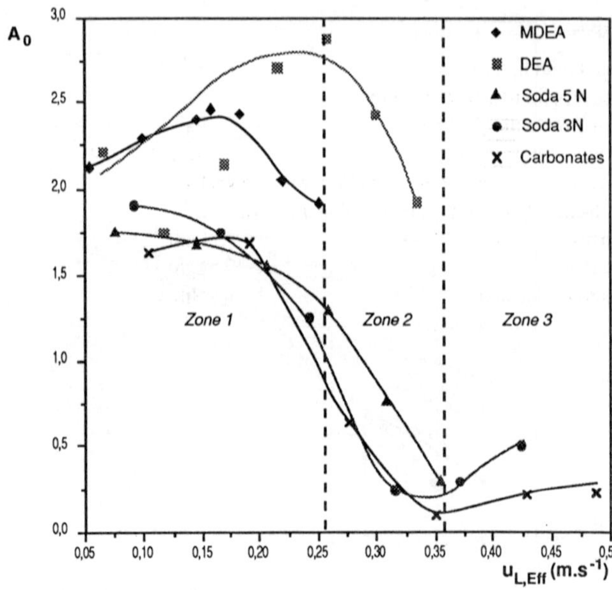

Figure 2 : Influence of hydrodynamic regimes upon A_0

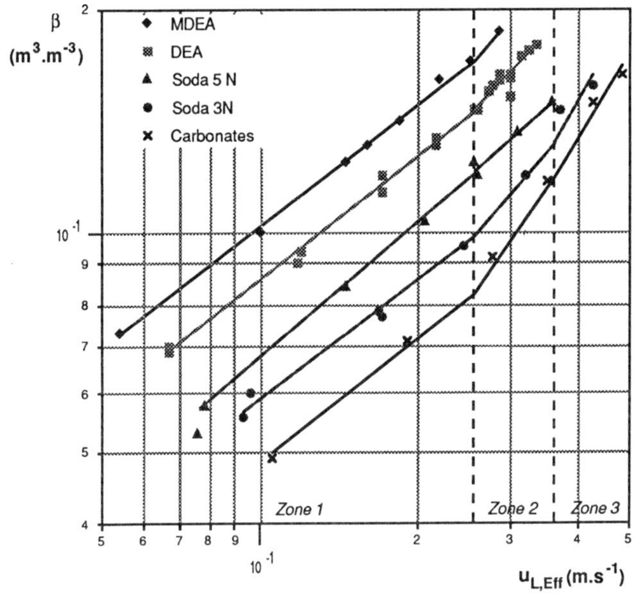

Figure 3 : Influence of hydrodynamic regimes upon β (F=0 √Pa)

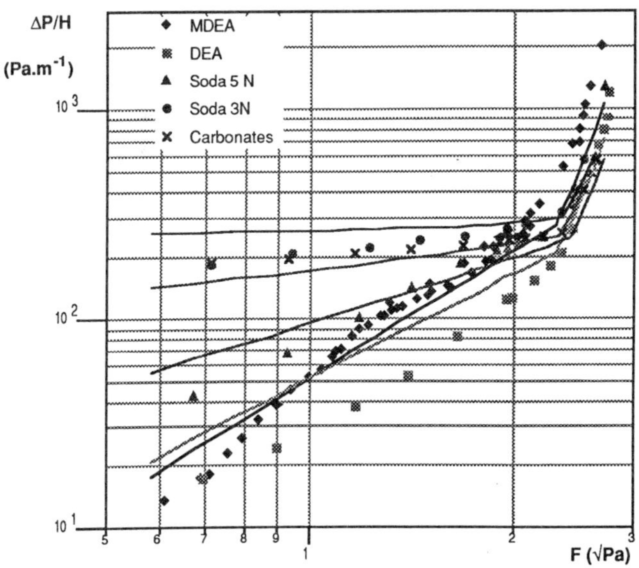

Figure 4 : Influence of gas kinetic factor upon ΔP/H (u_L=0.0226 $m.s^{-1}$)

SEPARATION OF ANHYDROUS ETHANOL FROM AQUEOUS SOLUTION BY ALTERNATIVES STRATEGIES

D.J. Gunn and H.N. Njenga
Department of Chemical Engineering, University College, Singleton Park, Swansea SA2 8PP

The common industrial process for the separation of anhydrous ethanol from ferments is the azeotropic separation of ethanol using benzene as an entraining agent. The process is less favoured today because of the known carcinogenic properties of benzene. In a recent experimental investigation the basic VLE data for two alternative processes have been obtained. It has been found that the ethanol-water azeotrope is removed by distillation at absolute pressures less than 80 mms Hg. In a second part of the investigation VLE data have been obtained for the water-ethanol system at several different concentrations of potassium acetate. Two processes for the separation of anhydrous ethanol based upon low pressure, and upon salt addition are described.

INTRODUCTION

In addition to substantial requirements of anhydrous ethanol for use in chemical manufacturing, there is a significant demand for the chemical as a supplement to conventional grades of motor fuel. For example the 'Gasohol' process allows the addition of up to 25% of anhydrous ethanol as a blend with motor fuel. This process can be of great economic importance when there are few indigenous sources of petroleum and there is a lack of foreign exchange, conditions that are frequently met in Third World countries.

The separation of anhydrous ethanol using benzene as an entraining agent is technically effective, but there is increasing concern about the carcinogenic properties of benzene. Alternative entraining agents such as pentane and furfural may be used, but each is a chemical of some toxicity, so that there is a clear need for the development of processes that are ecologically acceptable and financially viable. The question of economic viability can only be established on the basis of fully optimised designs, but the economics of application depend upon the site of the production plant, the site of plant fabrication, transport and raw material costs. Therefore the chosen form of the process will depend upon the location of the production plant, but in this paper we are concerned with the technical details of alternative processes for the separation of anhydrous ethanol.

SEPARATION OF ANHYDROUS ETHANOL AT LOW PRESSURES

An essential element of process feasibility is the dependence of the azeotropic composition upon pressure, and very little data are available. Wade and Merriman (1911) measured azeotropic compositions by equilibriating from either side of the azeotropic composition without estimating vapour-liquid (VLE) data. In the present investigation (Njenga, 1991) the azeotropic compositions were determined from VLE data at 70, 80, 100, 150, 300 and 500 mms Hg total pressure; the azeotropic compositions may be obtained from the condition,

$$\frac{\gamma_1}{\gamma_2} = \frac{p_2^o}{p_1^o} \qquad (1)$$

Figure 1 shows the data of both Wade and Merriman (1911) and Njenga (1991) illustrating that the azeotrope has been removed at absolute pressures of 80 mms Hg and below.

In comparing the VLE data with several types of correlating equations it was found that this particular system was best fitted to a two parameter Margules equation. The equations have the form,

$$\ln \gamma_1 = x_2^2 [1.596 - 1.474 x_1] \quad (2)$$

$$\ln \gamma_2 = x_1^2 [0.859 + 1.474 x_2] \quad (3)$$

The *yx* data calculated from these equations are shown in Figure 2. The problem of low pressure distillation is immediately evident, since the small separation between the *yx* line and the *y=x* line at high concentrations of ethanol shows that a large number of plates and a high reflux ratio are required. In view of the expense of low pressure distillation, it is probably economic to produce an alcohol-enriched stream of say, 0.85 mole fraction ethanol as feed for the low pressure still since VLE data at mole fractions are but little affected by pressure.

Even if a preliminary separation is carried out, distillation at a column top pressure of 70 mms Hg or less and the high reflux ratio required by the closeness of the *yx* line and the *y=x* line indicate a difficult and expensive operation. The requirement of a large number of trays means that the pressure drop across each tray will have to be low in a successful column design. Thus separation of anhydrous ethanol at low pressure will require a column of relatively large diameter operating at a high reflux ratio, difficult to design accurately and probably not economic if there are other contenders.

THE SYSTEM WATER-ETHANOL-POTASSIUM ACETATE

The addition of non-volatile salts to the liquid phase can be effective in enhancing the volatility of ethanol relative to water at high ethanol concentrations. Inorganic salts tend to be not effective because of low solubility in ethanol, but Johnson and Furter (1960) have shown that potassium acetate has a sufficiently high solubility to remove the azeotrope. Johnson and Furter studied VLE data for a range of solutions of ethanol and water saturated with potassium acetate. Although the strong effect of potassium acetate was shown, a direct application of their study is difficult because the solubility of the acetate increases as the concentration of ethanol in the solution falls. The addition of potassium acetate is most effective at high concentrations of ethanol, and therefore in Njenga's 1991 investigation the effect of potassium acetate was studied at several different but constant concentrations, limited to the solubility of potassium acetate in ethanol boiling at 760 mms Hg pressure. This value was found to be 0.11 mole fraction potassium acetate.

The addition of potassium acetate is most effective at higher concentrations so we need only be concerned with VLE data measured at an acetate concentration of 0.11 mole fraction. The VLE data are shown in Figure 3 as crosses. On this figure *y* represents the mole fraction of ethanol in the liquid, but *x* represents the molar ratio of ethanol to ethanol plus water. The line chosen to correlate the data is based upon an effective molar volume representation of the activity coefficients that is accurate except for a limited mid-range region where there is some overprediction.

The figure includes an example McCabe-Thiele construction for a given reflux ratio that shows an effective separation between 0.99 mole fraction ethanol and water containing very low concentrations of ethanol may be obtained from a column containing about 15 theoretical stages.

The *yx* line shows good separation from the *y=x* line at both high and low concentrations of ethanol, but the effect of salt addition is most marked at high concentration, since there is already a large separation at low concentrations of ethanol. A diagram for the separation with removal of acetate fom the liquid phase just above the feed location would be effectively the same as Figure 3.

The removal of potassium acetate from the liquid phase and its economic recovery are essential parts of this process for the production of anhydrous ethanol. To be effective the acetate has to be present in the reflux return to the column, but the anhydrate product has to be removed before the salt addition. The concentration of salt has to be at least maintained above the feed point although the number of trays required will be slightly reduced if the concentration of salt can be increased.

Below the feed point the dominant process requirement is to reduce the ethanol content of the process water for discharge as effluent or for possible recycle to the fermentation plant. This can be readily carried out with or without potassium acetate.

A process flowsheet for the separation of the anhydrate is shown as Figure 4. The ethanol-water feed from the ferment is filtered before entering the single distillation column. The overhead anhydrate is condensed, and split into two streams, one of which is the product anhydrate. The second stream is maintained at boiling temperature and fed to a mixer together with pelleted potassium acetate. The product is ethanol saturated with the salt at a concentration of 0.11 molar potassium acetate. This stream is returned as reflux to the distillation column. The saturation of the ethanol-water solutions decreases down the column because of the enhanced solubility of potassium acetate with increasing concentration of water, but the concentration of salt will be held fairly constant in the upper stages of the column.

The rate of liquid flow down the column is increased at the feed point for a feed at the boiling point thus diluting the acetate concentration and increasing the energy requirement for recovery, and therefore the liquid phase is diverted from the column just above the feed point. The arrangement for acetate recovery depends upon the scale of production, and to what extent additional capital plant may be installed to offset energy requirements. Here we have assumed that the appropriate policy is met by an evaporator followed by a spray dryer since any water in the acetate feed to the dissolver will reflect directly in an increased water content of the overhead product.

The condensate from the evaporator and spray dryer is returned to the column at the feed point through a preheater, and the product water is then available for further process use.

By comparison with the benzene process for the production of anhydrous ethanol the acetate process seems to be relatively free from environmental difficulties, and by comparison with the low pressure separation, both capital and energy requirements are likely to be significantly smaller. The acetate process appears to have environmental and economic potential in both Third World countries and elsewhere.

REFERENCES

Johnson, A.I. and Furter, W.F., 1960, *Can J Chem Eng*, 38: 78.

Njenga, H.N., 1991, Low pressure and salt effect on the ethanol-water vapour-liquid equilibrium, *PhD Thesis* (University of Wales).

Wade, J.D. and Merriman, R.W., 1911, Influence of water on the boiling point of ethyl alcohol at pressures above and below the atmospheric pressure, *J Chem Soc*, 99: 997.

ACKNOWLEDGEMENT
We wish to acknowledge financial support from the British Council and from the Government of Kenya for this study.

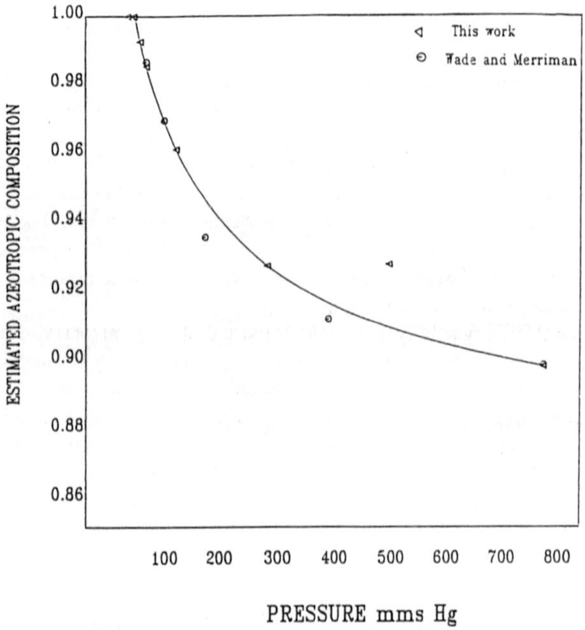

FIG. 1 ESTIMATED AZEOTROPIC COMPOSITIONS COMPARED WITH THE ESTIMATES OF WADE AND MERRIMAN (1911)

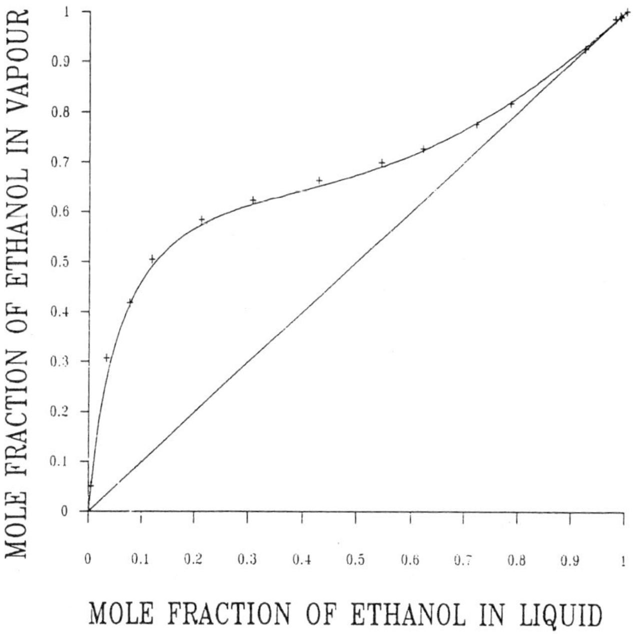

Fig 2. Vapour–liquid equilibrium data at 70 mms Hg

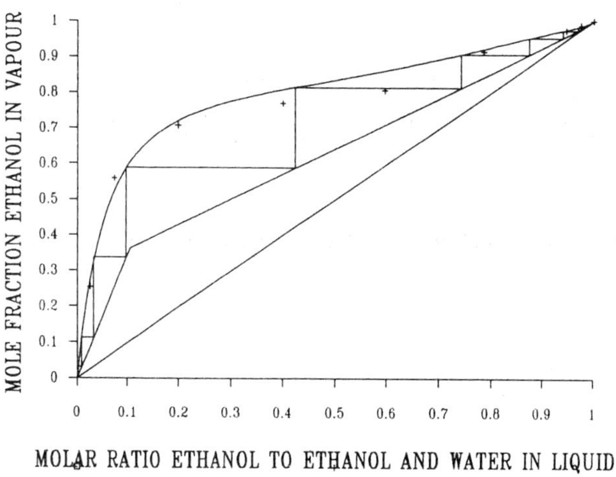

Fig. 3 VLE data and stage estimation at 0.11 mf Acetate

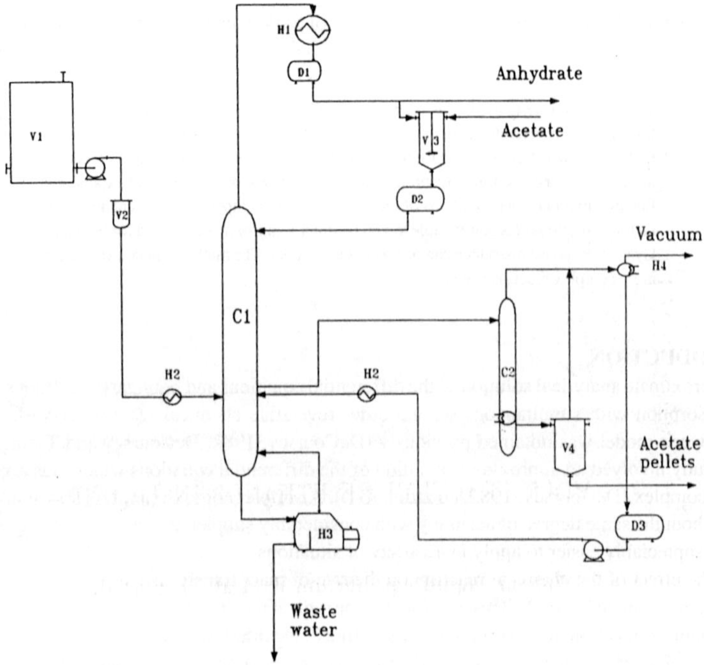

V1	Feed tank	C1	Distillation column	D1	Condensate drum	H1	Condenser
V2	Filter	C2	Evaporator	D2	Reflux Drum	H2	Preheater
V3	Mixer			D3	Evaporate Drum	H3	Reboiler
V4	Spray dryer					H4	Condenser

Fig.4 Flowsheet for the Production of Ethanol Anhydrate

A SIMPLER APPROXIMATION FOR ENHANCEMENT OF MASS TRANSFER BY SECOND-ORDER REVERSIBLE REACTION

W.J. DeCoursey
Department of Chemical Engineering, University of Saskatchewan, Canada, S7N 0W0

This paper presents a simpler method of predicting enhancement factors for mass transfer with reversible chemical reaction. This is an approximate analytical solution to the relevant differential equations and boundary conditions. Numerical methods are used to solve the resulting algebraic equations. The new method uses an exponential function to relate concentrations of the main diffusing component to the enhancement factor. This method is comparable to von Kármán's approximate method for boundary layers. Gas-phase resistance can be taken into account. The method can be extended to more complex reaction schemes.

INTRODUCTION

An approximate analytical solution of the differential equations and boundary conditions describing absorption with simultaneous second-order reversible chemical reaction according to the Danckwerts model was obtained previously (DeCoursey, 1982; DeCoursey and Thring, 1989). That study involved an approximate solution of the differential equations which was algebraicly rather complex (DeCoursey, 1982, equation (34)). A simpler approximation is now presented. It gives about the same degree of accuracy with considerably simpler algebra. The new approximation is appreciably easier to apply in a variety of situations.

The effect of the chemical reaction on the rate of mass transfer is expressed quantitatively by the enhancement factor, E. This is defined as the ratio of the time-mean mass flux at the interface with chemical reaction, to the time-mean mass flux at the interface without chemical reaction but with the same driving-force in the liquid phase and the same hydrodynamic conditions. As long as chemical equilibrium can be assumed in the bulk of the liquid (this is almost always true in practical applications of absorption), the definition of enhancement factor reduces to

$$E = \frac{k_L}{k_L^*} \tag{1}$$

where k_L is the mass transfer coefficient in the liquid phase with reaction, for driving-force in concentration units, and k_L* is the corresponding mass transfer coefficient without reaction. Then

$$\bar{N}_A = -D_A\left[\frac{d}{dx}\left([\bar{A}]-[A]_0\right)\right]_{x=0} = E\,k_L^*\left([\bar{A}]_i-[A]_0\right) \tag{2}$$

MATHEMATICAL FORMULATION

For the chemical equation

$$A + z_B B \Leftrightarrow z_c C + z_D D \tag{3}$$

the differential equations for conservation of mass with diffusion and reaction in absorption are:

$$D_A \frac{\partial^2 [A]}{\partial x^2} - \frac{\partial [A]}{\partial t} - R_A = 0 \tag{4}$$

$$D_B \frac{\partial^2 [B]}{\partial x^2} - \frac{\partial [B]}{\partial t} - z_B R_A = 0 \tag{5}$$

$$D_C \frac{\partial^2 [C]}{\partial x^2} - \frac{\partial [C]}{\partial t} - z_C R_A = 0 \tag{6}$$

$$D_D \frac{\partial^2 [D]}{\partial x^2} - \frac{\partial [D]}{\partial t} - z_D R_A = 0 \tag{7}$$

The initial condition is

$$\text{at } t = 0 \text{ and } x > 0,\ [A] = [A]_0,\ [B] = [B]_0,\ [C] = [C]_0,\ [D] = [D]_0 \tag{8a}$$

and these concentrations are in equilibrium with one another. The boundary conditions are

$$\text{at } x = \infty,\ [A] = [A]_0,\ [B] = [B]_0,\ [C] = [C]_0,\ [D] = [D]_0 \tag{8b}$$

again in equilibrium, and

$$\text{at } x = 0 \text{ and } t > 0,\ [A] = [A]_i,\ \frac{\partial [B]}{\partial x} = 0,\ \frac{\partial [C]}{\partial x} = 0,\ \frac{\partial [D]}{\partial x} = 0 \tag{8c}$$

BRIDGING RELATIONS

Bridging relations express concentrations at the liquid-gas interface, $x = 0$, in terms of corresponding concentrations in the bulk of the liquid, mathematically at $x = \infty$. They are determined by combining the differential equations (above) to eliminate the non-linear reaction terms, then integrating the resulting linear differential equations. For the Danckwerts model of mass transfer this procedure is simplified by the fact that time-mean concentrations are given by 's - multiplied' Laplace transforms of instantaneous concentrations (Danckwerts, 1970).

For example,

$$[\bar{A}] = s \int_{0}^{\infty} [A]\, e^{-st}\, dt \tag{9}$$

The resulting bridging relations have been developed previously (DeCoursey and Thring, 1989). The relation for component B is:

$$\beta = \frac{[\bar{B}]_i}{[B]_0} = 1 - \frac{(E^2 - 1)}{q'(E\sqrt{r_B} + 1)} \tag{10}$$

The corresponding bridging relations for components C and D are:

$$\frac{[\overline{C}]_i}{[B]_o} = \frac{[C]_o}{[B]_o} + \frac{z_c(E^2-1)}{z_B q'(E\sqrt{r_c}+1)} \tag{11}$$

$$\frac{[\overline{D}]_i}{[B]_o} = \frac{[D]_o}{[B]_o} + \frac{z_D(E^2-1)}{z_B q'(E\sqrt{r_D}+1)} \tag{12}$$

These relations are exact if all diffusivities are equal, so that $r_B = r_C = r_D = 1$. For unequal diffusivities an approximate expression for

$$\frac{d^2[\overline{A}]}{dx^2}$$

is required, based on the approximation that

$$[\overline{A}] - [A]_o = ([\overline{A}]_i - [A]_o)\ \exp\left(-x\,E\sqrt{\frac{s}{D_A}}\right) \tag{13}$$

For pseudo-first-order reactions this expression is exactly correct. For general second-order reactions it is correct in value and slope at the interface and bulk, but deviates from the true profile in between.

APPROXIMATE SOLUTION

One dimensionless quantity used in finding a solution is β. The other major dimensionless quantity is θ, defined as

$$\theta = \frac{[\overline{A}]_{ei} - [A]_o}{[\overline{A}]_i - [A]_o} \tag{14}$$

where

$$[\overline{A}]_{ei} = \frac{[\overline{C}]_i^{z_C}\,[\overline{D}]_i^{z_D}}{K\,[\overline{B}]_i^{z_B}} \tag{15}$$

(and some approximation is introduced by time-mean values) and

$$[A]_o = \frac{[C]_o^{z_C}\,[D]_o^{z_D}}{K\,[B]_o^{z_B}} \tag{16}$$

Now, two independent relations are found between θ and β. One is obtained by algebraic substitution of bridging relations in equations (15), (16) and (14). For 1:2:1:1 stoichiometry the result is

$$\theta = \frac{2q'}{K[B]_o}\left(\frac{1}{\beta^2}\left[\frac{[C]_o}{[B]_o} + \frac{E^2-1}{2q'(E\sqrt{r_c}+1)}\right]\left[\frac{[D]_o}{[B]_o} + \frac{E^2-1}{2q'(E\sqrt{r_D}+1)}\right] - \left(\frac{[C]_o}{[B]_o}\right)\left(\frac{[D]_o}{[B]_o}\right)\right) \tag{17}$$

Another relation is based on rate of reaction and rate of mass transfer. The time-mean rate of reaction is given by

$$\overline{R}_A = k_2 [\overline{B}]([\overline{A}] - [\overline{A}]_e) \tag{18}$$

An extension of the approximation of van Krevelen and Hoftijzer (1948) is to require that the Laplace-transformed version of equation (4),

$$D_A \frac{d^2[\overline{A}]}{dx^2} - s([\overline{A}] - [A]_o) - \overline{R}_A = 0 \tag{19}$$

be satisfied exactly at the interface but only approximately elsewhere. This is for three reasons. First, the reaction rate closest to the interface has the greatest influence on the enhancement factor. Second, the concentration of component A (from the gas phase) decreases rapidly with distance from the interface, so the reaction rate becomes negligible not far from the interface. Third, as the boundary conditions of equation (8c) express, the derivatives of concentrations of components other than A are zero at the interface because they are assumed not volatile. It is found that equation (13) satisfies equation (19) exactly at the interface. Substituting equation (13) and (18) in equation (19) at x = 0 gives

$$E^2 = 1 + \frac{k_2[B]_o}{s} \frac{[\overline{B}]_i}{[B]_o} \left(\frac{[\overline{A}]_i - [\overline{A}]_{ei}}{[\overline{A}]_i - [A]_o} \right)$$

Now introducing dimensionless quantities,

$$E^2 = 1 + Ha^2 \beta(1 - \theta)$$

or

$$\theta = 1 - (E^2 - 1)/(\beta Ha^2) \tag{20}$$

Equations (17) (or alternative for other stoichiometry) and (20) are the two desired independent relations between θ and β. It should be noted that equation (20) is much simpler in form than the following equation which it replaces (DeCoursey, 1982):

$$\theta = \left[\frac{\beta Ha^2 - (C^2 - 1)}{\beta Ha^2 - C(C+1)(1 - 1/\sqrt{1 + \beta Ha^2}} \right] \left[1 - E/\sqrt{(1 + \beta Ha^2)} \right] \tag{21}$$

Then θ and β are related independently by equation (17) and (20), and the two expressions for θ can be equated. The resulting non-linear algebraic equation can be solved numerically. Some limiting values of β and θ are useful in bounding the region for solution. For absorption with pseudo-first-order reaction (values of Hatta number approaching zero), we have β = 1 and θ = 0. For mass transfer with instantaneous reaction (values of Hatta number approaching infinity), β = β_a (asymptotic limit) and θ = 0. For irreversible reactions (K = ∞), θ = 0.

The numerical method I have used involves a few steps of bisection, followed by the secant method.

A complication arises when gas-phase resistance is present. That is that $[\bar{A}]_i$ depends on the relative resistance to mass transfer of the two phases, but $[\bar{A}]_i$ affects the value of q', which in turn affects the value of β and so the enhancement factor, and therefore relative resistances. This requires an outer loop to determine $[\bar{A}]_i$. An inner loop of iterative calculation is required to determine first β_a and E_a, then β and E.

DETAILS OF CALCULATION

A. CALCULATION OF $[\bar{A}]_i$

After appropriate parameters (eg bulk compositions, pressure, Henry's Law constant, equilibrium constant, chemical rate constant, mass transfer coefficients, diffusivities, etc) have been chosen, limiting possible values of $[\bar{A}]_i$ are found. Corresponding values of error, R, between assumed and calculated values of $[\bar{A}]_i$ are calculated.

One limit is for E = 1 (no enhancement). This gives an upper limit of $[\bar{A}]_i$, which is

$$[\bar{A}]_i = \left(\tilde{y}_{Ao} + \frac{k_L^*}{k_{\tilde{y}}}[A]_o\right) \Big/ \left(\frac{H_A}{P} + \frac{k_L^*}{k_{\tilde{y}}}\right) \tag{21}$$

With this assumed value, E_a and E are calculated by parts B and C below. Then a calculated value of $[\bar{A}]_i$ is found from

$$[\bar{A}]_{i,calc} = \left(\tilde{y}_{Ao} + \frac{Ek_L^*}{k_{\tilde{y}}}[\bar{A}]_o\right) \Big/ \left(\frac{H_A}{P} + \frac{Ek_L^*}{k_{\tilde{y}}}\right) \tag{22}$$

and the error is

$$R = [\bar{A}]_{i,calc} - [\bar{A}]_i \tag{23}$$

Another limiting condition corresponds to instantaneous irreversible reaction, for which an approximate value (Danckwerts, 1970) is

$$E_{a,irrev} = \frac{1}{\sqrt{r_B}} + \frac{[B]_o\sqrt{r_B}}{z_B[A]_i} \tag{24}$$

Substituting $E = E_{a,irrev}$ in equation (22) gives a quadratic expression for a lower limit of $[\bar{A}]_i$. Using this assumed value of $[\bar{A}]_i$, again E and E_a are calculated by parts B and C below. The new value of E is used to get $[\bar{A}]_{i,calc}$ by equation (22), and error is calculated by equation (23).

In general, values of the error R from the two limiting assumptions will be of opposite sign. The secant method is used to find successive new values of $[\bar{A}]$i, and for each, E_a and E are calculated, then $[\bar{A}]_{i,calc}$ and R. This is continued until R is smaller in absolute terms than a chosen tolerance, say 0.001 times the absolute value of $[\bar{A}]$i.

B. CALCULATION OF E_a, REVERSIBLE CASE
For each chosen value of $[\bar{A}]i$, q' is calculated, and asymptotic values of the enhancement factor, β_a, are calculated as follows. We know that β_a must be between 0 and 1. A lower limit is chosen, say 0.001 for absorption. An upper limit is chosen, say 0.99, again for absorption. The value of β_a must satisfy equation (17) with $\theta = 1$. About 4 bisections are used to localize the value of β_a. Then the secant method is used iteratively until a tolerance is satisfied. Once β_a is determined, the corresponding value of E_a for the reversible case is found by solving equation (10) for the enhancement factor.

C. CALCULATION OF E, REVERSIBLE CASE
One limit, the lower limit in the case of absorption with reaction, is β_a as calculated in Part B. This corresponds to instantaneous reaction. The other limit is $\beta = 1$, corresponding to absorption with pseudo-first-order reaction. Once again, a good strategy in numerical calculation of β is to start with bisection (say twice), and then to apply the secant method until the test for termination is satisfied. Once β is known, the enhancement factor can be calculated from the solution of equation (10) for E.

COMPARISON WITH NUMERICAL SOLUTION OF DIFFERENTIAL EQUATIONS

Results of the present study can be compared with numerical solutions of the differential equations corresponding to equations (4) to (8c). Solutions of the differential equations for a limited range of conditions were published by Secor and Beutler (1957) for the Higbie model of mass transfer. An important limitation is that the results of Secor and Beutler were for zero loading of the liquid solution with component A and its products of reaction. The data of Secor and Beutler for equal diffusivities were apparently in error for K = 1 (DeCoursey, 1982); aside from that the results of the present study agreed within 5% with the results of Secor and Beutler for stoichiometric ratios 1:1:1:1, for values of K from 0.1 to 20 and various values of Ha and q'. Agreement of results from the previous approximation (DeCoursey, 1982) with the same results of Secor and Beutler was within 5%. Secor and Beutler presented results for unequal diffusivities only at $K = 1$, $q' = 1$, and 1:1:1:1 stoichiometry. For those conditions the results of the present study agreed with Secor and Beutler within 4%, the same as for the previous approximation. Thus, agreement is very good.

Versteeg *et al* (1989) compared their numerical results, from numerical solution of the partial differential equations according to the Higbie model, with the previous approximation (DeCoursey, 1982). They found agreement within 10% over a wide range of parameters. On the other hand, they concluded that the method of Onda *et al* (1953) for reversible reactions can lead to erroneous results. They also compared their numerical results with experimental measurements. Their experimental results for absorption of CO_2 and H_2S (separately) in DEA solutions agreed with their numerical results within 20%, and the agreement in the case of absorption of CO_2 and H_2S in MEA solutions was within 40%. In all cases deviations were larger at higher absorption rates. The authors attributed deviations to uncertainties in physical constants, inaccuracy of the experimental method, and particularly Marangoni effects. Winkelman *et al* (1992) compared their numerical results with predictions by the method of DeCoursey and Thring (1989) and found agreement over an extremely wide range of parameters within 14%. They also confirmed inaccuracies in the results of Secor and Beutler. All these comparisons provide some indirect support to the methods of the present paper.

COMMENT ON APPROXIMATE SOLUTION

The exponential relationship for concentration of Component A from the gas phase, equation (13), is exact for absorption with pseudo-first-order reactions, and so at the lower limit of Hatta number equal to zero. Some approximation is involved at larger Hatta numbers. However, for mass transfer with instantaneous chemical reactions, corresponding to infinitely large Hatta numbers, it has been shown previously (DeCoursey and Thring, 1989) for irreversible reactions that enhancement factors by the present method agree very closely with exact relations. Thus, it is not surprising that the results of the present method seem to be acceptably accurate over the whole range of parameters.

The method used in this paper can be compared with the approximate solution for the boundary layer developed by von Kármán (1921) and discussed by Schlichting (1979). In that method von Kármán assumed an appropriate simple form for the mathematical relation and then determined constants from the boundary conditions. The same general approach has been used in this paper.

EXAMPLE OF CALCULATIONS

Absorption of CO_2 in MEA at low carbonation ratios involves principally the reaction

$$CO_2 + 2\,RNH_2 \Leftrightarrow RNH_3^+ + RNHCOO^-$$

At a point in the middle section of the absorber consider a 2 M MEA solution in water at 40°C, diffusivities of all reactants and products except CO_2 56% of the diffusivity of CO_2, pressure 15 atm. The carbonation ratio is 0.40 moles CO_2 absorbed per mole total MEA, and the bulk composition of the gas is 5 mole%. Calculations were done by a Fortran program according to the scheme discussed earlier. The limiting value of $[\bar{A}]_i$ for no enhancement was found to be 0.00752 moles/liter. The limiting value of $[\bar{A}]_i$ for instantaneous, irreversible reaction was found to be 0.00441 moles/liter. Two overall iterations gave $[\bar{A}]_i$ equal to 0.00570 moles/liter. The enhancement factor was found to be 28.4, very close to the instantaneous reversible enhancement factor. Approximately 50% of the total resistance to mass transfer was found to be in the gas phase.

CONCLUSION

A new method of approximate analytical solution of differential equations for absorption according to the Danckwerts model of mass transfer has been presented. It gives about the same accuracy as a previous approximation (DeCoursey and Thring, 1989) but involves much simpler algebraic relations. Results of an example calculation are given.

Perhaps the most important implication of the simpler method is that it now seems quite feasible to adapt the method to more complex reaction schemes, involving both parallel and series steps. That would allow calculations to be made for simultaneous absorption of CO_2 and H_2S, for example. Calculations for desorption would be very similar but with equations solved for a different set of boundary conditions.

REFERENCES

Danckwerts, P.V., 1970, *Gas Liquid Reactions* (McGraw-Hill, New York).

DeCoursey, W.J., 1974, *Chem Engng Sci*, 29: 1867–1872.

DeCoursey, W.J. and Thring, R.W., 1989, *Chem Engng Sci*, 44: 1715–1721.

Schlichting, H., 1979, *Boundary Layer Theory* (McGraw-Hill, New York).

Secor, R.M. and Beutler, J.A., 1967, *AIChEJ*, 13: 365–373.

van Krevelen, D.W. and Hoftijzer, P.J., 1948, *Rec Trav Chim*, 67: 563–586.

Versteeg, G.F., Kuipers, J.A.M., van Beckum, F.P.H. and van Swaaij, W.P.M., 1989, *Chem Engng Sci*, 44: 2295–2310.

von Kármán, T., 1921, *Z Angew Math Mech*, 1: 235.

Winkelman, J.G.M., Brodsky, S.J. and Beenackers, A.A.C.M., 1992, *Chem Engng Sci*, 47: 485–489.

NOMENCLATURE

A	dissolved gas
B	chemical reactant supplied by liquid phase
C,D	products of the forward reaction
D_A, D_B, D_C, D_D	diffusivities, m^2s^{-1}
E	enhancement factor
Ha	Hatta number, $\sqrt{k_2 [B]_o / s}$ for the Danckwerts model
k_L	mass transfer coefficient for liquid phase, m s^{-1}
k_L^*	mass transfer coefficient for liquid phase without chemical reaction m s^{-1}
$k_{\tilde{y}}$	mass transfer coefficient for gas phase, mol m^{-2} s^{-1}
k_2	second order chemical rate constant, m^3 mol^{-1} s^{-1}
P	pressure, Pa
q'	ratio of equivalents, $[B]_0/[z_B ([\overline{A}]_i - [A]_0)]$
R	difference between assumed and calculated values of $[\overline{A}]_i$, mol m^{-3}
R_A	reaction rate for component A, mol m^{-3} s^{-1}
r_B, r_C, r_D	diffusivity ratios (D_B/D_A, D_C/D_A, D_D/D_A respectively)
s	fractional rate of replacement of surface area for Danckwerts model, s^{-1}
t	time, s
x	distance from the interface, m
z_B, z_C, z_D	stoichiometric factors

GREEK LETTERS

β	$[\overline{B}]_i/[B]_0$
θ	dimensionless concentration (see equation (14))

SUBSCRIPTS

a	asymptotic value for instantaneous reaction
e	equilibrium
i	interface
0	bulk condition
Square bracket	concentration, mol m^{-3}
Overline	time-mean value

STATIC AND DYNAMIC BEHAVIOUR OF HETEROGENEOUS AZEOTROPIC DISTILLATION COLUMNS

R.G. Corrêa and S.B. Jørgensen
The Process Design, Dynamics and Control Group, Department of Chemical Engineering, Technical University of Denmark, DK–2800 Lyngby, Denmark

An investigation of steady-state and dynamic properties for control purposes of one type of heterogeneous azeotropic distillation is presented. The case considered is that of a minimum boiling ternary heterogeneous azeotrope and three binary azeotropes of which one is heterogeneous. The plant is described and investigated with the purpose ofcontrolling the two product purities; ie, the bottom products from the azeotropic and the dehydrating column, respectively. Three actuators are considered in the azeotropic column: the boilup rate V_A, the overhead split ratio R_A, and the entrainer feed flow rate F_E. The investigated plant has two degrees of freedom for fulfilling the primary control goal. Two actuator configurations R_AV_A and R_AF_E are investigated for control of the top conditions and bottom composition. A special distance measure is introduced for the top condition with the purpose of having one liquid phase in the column, yet ensuring phase split in the decanter. Steady-state operation is described by two regions, where it is shown that operation is statically optimal in one region. The dynamic characteristics vary with the operating regions and include the presence of overshoot and inverse response, in both regions. The multivariable dynamics also vary with the operating region. The R_AV_A configuration is shown to be much more well conditioned for control than the R_AF_E configuration

INTRODUCTION

The separation of binary azeotropic mixtures is a task commonly encountered in the chemical industry. This separation *can not* be performed by simple binary distillation. However, it is often possible to achieve the desired separation by distillation if a third component, the *entrainer*, is added to the system. The thermodynamic properties of the ternary mixture (binary azeotrope plus entrainer) determine if the separation can be performed either entirely by distillation *homogeneous* case or partly by decanting two liquid phases *heterogeneous* case. In the heterogeneous case, the action of the entrainer is to alter the relative volatility of the two azeotropic constituents and to induce a liquid-liquid phase separation.

Heterogeneous azeotropic distillation has been used to separate binary azeotropes since it was successfully applied for the first time by Young (1902). He prepared absolute alcohol from a binary mixture of ethanol and water using benzene as entrainer in a batch distillation.

Most of prior work on control of distillation have focused on ideal binary mixtures. The extensive literature on this subject illustrate that even simple binary distillation columns were hard to understand and control. Recent investigations in the field of homogeneous azeotropic distillation have shown that the choice of operating region significantly influences the possibility of effective operation and control (Andersen *et al*, 1989a–b). However, Jacobsen *et al* (1990) have shown that tight and robust control can be obtained with decentralized PI-controllers, by considering the high frequency behaviour of the column. The presence of two liquid phases in the heterogeneous case appears to increase the complexity further. The operation of such columns can be tricky, because disturbances can induce immiscibility inside the column, leading to loss of efficiency (Kovach and Seider, 1987). Some works on heterogeneous azeotropic distillation columns have focused on the calculation of liquid-liquid-vapour phase equilibria (Pham and Doherty, 1990a–b), dynamic simulation (Rovaglio and Doherty, 1990 and Wong *et al*, 1991), and upon selection of feasible heterogeneous entrainers and distillation sequences (Pham and Doherty, 1990a–c and Ryan and Doherty, 1989). Few papers relate to the control problems (Abu-Eishah and Luyben, 1985, Andersen, 1986, Bozenhardt, 1986 and 1988, Nisenfeld and Stravinski, 1968, Rovaglio *et al*, 1991 and 1992) with only a limited number of general results obtained.

The objective of this paper is to present steady-state and dynamic properties of a class of continuous heterogeneous azeotropic distillation column systems represented by the ternary systems ethanol-water-benzene (EWB) and isopropanol-water-cyclohexane (IWC). This class of systems is a minimum boiling heterogeneous ternary azeotrope with all three binary azeotropes (see Figures 2 and 3). First the selected plant structure is introduced. Then two control configurations for the azeotropic column are selected for further investigation. The modelling assumptions applied in this simulation study are given next. The steady-state properties of the two configurations are shown to depend uniquely upon the vapour flow rate and entrainer make-up flow rate relationship. This dependence gives rise to two different operating regions wherein the static and dynamic behaviour are investigated. Finally the multivariable properties are compared for the two configurations. This comparison shows that one configuration is much more well conditioned at high frequency than the other.

PLANT DESCRIPTION

The separation takes place in a plant consisting of two distillation columns and a decanter, as shown in Figure 1. A concentrated feed stream which has its composition close to the binary azeotrope is fed to the first (azeotropic) column. A stream which must condense into two liquid phases is taken overhead and sent to the decanter. The heavy phase from the decanter is fed to some intermediate tray of the second (dehydrating) column. A fraction of the condensed overhead stream from this column is recycled to the original feed stream. The light phase from the decanter is recycled to the top tray of the azeotropic column together with the reflux stream. A small make-up stream of entrainer is necessary to replace small losses mainly in the alcohol product stream. Four design issues in the sequence presented on Figure 1 should be stressed:

ENTRAINER-RICH PHASE

The returning of the entrainer-rich phase to the top plate, instead of to the feed tray, causes a marked decrease in the alcohol concentration in the vapour from the top tray and thus reduces the cost of alcohol recovery.

DECANTER BY-PASS

The azeotropic column should be refluxed with the correct amount of entrainer. An entrainer excess tends to be lost in the bottom product, in which case much more energy is required to perform the separation.

RECYCLE

To recover alcohol and entrainer which leave with the water-rich phase, part of the condensed overhead vapour from the dehydrating column can be recycled to the feed.

MAKE-UP
A make-up stream of entrainer is fed to replace small losses of entrainer in both product streams.

The early development of this process is described by Keyes (1929) with several modifications of the continuous process and complete details of patent numbers and dates.

DEGREES OF FREEDOM FOR CONTROL AND CONTROL CONFIGURATIONS

The heterogeneous azeotropic distillation schematically reported in Figure 1 may be viewed for operational purposes as having eight degrees of freedom. This number can be reduced by considering:

- the levels in both reboilers and the levels of both entrainer- and water-rich phases in the decanter are assumed perfectly controlled.
- the dehydrating column is designed with a sufficiently high number of trays and high reflux ratio to ensure a recycle stream with a composition close to the feed composition of the azeotropic column and without using a control loop in the top of the dehydrating column.
- the water composition in the bottom product of the dehydrating column is paired with the boilup of this column.

The remaining two degrees of freedom for control are then used to control the azeotropic column. That is, we have to pair 2 manipulated variables with 2 outputs. The two primary outputs are: d_{BIN} — for the top loop — and x_{BA1} — for the bottom loop. The output d_{BIN} is the minimum distance between the liquid composition on the top tray of the azeotropic column and the binodal curve. The idea behind the use of this measurement is to facilitate ensuring the presence of only one liquid phase on the top tray, while keeping the condensed overhead vapour composition inside the two-liquid phase region. Figure 2 presents how d_{BIN} is defined.

Two different control configurations are investigated in this paper: the $R_A V_A$ and $R_A F_E$ configurations (other possibilities are reported in Corrêa and Jørgensen, 1992). The names indicate the control inputs, eg, the $R_A V_A$ configuration refers to using split ratio R_A and boilup rate V_A to control d_{BIN} and x_{BA1}, respectively. Prokopakis and Seider (1983) and Rovaglio and Doherty (1990) suggested the use of the split ratio R_A as a manipulated variable. The split ratio has an interesting feature since it regulates part of the energy and material balances by regulating the amount of water recycled from the decanter. Although the make-up stream flow rate F_E is small when compared to the other flows and hence not recommendable for control (Prokopakis and Seider, 1983), it can be used to regulate the relative volatility in the azeotropic column by regulating the amount of entrainer in this column. The boilup V_A regulates the operating lines in the azeotropic column.

MODELLING CONSIDERATIONS

In the following sections a non-linear dynamic component and total mass balance model is used. Quasi-steady-state total mass balances are assumed for the dehydrating column and decanter. The azeotropic column flow dynamics are described by a Francis Weir formula

$$L_i = \frac{1.103\, L_w}{TMV_i}\left(\frac{1000}{A} M_i TMV_i - h_w\right)^{3/2} \tag{1}$$

where L_i is the liquid leaving stage i in mol/min, L_w is the length of the weir (35.0 cm), TMV_i is the total molecular volume in l/mol, A is the active area of the tray (1080.0 cm^2), M_i is the holdup in mol at tray i, and h_w is the height of the weir (5.5 cm). The vapour-liquid equilibrium thermodynamics are described using UNIFAC, while the liquid-liquid equilibrium is represented by an experimental correlation. The vapour-liquid-liquid equilibrium, which is checked at each tray, is calculated by both thermodynamic models. The energy balance is excluded as the reflux streams are not subcooled. All the algebraic and differential equations are solved simultaneously by using a semi-implicit Runge-Kutta method that handles the sparce analytical Jacobian matrix. The assumptions adopted in the development of the non-linear model were considered in order to obtain as simple a model as possible, yet retaining the characteristics important for control.

DEGREES OF FREEDOM FOR DESIGN

In the design of the azeotropic and dehydrating columns, and the decanter the number of trays will be considered fixed. Hence only the flow rates will be determined. Inspecting Figure 1, the total and component mass balances yield 25 variables and 15 equations and thus 10 degrees of freedom. Assuming x_{E1}, x_{E2}, F_A, x_{A1}, x_{A2}, R_A, and R_D given there remains 3 degrees of freedom for design. For given product specifications x_{BA1} and x_{BA2}, one degree of freedom still remains. Given the values of L_{BA}, L_{BD}, F_E, and V_A, the other variables can be calculated from the steady-state set of equations. For high product purity specifications ($x_{BA1} \approx 1$, $x_{BA2} \approx 0$, $x_{BD1} \approx 0$, and $x_{BD2} \approx 1$) and pure make-up stream ($x_{E1} = x_{E2} = 0$), it can be shown that relative changes in both product flows will be small

$$\frac{\Delta L_{BA}}{L_{BA}} \approx 0 \tag{2}$$

$$\frac{\Delta L_{BD}}{L_{BD}} \approx 0 \tag{3}$$

Therefore the remaining degree of freedom will be represented as a relation between, eg (F_E, V_A).

SPECIFICATION OF R_A FOR OPTIMAL OPERATING REGION

One important aspect to consider in the specification of R_A is to adopt a reflux policy to guarantee a configuration with the ability to handle process disturbances. With this point in mind, it will not be a good design specification to choose the split ratio $R_A = 0$, since the excess entrainer in the azeotropic column tends to be lost in the bottom product and, in order to avoid this situation, some of the water may be added to the reflux in the decanter by-pass stream, L_{RA} (Shinskey, 1984, Rovaglio and Doherty, 1990, and Prokopakis and Seider, 1983). This will be avoided with $R_A >$ 0. Another point that should be considered in the specification of R_A is related to the position of the condensed overhead vapour composition inside the two liquid phase region. The class of heterogeneous azeotropic systems represented by the IWC and EWB ternary mixtures presents a minimum boiling heterogeneous ternary azeotrope. This ternary azeotrope is ideally obtained by condensing the overhead vapour with an infinite number of trays in the azeotropic column. Since the number of trays is considered fixed and finite, the distance to ideal operating conditions is measured. The tie-lines and distillation regions for the EWB system given in Figure 3, show that the tie-line yielding the highest attainable purity, ie the 'best', that can be obtained in region I, where the azeotropic column operates, is the tie-line that passes through the heterogeneous ternary azetrope. The phrase 'highest available purity' refers to the split between the amount of entrainer in the entrainer-rich phase and the amount of water in the water-rich phase. To measure how close the condensed overhead vapour composition $\mathbf{y}_{NA}$ is to the heterogeneous ternary azeotrope $\mathbf{x}_{ter}$, the distance between these two points in the phase diagram has to be calculated.

$$d = \sqrt{(\mathbf{y}_{NA} - \mathbf{x}_{ter})^T (\mathbf{y}_{NA} - \mathbf{x}_{ter})} \tag{4}$$

Figure 4 presents the calculated d for the EWB ternary mixture, with 14 trays in the azeotropic column, by varying F_E and R_A. By inspection, it is decided to choose $R_{A_{EWB}}^{14trays} = 0.70$.

STEADY-STATE PROPERTIES

A continuous set of possible operating points can be obtained by varying the make-up stream, F_E, choosing a split ratio, R_A (let us say, $R_{A_{EWB}}^{14trays} = 0.70$) and adjusting the other flows accordingly for a given separation. As stated above, we will describe this set as boilup, V_A, versus make-up stream, F_E. Figure 5 presents the set of operation points for the EWB-column at $R_{A_{EWB}}^{14trays} = 0.70$ and product specifications equal to 0.998 mole fraction ethanol in the azeotropic column and 0.99 mole fraction water in the dehydrating column. Two regions with significantly different steady-state properties are shown in Figure 5. If the make-up flow rate in region I is increased, the increasing amount of entrainer will improve the relative volatility between alcohol and water and less boilup is needed to keep the specifications. Further increase of F_E will move the operating point to operating region II where entrainer excess leaves the azeotropic column in the bottom stream, and in order to keep the specifications, the boilup rate must increase. This observation explains the minimum boilup rate point observed near operating point 2. Economically optimal operation will result in a design in region I. The operating point will move from point 2 towards point 1 if the cost of entrainer dominates the cost of boilup (energy). Consequently it is relevant to investigate the static and dynamic properties at the three selected operating points 1, 2, and 3. A similar relationship was also observed for the IWC column (Corrêa and Jørgensen, 1992).

The steady-state gain matrices and condition numbers γ presented in Figure 5 were obtained at operating points 1, 2, and 3 from a linear state-space model, for the $R_A V_A$ configuration (Figure 5a) and for the $R_A F_E$ configuration (Figure 5b):

REGION I

An increase in boilup rate improves the purity in the bottom product of the azeotropic column (change in operating lines) and moves $\mathbf{x}_{NA}$ away from the binodal curve (change in operating lines). An increase in the make-up flow rate improves the purity in the bottom product of the azeotropic column (change in relative volatility) and moves $\mathbf{x}_{NA}$ closer to the binodal curve (change in entrainer inventory).

REGION II
An increase in boilup rate improves the purity in the bottom product of the azeotropic column (change in operating lines) and also moves $\mathbf{x}_{NA}$ away from the binodal curve (change in operating lines). An increase in the make-up stream decreases both the bottom product in the azeotropic column (change in entrainer inventory) and d_{BIN} (change in entrainer inventory).

These two regions are separated by an operating point where the entrainer flow rate (make-up stream) has almost no effect on the bottom product of the azeotropic column, since improving relative volatility and increase of load (change in entrainer inventory) have an equal impact. Heterogeneous azeotropic distillation does not present a third region as observed in homogeneous azeotropic distillation (Andersen *et al*, 1989a), since the heterogeneous column does not have an extractive section, as homogeneous azeotropic columns do.

The $R_A V_A$ configuration does not present any steady-state control problem and it has the steady-state gains at operating points 2 and 3 well conditioned. From a steady-state point of view, the control problems resemble those of simple binary distillation. However, the $R_A F_E$ configuration presents a change in the steady-state gain sign between x_{BA1} and F_E, around operating point 2. Thus control properties in region II are different from those in region I and the control design adopted for the former region can not be used for the latter. At operating point 1, the EWB-column is running almost without entrainer feed which leads to singularities and thus higher condition numbers for both configurations, as observed in Figure 5.

The liquid composition profiles and relative volatilities for points 2, 1, and 3 are shown in Figures 6a,c,e respectively, together with the same liquid composition profile in a phase diagram (Figures 6b,d,f) for the EWB column. Observe that the alcohol profile starts to decrease and the entrainer profile starts to increase for the EWB column close to tray no. 5 for point 2 (Figure 6a). Around this tray the relative volatilities between alcohol and water α_{12} and between alcohol and entrainer α_{13} change significantly. We will call this region *the front of the profile*. The phase diagram profiles (Figure 6b) show how close the top liquid composition in the azeotropic column is to the binodal curve, although outside the two-liquid phase region. At point 1 the front of the profile is moved upwards to tray no. 7 in the azeotropic column since less entrainer is present in this column. Consequently, water moves downwards in the azeotropic column (Figure 6c). The column profile, in the phase diagram (Figure 6d), is shifted slightly towards the water vertex compared to Figure 6b, but the liquid composition profile on the top tray remains outside the two-liquid phase region, as at point 2. Figure 6e shows that the front of the profile in the azeotropic column is now moved downwards to tray no. 4 in the EWB column. The increasing presence of entrainer drives water out of the column by its top. Figure 6f presents the column profile in the phase diagram, which is shifted to the alcohol-entrainer side and again presents no trays inside the two liquid phase region. It is seen from the relative volatilities α_{12} and α_{13} that the stripping section removes entrainer from the bottom, while the rectifying section is responsible for removing water from the top of the column. Tray 12 is the feed tray.

The analysis presented above can also be applied to other separations belonging to the same class of heterogeneous azeotropic systems, eg an IWC column (Corrêa and Jørgensen, 1992).

DYNAMIC PROPERTIES

The steady-state properties reported above are here extended to the dynamic behaviour for the Single-Input-Single-Output (SISO) and for the Multi-Input-Multi-Output (MIMO) cases. The transient responses of d_{BIN} and x_{BA1} to step changes in V_A, R_A, and F_E will be considered in the analysis. The transient simulations are obtained from the non-linear dynamic model outlined above. The steps were performed at each operating point 1, 2, and 3 reported in the steady-state analysis.

SINGLE VARIABLE CASE

Figure 7 presents the step responses of d_{BIN} and x_{BA1} for the ethanol-water-benzene (EWB) column to positive step changes in boilup V_A, split ratio R_A, and make-up stream F_E of 1% of their steady-state values at operating points 1, 2, and 3.

Boilup V_A

Inverse response is present in x_{BA1} for disturbances in V_A, but just barely observed in Figures 7b,d,f. No inverse response is observed for disturbances in d_{BIN}. The non-minimum phase behaviour observed for x_{BA1} is the result of an initial increase of the entrainer composition x_{BA3} in the bottom product, which is a consequence of an initial increase of the liquid holdup on trays, since an increase in V_A initially increases the liquid flowrate. Subsequently this increased flowrate moves liquid out of the trays down the column. Afterwards, the improvement of operating lines dominate the change in the entrainer inventory and water is further removed from the bottom product up in the column with a consequent improvement in the bottom purity (Figure 7b). The liquid holdup is then increased again and will be larger than the initial steady-state value, as the reflux ratio $\left(\frac{R_A V_A + L_{DL}}{L_{DH}}\right)$ at the top of the azeotropic column is increased. Since the improvement in operating lines dominates in both regions I and II, inverse response will be observed at all three operating points.

Split ratio R_A

An increase in R_A will initially move water into the column bypassing the decanter which will result in a decrease of both the amount of alcohol in the bottom product of the azeotropic column x_{BA1} and d_{BIN}. Since this effect dominates the slow increase of alcohol and entrainer from the decanter, we will observe overshoot but not inverse response at all operating points.

Entrainer make-up stream F_E

In region I where the effect of improved relative volatility dominates the load increase, improvement in the bottom composition of alcohol in the azeotropic column is observed with no overshoot or inverse response. The reverse effect is observed in region II, where the bottom composition of alcohol is reduced due to an increase of the make-up stream. Also no overshoot or inverse response is observed. An increase in F_E reduces d_{BIN} in all regions again without overshoot or inverse response the time constant is however very large due to the relatively small entrainer flowrate, therefore this effect is not visible in Figures 7b,f.

Multivariable case

The previous results regarding the Single Variable dynamics will now be extended to multivariable dynamics for 2×2 dual composition control. Both the $R_A V_A$ and $R_A F_E$ configurations are investigated:

$$\begin{pmatrix} \Delta^s d_{BIN}(s) \\ \Delta^s x_{BA1}(s) \end{pmatrix} = G^s_{R_A V_A}(s) \begin{pmatrix} \Delta^s R_A(s) \\ \Delta^s V_A(s) \end{pmatrix} \tag{5}$$

$$\begin{pmatrix} \Delta^s d_{BIN}(s) \\ \Delta^s x_{BA1}(s) \end{pmatrix} = G^s_{R_A F_E}(s) \begin{pmatrix} \Delta^s R_A(s) \\ \Delta^s F_E(s) \end{pmatrix} \tag{6}$$

where the superscript s denotes scaled variables (Corrêa and Jørgensen, 1992).

The Singular Value Decomposition (SVD) is a frequency domain technique which is a convenient tool for analysis of multivariable systems (Skogestad and Morari, 1988a and 1988b). It is a natural extension of the SISO frequency analysis. In the MIMO case, the effect of inputs upon outputs depends strongly on the directions of the inputs; ie, how the inputs are aligned with the maximum ($\overline{\sigma}$) and the minimum ($\underline{\sigma}$) singular values of the system and how the outputs are aligned with the singular values. For a 2×2 system the alignment with the minimum singular value ($\underline{\sigma}$) is obtained by reversing the indices of the alignment with the maximum singular value ($\overline{\sigma}$).

Region I

The maximum and minimum singular values for the EWB-column at point 1 in the $R_A V_A$ configuration are shown in Figure 8. The high gain direction resembles the single variable gain $G_{x_{BA1} V_A}(iw)$. The V_A actuator is aligned with the maximum singular value up to a frequency of 4 $\times 10^{-4}$ min^{-1}, from where R_A starts to be aligned with $\overline{\sigma}$ (Figure 9a). The output alignment shows that the output x_{BA1} is easier to change than d_{BIN} (Figure 9).

The input alignment for the EWB-column in the $R_A F_E$ configuration (Figure 9a) shows that the maximum singular value is obtained for changes in R_A up to a frequency of 10^{-3} min^{-1}, where the alignment changes to F_E. The output alignment (Figure 9b) shows that x_{BA1} at high frequencies is mainly aligned with the high gain direction. Within the frequency range 6×10^{-2} and 10^{-3} min^{-1} this alignment is changed significantly. This behaviour seems related to the increased presence of entrainer which improves the relative volatility between alcohol and water α_{12}, this effect will take some time before becoming effective. The input alignment for the EWB column is very similar to that of the IWC column, while the output alignment is very different from that reported in Corrêa and Jørgensen (1992).

Operating Point 2

At this operating point the high gain direction for the EWB-column in $R_A V_A$ configuration (Figure 10) resembles the gain $G_{x_{BA1} V_A}(iw)$, as at point 1 (Figure 8), over the whole frequency range, while the low gain direction resembles the gain $G_{d_{BIN} R_A}(iw)$. The input V_A is almost entirely aligned with the maximum singular value (Figure 11a), which makes x_{BA1} the easier output to change (Figure 11b).

The input alignment for the EWB-column in the $R_A F_E$ configuration (Figure 11a) shows that R_A affects the high gain direction and F_E the low gain direction. From a frequency of 10^{-2} min^{-1} up to high frequencies the output alignment is very similar to the output alignment at operating point 1 of the EWB-column (Figure 9b). But the the output alignment in the low frequency range; ie, up to 10^{-2} min^{-1} differs from that of operating region 1, in that the small effect of F_E on the outputs dominates. The input alignment and the low frequency behaviour of the output alignment is again very similar to that of the IWC-column (Corrêa and Jørgensen, 1992), however the high frequency behaviour of the output alignment is different.

Region II

The high gain direction for the EWB-column in $R_A V_A$ configuration (Figure 12) resembles the gain $G_{x_{BA1} V_A}$ (iw), as at point 2 (Figure 10), while the low gain direction resembles the gain $G_{d_{BIN} R_A}$ (iw). The input alignment (Figure 13a) is very similar to the input alignment at point 1 (Figure 9a). The same can be said for the output alignment (Figure 9b), except for a small difference at low frequencies, where d_{BIN} improves its alignment with the high gain direction.

The input and the output alignments (Figures 13a and b) for the EWB-column in the $R_A F_E$ configuration are very similar to those at operating point 1 (Figures 9a and b), with two exceptions: (1) in the input alignment (Figure 13a), R_A is mainly aligned with the high gain direction. No change of alignment is seen at low frequencies as observed at point 1 (Figure 9a) and; (2) the change in the output alignment observed at this operating point (Figure 13b) at low frequencies is not so drastic as the one reported for point 1 (Figure 9b). In fact, both outputs are nearly equally easy to change at frequencies below 2×10^{-4} min^{-1}.

It is interesting to note that the multivariable behaviour for the $R_A V_A$ configuration reveals no right half plane transmission zeroes.

SUMMARY

For the selected type of heterogeneous azeotropic distillation, it is shown that the presence of entrainer complicates the dynamic properties compared to simple binary distillation. The significance of these properties depends upon the operating point chosen. For the EWB-column (and IWC-column, Corrêa and Jørgensen, 1992) in region I for positive changes in F_E an improvement in the bottom composition of alcohol in the azeotropic column is observed with no overshoot or inverse response, since the effect of improved relative volatility dominates the increase load. In region II, the bottom composition of alcohol is reduced due to an increase of the make-up stream, since now the effect of improved relative volatility is dominated by the load increase. Also no overshoot or inverse response is observed. At operating point 2, the effect of improved relative volatility and load increase are of equal importance and for that reason F_E has almost no impact on x_{BA1}. The existence of no overshoot or inverse response is related to the fact that both effects have large time constants. The presence of overshoot and inverse response is observed in all operating regions for changes in V_A. For the output d_{BIN}, early in the response an improvement in the operating lines is obtained, which increases d_{BIN}. With a larger time constant a decrease in d_{BIN} is observed due to the increased presence of entrainer, which comes from the decanter. Since the improved operating lines dominate, only overshoot is observed. However, for the x_{BA1} output, the response is initially characterized by moving entrainer into the bottom product with consequent decrease of x_{BA1}. With a longer time constant, the dominant effect of improved operating lines increases x_{BA1}. The different time constants and nearly equal gains of the two opposing effects result in an non-minimum phase behaviour of x_{BA1}.

For changes in R_A it is observed that in both operating regions there is an initial effect of decreasing d_{BIN} and x_{BA1} due to an increased water flow into the column top. With a longer time constant, an increase of both outputs is observed due to an increase of the relative amount of alcohol and entrainer in the column by going through the decanter. Since the first effect dominates the slow increase of alcohol and entrainer, an overshoot is observed but no inverse response.

Finally, the Single Variable nonminimum phase zeros observed at all operating points in the $R_A V_A$ configuration are removed by the multivariable dynamics, where no RHP transmission zeros are founded for the EWB-column. This was also observed for the IWC-column (Corrêa and Jørgensen, 1992). The EWB-column presented larger overshoots than the IWC-column in both the multiloop and multivariable cases.

CONCLUSIONS

Steady-state operation is economically most feasible in operating region I.

The dynamic properties of the azeotropic column when it is operated with only one liquid phase in the column are investigated for three different inputs: vapour flow rate V_A, overhead split ratio R_A, and entrainer feed flow F_E. The two latter actuators do not have analogues in simple binary distillation. The qualitative dynamics for disturbances in V_A are analogous to those of simple binary distillation except that non-minimum phase behaviour is observed for positive changes in V_A, due the increased downwards transport of entrainer. The response to disturbances in F_E are very slow, thus this input is not suitable as a fast actuator.

For multivariable dynamics no RHP transmission zeros are present in the $R_A V_A$ configuration for control of this type of azeotropic distillation. If the performance limitations imposed by the Single Variable inverse response of V_A upon a multiloop configuration are significant, then a multivariable regulator should be used.

ACKNOWLEDGMENT

Financial support was provided to Ronaldo G. Corrêa by Conselho Nacional de Desenvolvimento Científico e Tecnológico (CNPq) of Brazil, and is appreciated.

REFERENCES

Abu-Eishah, S.I. and Luyben, W.L., 1985, Design and control of a two-column azeotropic distillation system, *Ind Eng Chem Process Des Dev*, 24 (1): 132–140.

Andersen, H.W., Laroche, L. and Morari, M., 1989a, *Effect of Design on the Control of Homogeneous Azeotropic Distillation, Internal Report* (California Institute of Technology, Pasadena, California, USA).

Andersen, H.W., Laroche, L. and Morari, M., 1989b, *Dynamics of Homogeneous Azeotropic Distillation Columns, Internal Report* (California Institute of Technology, Pasadena, California, USA).

Andersen, J.E., 1986, Multivariable control of an azeotropic distillation column, *Advances in Instrumentation*, 41, Part 2: 469–476.

Bozenhardt, H.F., 1986, Modern control 'tricks' to solve an old problem azeotropic distillation, *Advances in Instrumentation*, 41, Part 2: 661–674.

Bozenhard, H.F., 1988, Modern control tricks solve distillation problems, *Hydrocarbon Processing*, 67 (6): 47–50.

Corrêa, R.G. and Jørgensen, S.B., 1992, Understanding the dynamic characteristics of heterogeneous azeotropic distillation columns for effective operation and control, Presented at *European Symposium on Computer Applications in Process Engineering (ESCAPE-1), May 24–28, 1992, Elsinore, Denmark.*

Jacobsen, E.W., Laroche, L., Morari, M., Skogestad, S. and Andersen, H.W., 1991, Robust control of homogeneous azeotropic distillation columns, *AIChE Journal*, 37 (12): 1810–1824.

Keyes, D.B., 1929, The manufacture of anhydrous ethyl alcohol, *Industrial and Engineering Chemistry*, 21 (11): 998–1001.

Kovack III, J.W. and Seider, W.D., 1987, Heterogeneous azeotropic distillation: experimental and simulation results, *AIChE Journal*, 33 (8): 1300–1314.

Nisenfeld, A.E. and Stravinski, C.A., 1968, Feedforward control for azeotropic distillations, *Chemical Engineering*, 75 (20): 227–236.

Pham, H.N. and Doherty, M.F., 1990a, Design and synthesis of heterogeneous azeotropic distillation. I. Heterogeneous phase diagrams, *Chemical Engineering Science*, 45 (7): 1823–1836.

Pham, H.N. and Doherty, M.F., 1990b, Design and synthesis of heterogeneous azeotropic distillation. II. Residue curve maps, *Chemical Engineering Science*, 45 (7): 1837–1843.

Pham, H.N. and Doherty, M.F., 1990c, Design and synthesis of heterogeneous azeotropic distillation. III. Column sequences, *Chemical Engineering Science*, 45 (7): 1845–1854.

Prokopakis, G.J. and Seider, W.D., 1983, Dynamic simulation of azeotropic distillation towers, *AIChE Journal*, 29 (6): 1017–1029.

Rovaglio, M. and Doherty, M.F., 1990, Dynamics of heterogeneous azeotropic distillation columns, *AIChE Journal*, 36 (1): 39–52.

Rovaglio, M., Faravelli, T. and Doherty, M.F., 1991, Operability and control of azeotropic heterogeneous distillation sequences, *Proceedings of the 4th International Symposium on Process Systems Engineering — PSE'91, August 5–9, 1991, Montebello, Quebec, Canada.*

Rovaglio, M., Faravelli, T., Biardi, G., Gaffuri, P. and Soccol, S., 1992, Precise composition control of heterogeneous azeotropic distillation towers, to be presented at *European Symposium on Computer Applications in Process Engineering (ESCAPE-1), May 24–28, 1992, Elsinore, Denmark.*

Ryan, P.J. and Doherty, M.F., 1989, Design/optimization of ternary heterogeneous azeotropic distillation sequences, *AIChE Journal*, 35 (10): 1592–1601.

Shinskey, F.G., 1984, *Distillation Control For Productivity and Energy Conservation* (McGraw-Hill Book Company, New York, USA).

Skogestad, S. and Morari, M., 1988a, Understanding the dynamics behavior of distillation columns, *Ind Eng Chem Res*, 27 (10): 1848–1862.

Skogestad, S. and Morari, M., 1988b, LV-control of a high-purity distillation column, *Chemical Engineering Science*, 43 (1): 33–48.

Wong, D.S.H., Jang, S.S. and Chang, C.F., 1991, Simulation of dynamic and phase pattern changes for an azeotropic distillation column, *Comput Chem Engng*, 15 (5): 325–335.

Young, S., 1902, The preparation of absolute alcohol from strong spirit, *Journal of Chemical Society*, 81, part I: 707–717.

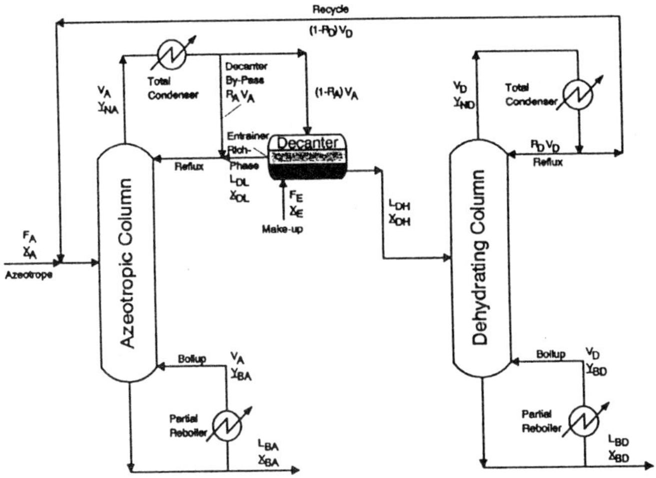

Figure 1: Schematic representation of a heterogeneous distillation sequence.

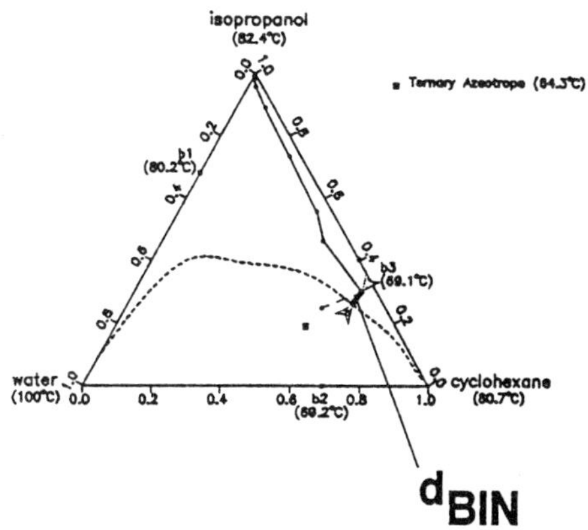

Figure 2: Minimum distance between $\mathbf{x}_{NA}$ and binodal curve measured in a isopropanol-water-cyclohexane azeotropic column.

Solid lines connecting '*' are the liquid composition profile in the azeotropic column.

b1, b2, and b3 are binary azeotropes.

.... binodal curve.

- - - - - - - - - - - - $\mathbf{y}_{NA}$ in equilibrium with $\mathbf{x}_{NA}$.

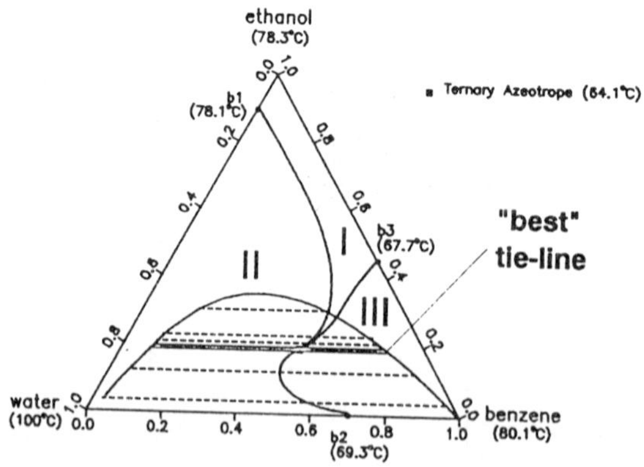

Figure 3: 'Best' tie-line for the ternary ethanol-water-benzene at 1 atm. pressure.
I, II, and III are distillation regions.
Solid curves connecting binary azeotropes are simple possible distillation boundaries.
Solid curve connecting tie-lines is binodal curve.
b1, b2, and b3 are binary azeotropes.
- - - - - - - - - - - - tie-lines.

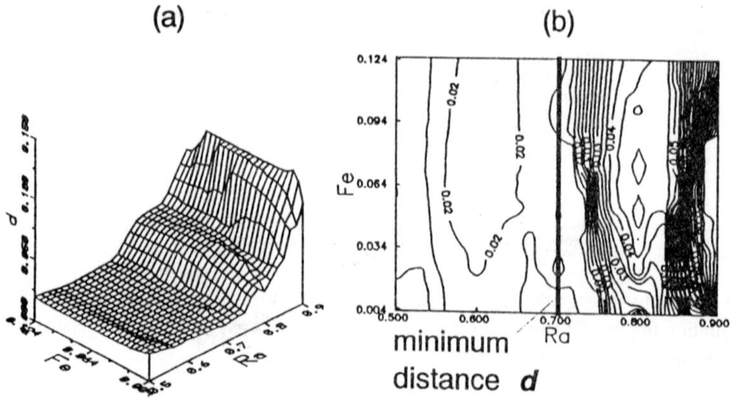

Figure 4: Distance d as defined by equation (4) for the ternary ethanol-water-benzene with 14 trays in the azeotropic column. (a) surface. (b) contour.
(b) — Solid vertical line indicates $R_{A_{EWB}}^{14\ \text{trays}} = 0.70$.

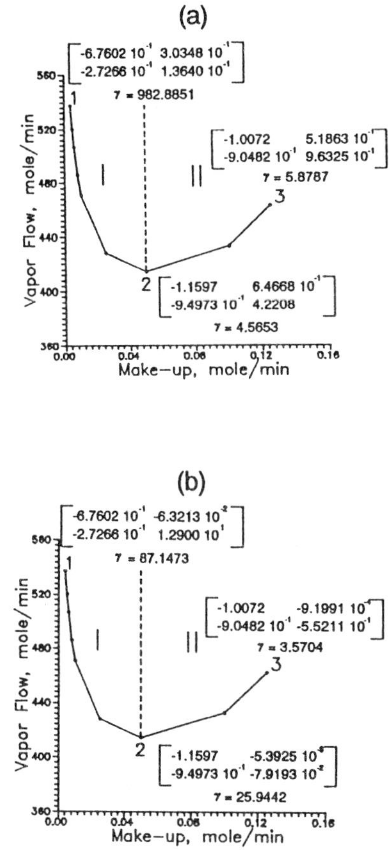

Figure 5: Steady-state gains and condition number for (a) $R_A V_A$ configuration and (b) $R_A F_E$ configuration for the EWB-column. Steady-state gains and condition number depend uniquely upon the vapour flow rate-make-up flow rate relationship as shown.

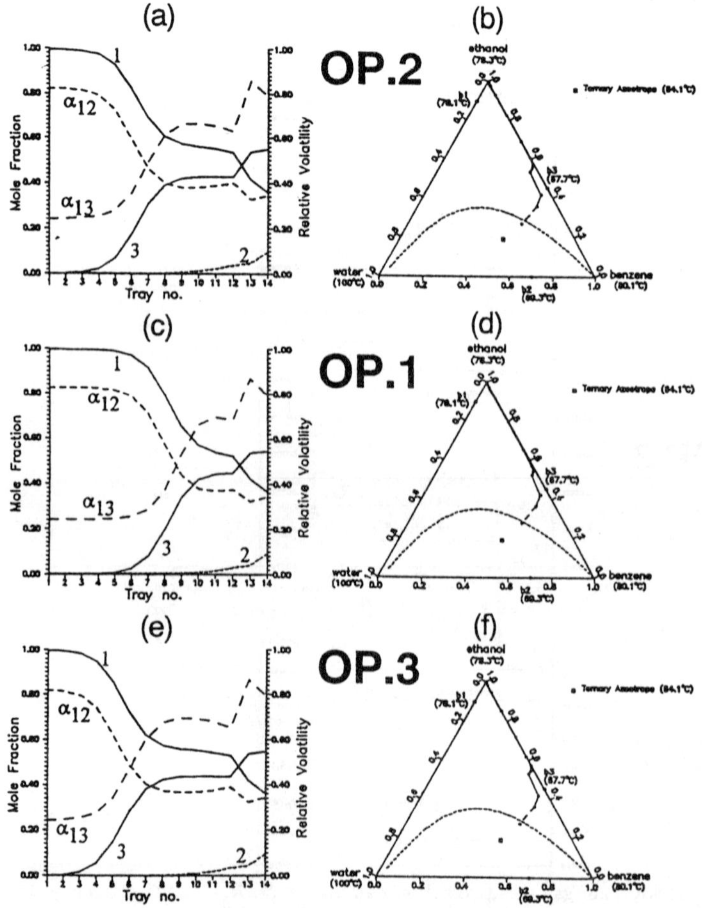

Figure 6 (a)(c)(e): Composition and relative volatility profiles and (b)(d)(f) composition profile in the phase diagram of a 14-tray azeotropic column for dehydration of aqueous ethanol using benzene as entrainer at operating points 2, 1 and 3.

(a)(c)(e) — 1-ethanol, 2-water and 3-benzene.

α_{12} — relative volatility between 1 and 2.

α_{13} — relative volatility between 1 and 3.

(b)(d)(f) — b1, b2 and b3 are binary azeotropes.

Solid lines connecting '*' are the liquid composition profile in the azeotropic column.

- - - - - - - - - $\mathbf{y}_{NA}$ in equilibrium with $\mathbf{x}_{NA}$.

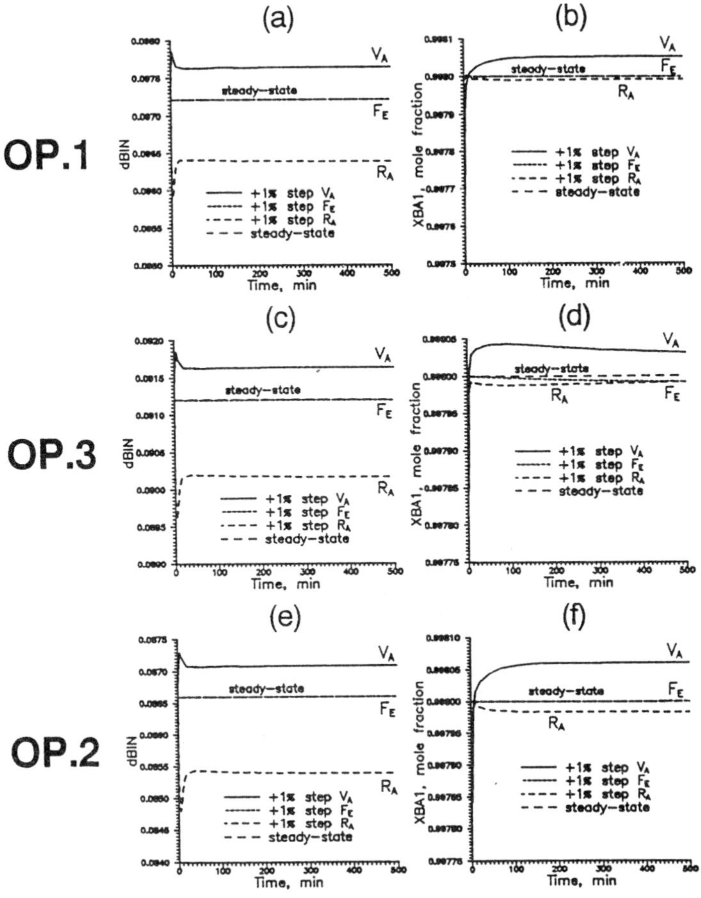

Figure 7: Non-linear open-loop responses of d_{BIN} and x_{BA1} towards +1% step changes in V_A, F_E, and R_A at (a), (b) op. point 1, (c), (d) op. point 3, and (e), (f) op. point 2 for the EWB-column.

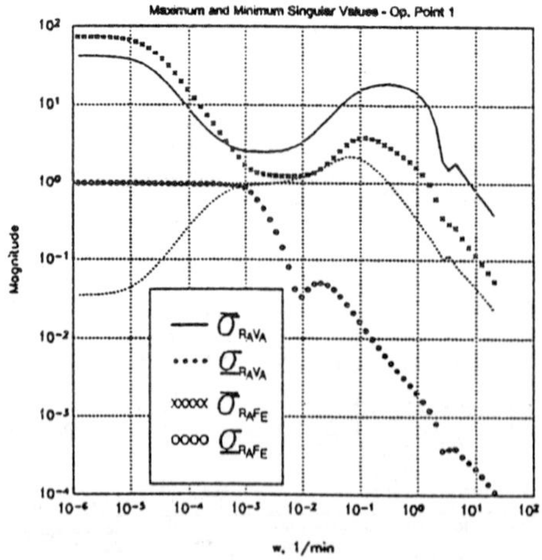

Figure 8: Multivariable open-loop dynamics for R_AV_A and R_AF_E configurations in region I (op. point 1) and EWB-column: maximum and minimum singular values.

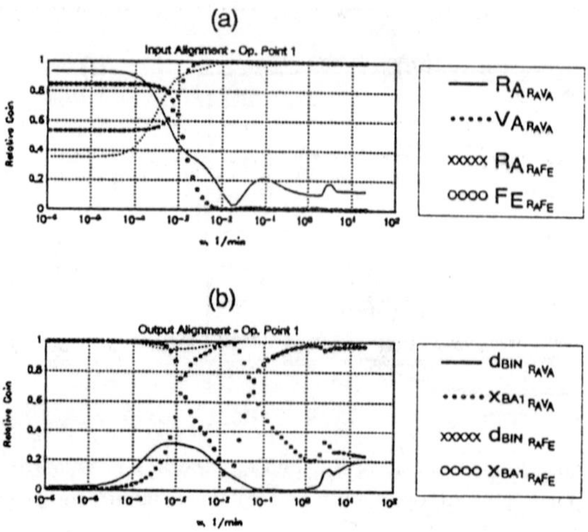

Figure 9: Multivariable open-loop dynamics for R_AV_A and R_AF_E configurations in region I (op. point 1) and EWB-column: (a) input alignment with the maximum singular value. (b) output alignment with the maximum singular value.

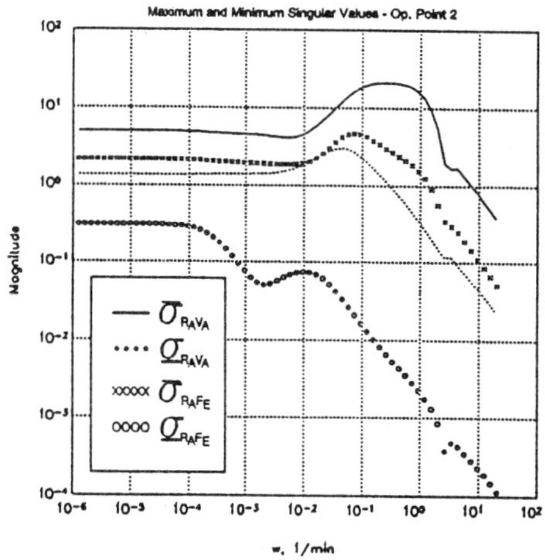

Figure 10: Multivariable open-loop dynamics for R_AV_A and R_AF_E configurations at operating point 2 and EWB-column: maximum and minimum singular values.

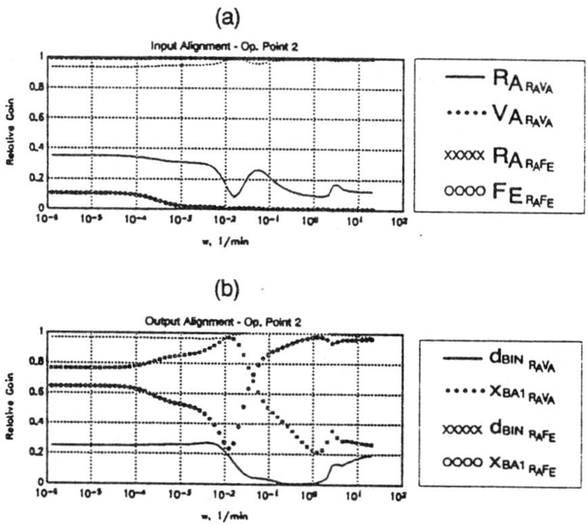

Figure 11: Multivariable open-loop dynamics for R_AV_A and R_AF_E configurations at operating point 2 and EWB-column: (a) input alignment with the maximum singular value. (b) output alignment with the maximum singular value.

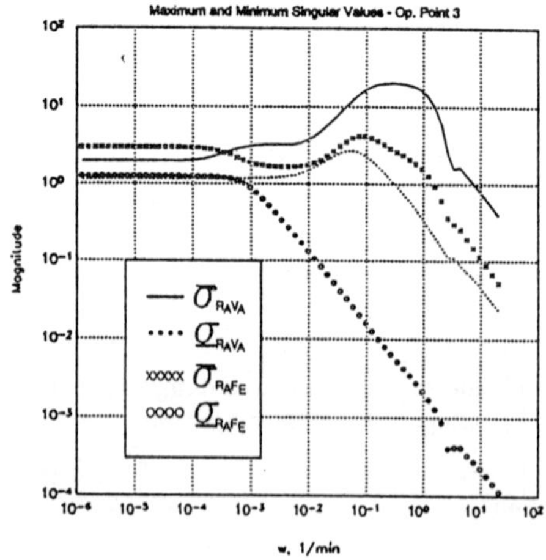

Figure 12: Multivariable open-loop dynamics for $R_A V_A$ and $R_A F_E$ configurations in region II (op. point 3) and EWB-column: maximum and minimum singular values.

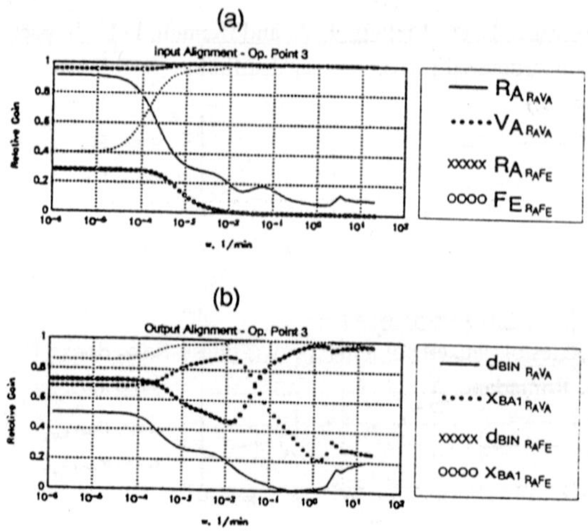

Figure 13: Multivariable open-loop dynamics for $R_A V_A$ and $R_A F_E$ configurations in region II (op. point 3) and EWB-column: (a) input alignment with the maximum singular value. (b) output alignment with the maximum singular value.

INDEX

A

D

E

F

G

H

I

J

M

N

O

P

R

S

T

U

V

W

Y

Z

B314